I0056210

Modern Optics

Modern Optics

Editor: Roderick Swayne

NY RESEARCH
PRESS

New York

Published by NY Research Press
118-35 Queens Blvd., Suite 400,
Forest Hills, NY 11375, USA
www.nyresearchpress.com

Modern Optics
Edited by Roderick Swayne

© 2018 NY Research Press

International Standard Book Number: 978-1-63238-584-0 (Hardback)

This book contains information obtained from authentic and highly regarded sources. Copyright for all individual chapters remain with the respective authors as indicated. All chapters are published with permission under the Creative Commons Attribution License or equivalent. A wide variety of references are listed. Permission and sources are indicated; for detailed attributions, please refer to the permissions page and list of contributors. Reasonable efforts have been made to publish reliable data and information, but the authors, editors and publisher cannot assume any responsibility for the validity of all materials or the consequences of their use.

Trademark Notice: Registered trademark of products or corporate names are used only for explanation and identification without intent to infringe.

Cataloging-in-Publication Data

Modern optics / edited by Roderick Swayne.
    p. cm.
Includes bibliographical references and index.
ISBN 978-1-63238-584-0
1. Optics. 2. Physics. I. Swayne, Roderick.
QC355.3 .M63 2018
535--dc23

Contents

Permissions

List of Contributors

Index

Preface

The study of properties, characteristics, and behavior of light and its relation with matter is known as optics. This field includes learning about the characteristics of infrared, ultraviolet and visible light. It also incorporates the construction of instruments and machines that use and detect light. This book outlines the processes and applications of optics in detail. It lays special attention on the theories and techniques related to the field. The book explores all the important aspects of this area in the present day scenario like quantum optics, physical optics, and fiber optics, etc. It presents the complex subject of optics in the most comprehensible and easy to understand language. Students, researchers, experts and all associated with this field will benefit alike from the book.

Significant researches are present in this book. Intensive efforts have been employed by authors to make this book an outstanding discourse. This book contains the enlightening chapters which have been written on the basis of significant researches done by the experts.

Finally, I would also like to thank all the members involved in this book for being a team and meeting all the deadlines for the submission of their respective works. I would also like to thank my friends and family for being supportive in my efforts.

Editor

Corollaries of Point Spread Function with Asymmetric Apodization

Naresh Kumar Reddy Andra,[1,2] Udaya Laxmi Sriperumbudur,[3] and Karuna Sagar Dasari[2]

[1]*Department of Physics, CMR Institute of Technology, Medchal Road, Kandlakoya, Hyderabad, Telangana 501401, India*
[2]*Optics Research Group, Department of Physics, University College of Science, Osmania University, No. 49, Hyderabad, Telangana 500007, India*
[3]*Department of Physics, Keshav Memorial Institute of Technology, Narayanguda, Hyderabad, Telangana 500029, India*

Correspondence should be addressed to Naresh Kumar Reddy Andra; naarereddy@gmail.com

Academic Editor: Ivan Moreno

Primary energy based corollaries of point spread function with asymmetric apodization using complex pupil function have been studied in the case of three-zone aperture. Merit function like semicircled energy factor, excluded semicircled energy, and displaced semicircled energy were analyzed with respect to Airy case in terms of phase and amplitude apodization. Analytical results have been presented for the optimum parameters of phase and amplitude asymmetric apodization.

1. Introduction

The principal corollaries measure the fraction of the total energy enclosed in the PSF of specific radius "δ" in the focal plane of optical imaging systems. Flux in a circle of specific radius (encircled energy) is the significant parameter in quantifying the performance of optical without aberrations. A lot of work has been done on the energy corollaries, but all the studies were pertained to symmetrically apodized PSF where the energy distribution on both sides of the diffraction centre is symmetric, whereas current study is dealing with asymmetric PSF. Asymmetric apodization is defined as the modification in the diffraction pattern by suppressing the optical side lobes on one side of the PSF at the cost of enhancing the side lobes on the other side while decreasing the width of the central maximum. For the obtained asymmetric PSF, where the central disc is not an exact circle, the corollaries based on the energy of the PSF, namely, the semicircled energy, excluded semicircled energy, displaced semicircled energy, and mean apodization ratio on both sides of the PSF, are investigated for the amplitude and phase filters.

Initially the idea of asymmetric apodization was introduced by Cheng and Siu [1, 2]. Later Siu et al. continued their work and successfully applied asymmetric apodization to one-dimensional arrays [3]. All these studies are significant breakthrough in asymmetric apodization studies. Based on their studies Reddy and Sagar [4] applied the concept of asymmetric apodization to semicircular arrays of circular pupil functions and they achieved improved side-lobe suppression. Reddy and Sagar [5] introduced semicircled energy factor for analyzing the flux distribution in the image plane of optical systems with phase-only pupil. Naresh Kumar Reddy et al. [6] studied Strehl ratio, total transmission factor, and half power diameter for point spread function of imaging systems with asymmetric apodization. Rayleigh [7] discussed importance of the encircled factor to determine the illuminations in the edge ringing effect. Luneberg [8] studied third apodization problem to find the optimum pupil function for maximum encircled energy. Ueno and Asakura [9] investigated optimum pupil apodization for maximum encircled energy with specified overall transmittance. Mondal [10, 11] studied corollaries of PSF apodized with various amplitude filters. A few earlier studies in this case are limited [12–15]. Earlier studies [1–6] are the basis for current investigation. The present study on corollaries of the PSF of an asymmetric optical imaging system enables us to know the effects of asymmetric apodization on the consequences like increase in resolution in the Sparrow criterion, decrease in full width half maximum (FWHM), and so forth. Finally a comparison study can be made between Airy PSF corollaries and APSF corollaries from which important conclusions are presented.

2. Theory

Based on the Fresnel-Kirchhoff diffraction theory, amplitude impulse response of phase and amplitude apodized optical imaging systems can be given as

$$A_F(0, u) = 2 \int_0^1 f(r) J_0(ur) r \, dr, \qquad (1)$$

where $J_0(ur)$ is the Bessel function of the first kind and zero order with argument (ur). $f(r)$ is the pupil function, where $u = (2\pi/\lambda)\sin\theta$. Here r is the radial coordinate in the complex pupil function. $f(r)$ for the three-zone amplitude and phase filter is given by

$$f(r) = \begin{cases} -i \exp(iur\cos(\phi - \varphi)), & 1 - b \leq r \leq 1, \ -\dfrac{\pi}{2} \leq \varphi \leq \dfrac{\pi}{2}, \\ \left(1 - 4\beta r^2 + 4\beta r^4\right) \exp(iur\cos(\phi - \varphi)), & 0 \leq r \leq 1 - b, \ 0 \leq \varphi \leq 2\pi, \\ i \exp(iur\cos(\phi - \varphi)), & 1 - b \leq r \leq 1, \ \dfrac{\pi}{2} \leq \varphi \leq \dfrac{3\pi}{2}. \end{cases} \qquad (2)$$

On introducing the amplitude apodization parameter β to control the degree of nonuniformity of transmission in the central region of the pupil of radius $(1 - b)$ and introducing semicircular edge ring of width b to control the phase apodization of the pupil function, complex pupil function $f(r)$ consists of three zones with two semicircular edge rings of equal width "b" with opposite phase transmittances of the form $+i$ and $-i$ and the central circular zone is with amplitude apodizer and its corresponding phase transmittance is zero. For amplitude apodizer, the amplitude transmittance at the center $(r = 0)$ of aperture $f(r)$ is equal to unity, that is, maximum. It decreases towards the edge of the aperture as "r" goes to 0.5 for all values of β. From Figure 2, it is clear that $f(r)$ is minimum at edges for any degree of amplitude apodization. β is the apodization parameter controlling the degree of nonuniformity of the transmittance over zone of radius $(1 - b)$. The range it takes is $0 \leq \beta \leq 1$. It is clear that, for $\beta = 0$, the transmittance of this zone is uniform. Transmittance over the rest of the two zones of circular aperture of unit radius is unity. Here b is the certain width of semicircular edge ring. The range of values it takes is $0 \leq b \leq 0.1$. It is obvious that, for $b = 0$, the PSF obtained is symmetric

in nature. The asymmetry in the PSF increases with the values of b (degree of phase apodization) and further improved by increase in amplitude apodization parameter β. The design of pupil function can be seen in Figure 4.

The optics term semicircled energy refers to a measure of concentration of energy over one side of the PSF from the diffraction centre. Calculation of the semicircled energy factor (SCEF) on both sides of the PSF gives the total distribution of energy in the asymmetric PSF. In the case of two-dimensional aperture, the side-lobe region is specified not only by "u," which related to the orientation θ, but also by azimuth angle Φ, which determines how large the good side "window" is, taken as $-\pi/3 < \Phi < \pi/3$. By the minimum of intensity square, the semicircular edge ring width is determined to be $b = 0.1$. $A_F(0, u)$ is the amplitude in the image plane at point "u" (reduced dimension less diffraction coordinate), units away from the diffraction head due to the pupil function. Hence, the integration over Φ introduces just the same constant in both the numerator and the denominator. Amplitude contains Bessel functions of the first kind, which oscillate from positive to negative values very rapidly and become zero at a finite distance from the centre of the diffraction image ($u = 0$). Thus

$$\text{SCEF}(\delta)_{\text{Good Side}} = \frac{\int_0^\delta \left[\int_0^{1-b} f(r) J_0(ur) r \, dr\right] u \, du - i \int_{1-b}^1 \int_{-\pi/2}^{\pi/2} \exp(iur\cos(\phi - \varphi)) r \, dr \, d\varphi}{\int_0^1 |f(r)|^2 r \, dr},$$

$$\text{SCEF}(\delta)_{\text{Bad Side}} = \frac{\int_0^{-\delta} \left[\int_0^{1-b} f(r) J_0(ur) r \, dr\right] u \, du + i \int_{1-b}^1 \int_{\pi/2}^{3\pi/2} \exp(iur\cos(\phi - \varphi)) r \, dr \, d\varphi}{\int_0^1 |f(r)|^2 r \, dr}. \qquad (3)$$

Excluded semicircled energy (ESCE) is complementary quantity of the semicircled energy. In order to examine the

outer ring structure in detail, this factor is convenient as compared to the semicircled energy. The function of an asymmetric apodizer is best fulfilled if this factor is maximum on good side and minimum on bad side of the obtained asymmetric PSF.

The expression for excluded semicircled energy is as follows:

$$\text{ESCE}(\delta) = 1 - \text{SCEF}(\delta),$$

$$\text{ESCE}(\delta) = \frac{\int_R^{\pm\delta} |A_F(0,u)|^2 u\,du}{\int_0^\infty |A_F(0,u)|^2 u\,du}. \quad (4)$$

This energy corollary is useful in some photometric situations. For example, the contrast at the center of the image of a black disk seen against a uniform incoherent background is semicircled energy, so excluded semicircled energy gives the residual intensity. Such situation exists in the transit of planet across the Sun. this parameter is useful in evaluating apodization for suppressing the ring structure. Displaced semicircled energy (DSCE) is the difference between the semicircled energy (SCEF) of the perfect system (Airy) and given case (asymmetrically apodized). It is useful to compare the energy distribution in the actual optical system to its perfect counterpart. Thus

$$\text{DSCE}(\delta) = \text{SCEF}_A(\delta) - \text{SCEF}_F(\delta), \quad (5)$$

where $\text{SCEF}_A(\delta)$ gives the semicircled energy for Airy case and $\text{SCEF}_F(\delta)$ gives the semicircled energy for asymmetrically apodized case. If the displacement semicircled energy is positive, there is an outward displacement of energy, and if it is negative, there is an inward displacement of energy. This parameter is more sensitive to aperture obscuration and less sensitive to image motion.

3. Results and Discussions

The improved resolution of asymmetric point spread function with amplitude and phase filters has been analyzed in terms of energy corollaries on both sides of the main lobe. However, we reported the majority of the results for the good side which is the half part of the complete pattern. The results on investigations on semicircled, excluded, and displaced energies have been studied. In order to investigate the flux enclosed in the detector plane of optical imaging, systems have been obtained as function of dimension less diffraction parameter "u" by employing the standard numerical method of integration. An iterative method has been developed and applied to determine the asymmetric radial distribution of energy within a specific radius "δ" in the plane of observation or detection.

Semicircled energy factor of the APSF as function of "δ" for various combinations of b and β is obtained for Airy case ($b = 0$ and $\beta = 0$), symmetric apodized case ($b = 0$ and $\beta \neq 0$), and asymmetric apodized case ($b \neq 0$). In our analysis, it has been observed that, for Airy case, SCEF(δ) increases rapidly for first few values "δ" in the receiver plane and later on slows down considerably before finally approaching half of the unity asymptotically.

In this case, there is quick increase in SCEF up to a value of δ around 3.0 which results in a fast growth in the peak

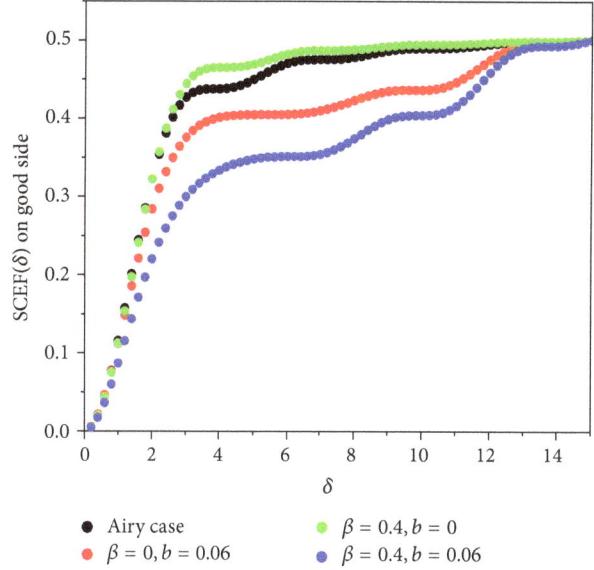

FIGURE 1: Semicircled energy on good side for different values of b and β.

intensity of the main lobe. For $\beta = 0.4$ and $b = 0$, it has been observed that with the introduction of three-zone amplitude filter in the central region of the aperture there is a variation in SCEF compared to Airy case. Till δ is equal to 3.75 the semicircled energy is equal to that of Airy case and beyond 3.75 the semicircled energy is greater than the Airy case. Two equal semicircular edge rings of width b (0.06) with opposite transparencies have been distributed over the central circular region of the aperture with which an asymmetric PSF has been obtained with unequal energies on both sides of the central lobe. The side on which the side lobes are enhanced is referred to as bad side and the other is referred to as good side. From Figure 1, it is clear that for all the values of δ varying from 0 to 15 the semicircled energy is found to be less than other cases (Airy case and also symmetrically apodized case).

From Figure 1, it has been observed that the value of SCEF is not as much of Airy case. It is also noticed that with the introduction of amplitude apodizer in the central region of the aperture the semicircled energy factor is further decreased comparatively with the above-mentioned cases. Figure 2 shows the variation in amplitude transmittance for different values of amplitude apodization parameter β as a function of the radial coordinate "r" of the pupil. It shows a monotonic decrease of amplitude transmittance $t(r)$ with the normalized distance r. At the center of the pupil ($r = 0$) transmittance is unity, that is, highest for all values of β, and it decreases towards the edges as r goes to 0.5. Transmittance is minimum at the edges for all values of β. Figure 3 explains the study of SCEF on both sides of the PSF for $b = 0$, while β varies from 0 to 1. In this case the semicircled energy on both sides of the central lobe is symmetric or uniform. Hence the results obtained on the bad side are similar to those of the good side. It has been observed that the lowest values of semicircled energy are obtained for $\beta = 0$ and the highest values have been obtained for $\beta = 1$. It is clear that

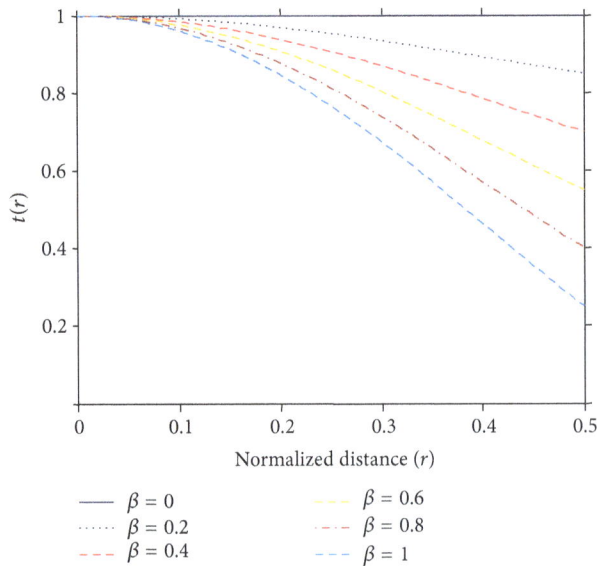

FIGURE 2: Amplitude transmittance of the amplitude apodizer decreases with radial coordinate.

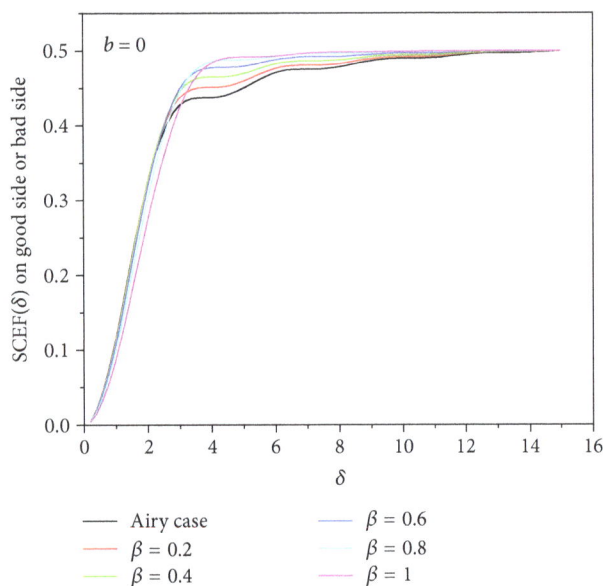

FIGURE 3: Semicircled energy on good side or bad side for various values of β when $b = 0$.

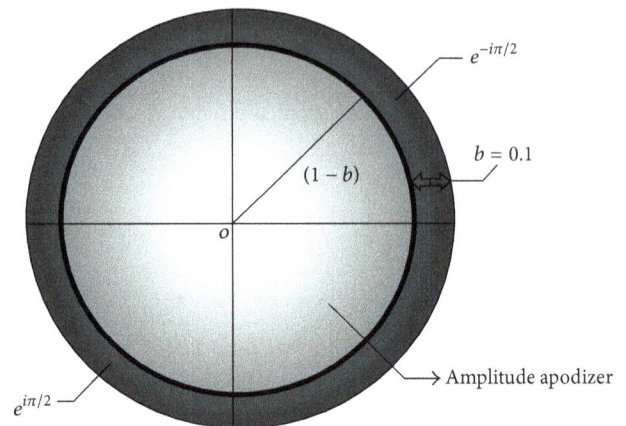

FIGURE 4: Generalized scheme of complex pupil function.

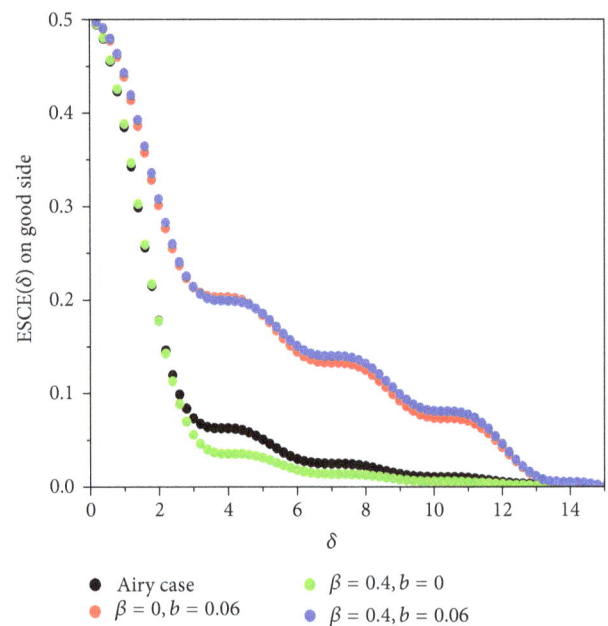

FIGURE 5: Excluded energy on good side for different values of b and β.

the semicircled energy vanishes at the origin ($\delta = 0$) and increases monotonically, approaching half of the unity. The semicircled energy on good side is decreasing with increasing the value of b. For all values of δ varying from 0 to 15 it is observed that the semicircled energy value is less than Airy case, whereas on bad side semicircled energy obtained its maximum factor which is more than Airy case for $b = 0.02$ and 0.04. On further rise in semicircular edge ring width above 0.04 SCEF (δ) on bad side decreases.

The values of the excluded semicircled energy for various values of δ as function of b and β are obtained. It has computed quite easily from SCEF. For $b \neq 0$ (0.06) and $\beta = 0$ or $\beta \neq 0$ (0.4), more excluded energy has been recorded than that of Airy case, whereas for symmetrically apodized case ($\beta \neq 0$ and $b = 0$) the excluded energy is detected less than Airy case. It has been clearly illustrated in Figure 5. It is concluded that with increase in β value the excluded energy is decreasing. It is depicted clearly in Figure 6. With increase of degree of amplitude apodization in the central region of the aperture, the central disc widens; more energy is accumulated in the central region compared to the energy concentrated in the near vicinity of the diffraction centre, whereas in the case of asymmetric apodization the energy is shifting to one side of the diffraction centre (bad side) and simultaneously narrowing the central lobe of the diffraction pattern. The side on which the optical lobes are suppressed is termed as good side and the energy from this side has been excluded more to the outer rings.

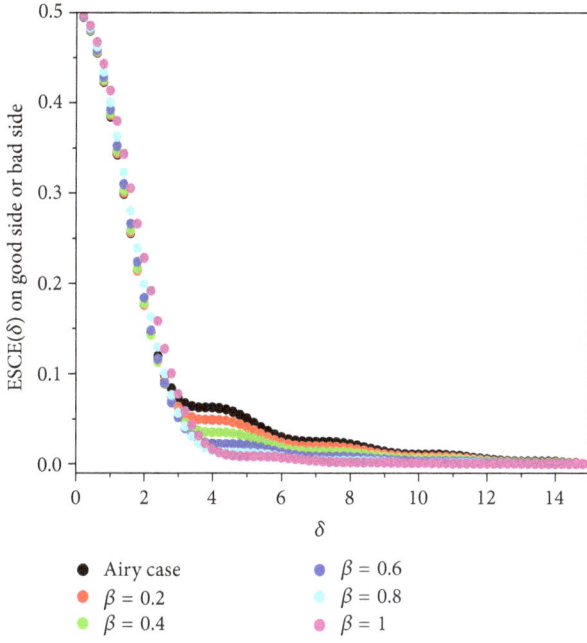

FIGURE 6: Excluded energy on good side or bad side for various values of β when $b = 0$.

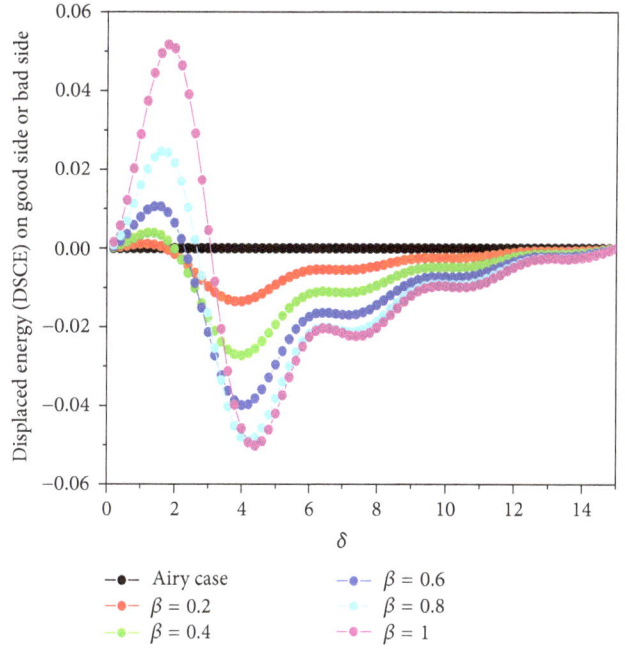

FIGURE 8: Displaced energy on good side or bad side for various values of β when $b = 0$.

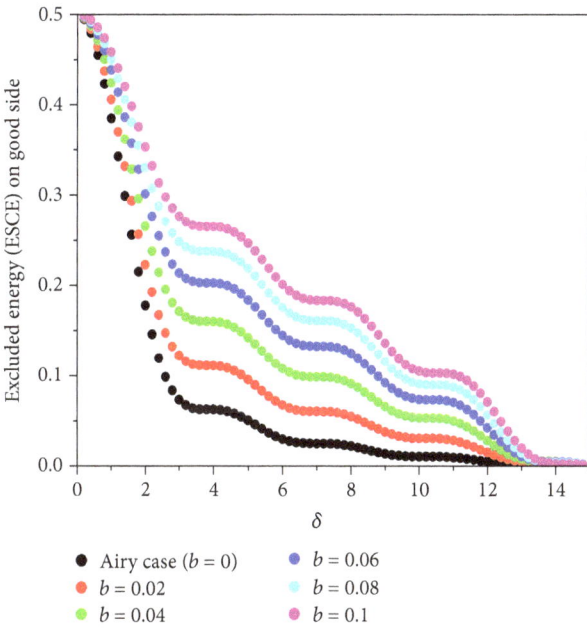

FIGURE 7: Excluded energy on good side for various amount of asymmetric apodization (b) when $\beta = 0$.

The excluded energy on the good side increases with increase in the value of "b." It is observed that it is minimum for $b = 0$ and maximum for $b = 0.1$. It is observed in more detail in Figure 7.

The values of displaced semicircled energy on good side for various combinations of b and β are studied. There is no displaced energy for Airy case by definition. It has been observed that for symmetrically apodized optical system

($\beta \neq 0$ and $b = 0$) the displaced energy is negative, which means that the semicircled energy possesses higher values. Hence the displaced energy is lying on the negative axis. For an optical system ($\beta = 0$ and $b \neq 0$) the semicircled energy on the good side is less than Airy case; hence by definition the values of displaced energy are positive for all the values of δ. For $\beta \neq 0$ and $b \neq 0$, the semicircled energy of the system is very low compared to the semicircled energy in Airy case; hence the displaced energy values are high compared to the other cases. Figure 8 illustrates that in absence of phase apodization ($b = 0$) for all values of β the displaced energy is lower than Airy case for lower values of δ and on further rise in δ the displaced energy is on negative axis. Negative values of displaced energy in Figure 8 indicate that the semicircled energy increases with degree of amplitude apodization (β). Figure 9 shows the corresponding distribution of diffraction filed, where the Airy PSF is presented as a solid black curve. It is clearly seen that the complex pupil under different considerations obtains the optical side lobes much lower than Airy ones and the central lobe is narrowed.

4. Conclusions

The following explanation for the energy based corollaries of asymmetric point spread function can be given. With introduction of amplitude apodizer in the central region gives modified PSF with equal energies (equal intensity in side lobes or secondary maxima) on both sides of the diffraction centre. Hence, as β increases, the semicircled energy on both sides of the PSF increases. In the case of asymmetric

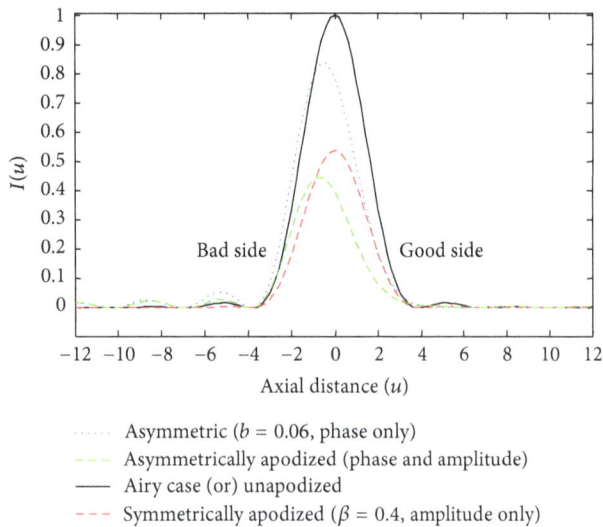

FIGURE 9: The point spread function of a complex pupil filter under different conditions.

apodization, as b increases, the energy distribution on both sides of the diffraction centre is not equal (i.e., asymmetric). With increase in b value the semicircled energy decreases on the good side and at the same time for $b = 0.02$ and 0.04 the semicircled energy on bad side is more than that of Airy case. Hence the excluded energy increases on good side of the APSF as degree of asymmetric apodization increases. Increase in the semicircled excluded energy results in the enhancement of resolution of the optical system. On the contrary, as β increases, the amount of excluded energy decreases. Hence the highest excluded energy is recorded for maximum amplitude apodization ($\beta = 1$). For lower values of δ, the positive values of displaced energy are obtained which indicates more energy in the Airy pupil than that of the apodized pupil with complex pupil filters. The displaced energy on good side of the diffraction centre increases as b increases rendering the asymmetrically apodized pupils energy lower than Airy pupils. In this case, the energy is shifting to one side of the diffraction centre (i.e., bad side of the PSF) and simultaneously narrowing the central lobe of the diffraction pattern. The side on which the optical lobes are suppressed is termed as good side and the energy from this side has been excluded more to the outer rings. This asymmetry increases with the width (b) of the edge rings and is promoted by the increase in the degree of apodization (β) of the central circular region of the aperture. Thus, present study has many applications in diverse potential fields such as confocal microscopy, spectroscopy, astronomy, communication, and medical imaging.

Conflict of Interests

The authors declare that there is no conflict of interests regarding the publication of this paper.

Acknowledgment

The authors are thankful to Anup L. Shah, Scientist-H (Project Director, CHESS, RCI campus, DRDO), Hyderabad, for technical support and some interesting suggestions concerning the bibliographic background.

References

[1] L. Cheng and G. G. Siu, "Asymmetric apodization," *Measurement Science and Technology*, vol. 2, no. 3, pp. 198–202, 1991.

[2] G. G. Siu, L. Cheng, D. S. Chiu, and K. S. Chan, "Improved sidelobe suppression in asymmetric apodization," *Journal of Physics D: Applied Physics*, vol. 27, no. 3, pp. 459–463, 1994.

[3] G. G. Siu, M. Cheng, and L. Cheng, "Asymmetric apodization applied to linear arrays," *Journal of Physics D: Applied Physics*, vol. 30, no. 5, pp. 787–792, 1997.

[4] A. N. K. Reddy and D. K. Sagar, "Point spread function of optical systems apodized by semi circular arrays of 2D aperture with asymmetric apodization," *Journal of Information and Communication Convergence Engineering*, vol. 12, no. 2, pp. 83–88, 2014.

[5] A. N. K. Reddy and D. K. Sagar, "Semi circled energy of asymmetrically apodized optical systems," *Advances in Applied Science Research*, vol. 5, no. 2, pp. 42–48, 2014.

[6] A. Naresh Kumar Reddy, R. Komala, M. Keshavulu Goud, and S. Lacha Goud, "A few PSF-based corollaries of optical systems apodised asymmetrically with two-dimensional complex pupil filters," *Armanian Journal of Physics*, vol. 4, no. 4, pp. 200–205, 2011.

[7] L. Rayleigh, "Investigations in optics, with special reference to the spectroscope," *Philosophical Magazine Series 5*, vol. 8, no. 49, pp. 261–274, 1879.

[8] P. Luneberg, "Measurement of contrast transmission characteristics in optical image formation," *Optica Acta*, vol. 1, no. 2, pp. 80–89, 1954.

[9] T. Ueno and T. Asakura, "Apodization for maximum encircled energy with specified over-all transmittance," *Journal of Optics*, vol. 8, no. 1, pp. 15–31, 1977.

[10] K. P. Rao, P. K. Mondal, and T. Seshagiri Rao, "Diffracted field characteristics of Straubel class of apodisation filters," *Pramana*, vol. 7, no. 6, pp. 389–396, 1976.

[11] K. Surendar, S. L. Goud, and P. K. Mondal, "PSF corollaries of apodized optical systems," *Acta Ciencia Indica*, vol. 18, no. 1, pp. 91–96, 1992.

[12] E. C. Kintner, "Calculating the encircled energy in the point-spread function," *Optica Acta: International Journal of Optics*, vol. 24, no. 10, pp. 1075–1076, 1977.

[13] J. Campos, F. Calvo, and M. J. Yzuel, "Resolving power and encircled energy in aberrated optical systems with filters optimized for the Strehl ratio," *Journal of Optics*, vol. 19, no. 3, pp. 135–144, 1988.

[14] A. A. Dantzler, "Encircled energy correction method for ray-trace programs," *Applied Optics*, vol. 27, no. 24, pp. 5001–5002, 1988.

[15] S. J. Park and C. S. Chung, "Influence of Bessel and Bessel-Gauss beams on the point spread function and the encircled energy," *Journal of the Korean Physical Society*, vol. 30, no. 2, pp. 194–201, 1997.

Visible Light Communication System Using Silicon Photocell for Energy Gathering and Data Receiving

Xiongbin Chen,[1,2] Chengyu Min,[1] and Junqing Guo[1]

[1]*State Key Laboratory of Integrated Optoelectronics, Institute of Semiconductors, Chinese Academy of Sciences, Beijing, China*
[2]*School of Electronic, Electrical and Communication Engineering, University of Chinese Academy of Sciences, Beijing, China*

Correspondence should be addressed to Xiongbin Chen; chenxiongbin@semi.ac.cn

Academic Editor: Liang Wu

Silicon photocell acts as the detector and energy convertor in the VLC system. The system model was set up and simulated in Matlab/Simulink environment. A 10 Hz square wave was modulated on LED and restored in voltage mode at the receiver. An energy gathering and signal detecting system was demonstrated at the baud rate of 19200, and the DC signal is about 2.77 V and AC signal is around 410 mV.

1. Introduction

Solar cell has drawn great interest over the past 30 years, and there is a tendency to use it more widely and practically. Visible light communication is also very amazing [1] as a new kind of wireless communication technology with less energy consumption, higher response speed, and more privacy.

Energy gathering and signal detecting system is a new idea. Energy harvesters are widely used in sensor networks. But energy gathering can be hardly seen in the VLC. We noticed that the silicon-based solar panels could receive VLC data and gather energy at the same time.

Research works in this area can be found in [2]; the researchers from Korea used a solar cell as a simultaneous receiver of solar power and visible light communication (VLC) signals. Some research on the efficiency and frequency response of solar cell had been launched.

In our works, solar cell was studied totally under visible light. We set up models similar to the real lighting conditions and run simulations in Matlab/Simulink. Simulation results indicate that it is possible to gather energy and receive data through the same solar panels.

We implement the system using commercial components. Our experiments based on the prototype show that the solar panels can gather energy for low power circuit and detect the VLC signal at the same time.

2. Model Analysis

In this section, we analyzed the model of LED and solar cell and then formulated their relationship with some approximations.

2.1. Model of LED Light Source. The LED conforms to Lambert emission rule. When the transmitted optical power is P_t, the received power P_r (w/m^2) is expressed as [3]

$$P_r = \begin{cases} P_t \dfrac{m+1}{2\pi d^2} \cos^m(\phi) \, Ts(\psi) \cdot g(\psi) \cos(\psi) & 0 \le \psi \le \psi_c \\ 0 & \psi > \psi_c, \end{cases} \quad (1)$$

where d is the distance between LED and PD, ϕ is the irradiance angel, ψ is the incidence angel of PD, $Ts(\psi)$ is the optical filter gain, $g(\psi)$ is the optical concentrator gain, ψ_c is the field of view of PD, and m is Lambert emission order.

The SNR for VLC and the illuminance value on PD are given as follows:

$$\text{SNR} = \frac{\gamma^2 P_{r\text{signal}}^2}{\sigma_{\text{shot}}^2 + \sigma_{\text{thermal}}^2 + \gamma^2 p_{r\text{ISI}}^2} \quad (2)$$

$$E_{\text{hor}} = \frac{I(0)\cos^m(\Phi)\cos(\psi)}{d^2}.$$

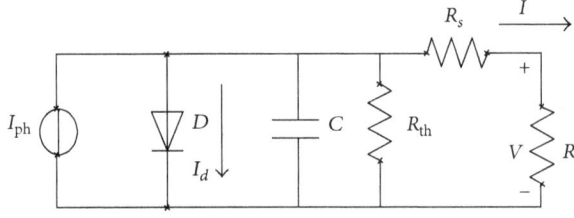

FIGURE 1: The equivalent circuit diagram of a typical solar cell.

2.2. Model of Solar Cell. The equivalent circuit diagram of a typical solar cell is as shown in Figure 1 [4].

It can be formulated as

$$I = N_1 I_P - I_0 \left\{ \exp\left[\frac{q(V + IR_s)}{N_2 AKT} \right] - 1 \right\} - C\frac{dV}{dt} - \frac{V + IR_s}{R_{th}}, \tag{3}$$

where N_1 is the number of solar cells in parallel, N_2 is the series number, I_P is the light current, I_0 is the diode saturation current, V is the output voltage of solar cell, I is the output current, and A is a constant which is typically in the rang 1 to 3. As $R_{th} \gg R_s$, if set

$$K_0 = \frac{AKT}{q} \tag{4}$$

then (3) can be written as

$$I = N_1 I_P - I_0 \left\{ \exp\left[\frac{(V + IR_S)}{N_2 K_0} \right] - 1 \right\} - C\frac{dV}{dt}. \tag{5}$$

For solar cell, the light current is positively proportional with received illuminance power:

$$I_P = \frac{S}{1000} I_{SC}. \tag{6}$$

$$I = \frac{U}{R_h} \tag{10}$$

S is the illuminance power of solar cell. The standard sun light illuminance power at normal room temperature is 1000 w/mm^2. I_{SC} is the short circuit current.

We can set $I = 0$, so the solar cell works in the open state; then (5) can be expressed as

$$N_1 I_P = I_0 \left[\exp\left(\frac{U_{OC}}{k_0 N_2} \right) - 1 \right]$$
$$I_0 = \frac{N_1 I_P}{\exp(U_{OC}/k_0 N_2) - 1}. \tag{7}$$

2.3. Model of the System. For our system, solar cell is used as the PD. The two models can be connected by making $P_r = S$. In this way, (3) can be expressed as follows:

$$I = N_1 \frac{P_r}{1000} I_{SC} - \frac{N_1 P_r I_{SC}}{\exp(U_{OC}/N_2 k_0) - 1} \cdot \left[\exp\left(\frac{U + U/R_h \cdot R_S}{N_2 k_0} \right) - 1 \right]. \tag{8}$$

Combine (5), (6), and (7) together:

$$I = \frac{N_1 P_r I_{SC}}{1000} \left\{ 1 - \frac{1}{1000 \left[\exp(U_{OC}/N_2 k_0) - 1 \right]} \cdot \left[\exp\left(\frac{U + (U/R_n) R_S}{N_2 k_0} \right) - 1 \right] \right\}. \tag{9}$$

R_h is the load resistance of solar cell. In conclusion, k_1, k_2, and k_3 are constants related to N_1, N_2, I_{SC}, and U_{OC}, so the relationship between U of solar cell and the LED power can be formulated as (11):

$$U = \begin{cases} P_t \dfrac{m+1}{2\pi d^2} \cos^m(\phi) Ts(\psi) \cdot g(\psi) \cos(\psi) \dfrac{1}{k_1} \left[1 - \dfrac{1}{k_2 R_h} \exp\left(\dfrac{U}{k_3} \right) \right] & 0 \leq \psi \leq \psi_c \\ 0 & \psi > \psi_c. \end{cases} \tag{11}$$

3. Results and Discussions

We set up the two models in Matlab/Simulink and combined them for simulation.

Solar cell model was simulated separately first. The model is based on the equations of (5), (6), and (7). Assuming that it works in the stable room temperature at 298 K, we chose solar cell AM-5308 for our experimental study. Parameters are set in Table 1.

The LED illumination model and Si photocell array model were combined to simulate the practical system. Figure 2 shows that U_{OC} for 4×4 and I_{SC} for 2×8 are half of values for 4×8 arrays individually. U_{OC} of 2×8 and 4×8 are 3~3.5 V, which possibly charge lithium battery. In Figure 3, we got the 2×8 arrays solar cell's I-V curves through the simulations under different illumination from 300 Lx to 1000 Lx. These numbers represent the daily scene illumination value, including living room, library, hospital

TABLE 1: Parameters for solar cell.

Parameter	Value
Areas	$3 \times 36 \text{ mm}^2$
Open circuit voltage	$U_{OC} = 0.3 \text{ V}$
Short circuit current	$I_{SC} = 15 \text{ uA}$
Series resister	$R_s = 0.0052 \, \Omega$
Standard condition	$E_v = 100 \text{ Lx}$
Parallel number of solar cells	$N_1 = 2$
series number of solar cells	$N_2 = 8$
Load resistance	$R_h = 0 \sim 5000 \, \Omega$
Capacity	20 nF
Temperature	$T = 298 \text{ K}$
Electronic charge	$q = 1.62e - 19c$
Boltzmann's constant	$K = 1.38066e - 23$
Gain of optical filter	$T_s(\psi) = 1$
Field of view	$\Psi = 60°$
Half angel of LED	$\Phi = 60°$
Lambert constant	$m = 1$
Distance	$d = 1.0 \text{ m}$
Optical concentrator gain	1
Power of LED	15 W
Illuminance	68 lm/W

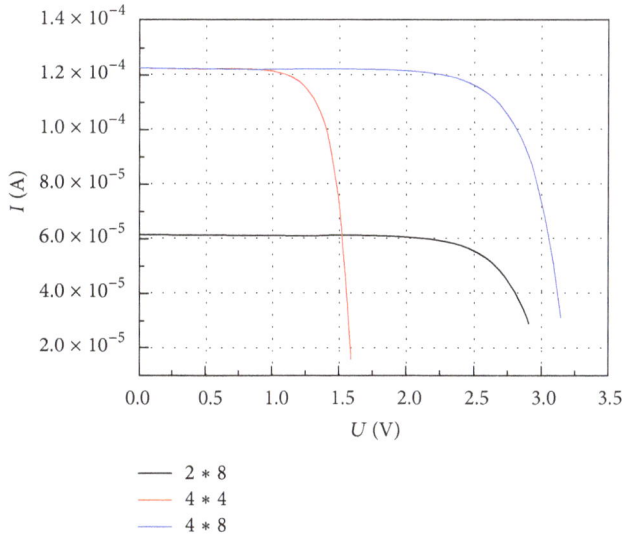

FIGURE 3: I-V curves of 2×8 arrays under different illumination.

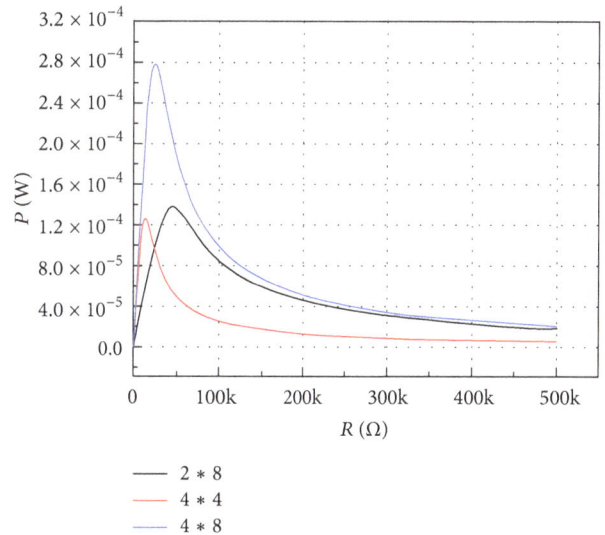

FIGURE 4: P-R curves for different solar cell arrays ($E_v = 300$ Lx).

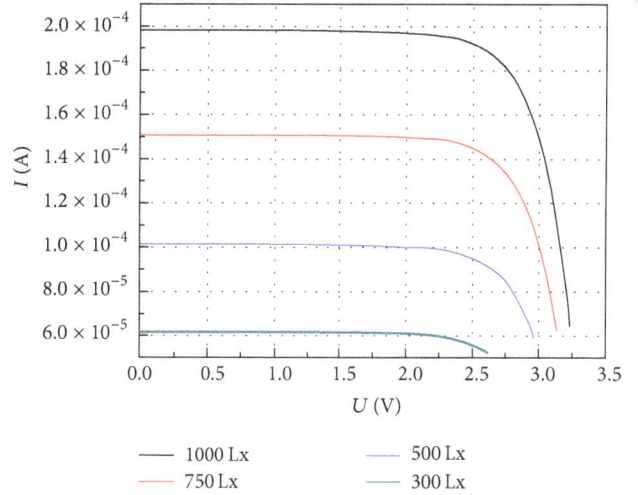

FIGURE 2: I-V curves for different solar cell arrays ($E_v = 300$ Lx).

operating room, and sports venue. Power properties of different arrays under different illumination values are also simulated in Figures 4 and 5. The output power of 2×8 arrays under 300 Lx and 50 KΩ is 1.4×10^{-4} W. The single receiving area of Si photocell chip is $3 \times 36 \text{ mm}^2$. For 2×8 arrays, the area is 1728 mm². So the efficiency of the 2×8 array is 8.1%. The spectral response of Si photocell chip made influence on the received light power as our LED light is mainly made up of blue and yellow.

The output voltage in simulation and experiment is among 2.7 V to 3.5 V. It increased to saturation state when the illumination value is above 500 Lx. It is stable for supplying power. The simulation value and experiment matched perfectly in Figure 6.

Then, a square signal as in Figure 7 is modulated on LED as the transmitting data. The period of the square signal is 0.1 s. The duty cycle is 50%.

The output power of the solar cell depends on the load resistance. The maximum output power about 1.2×10^{-3} W can be achieved, when load resistance is 4 kΩ, under illumination at 300 lx. The output power of solar cell with different load resistance is shown in Figure 8; the x-axis unit is 10 KΩ.

The output voltage of solar cell rises to 2.5 V after several pulses. The waveform of output voltage of solar cell under continuous pulse modulation is shown in Figure 9; the x-axis unit is 1 S.

An energy gathering and signal detecting system was demonstrated as Figure 10. To fit the working condition of solar cell, we used a 15 W LED which could simulate the

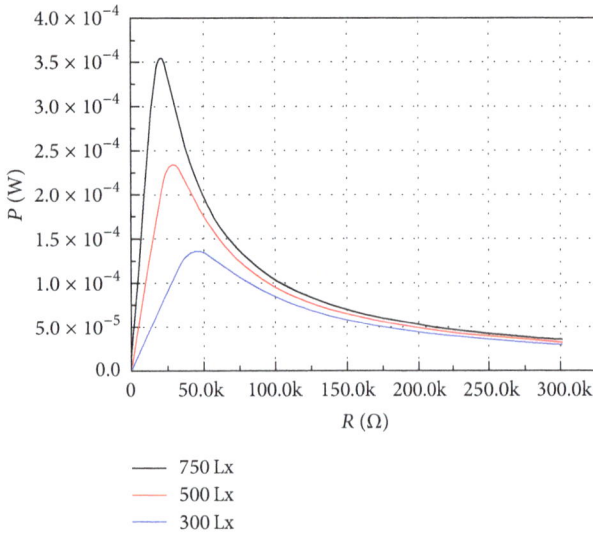

FIGURE 5: *P-R* curves under different illumination.

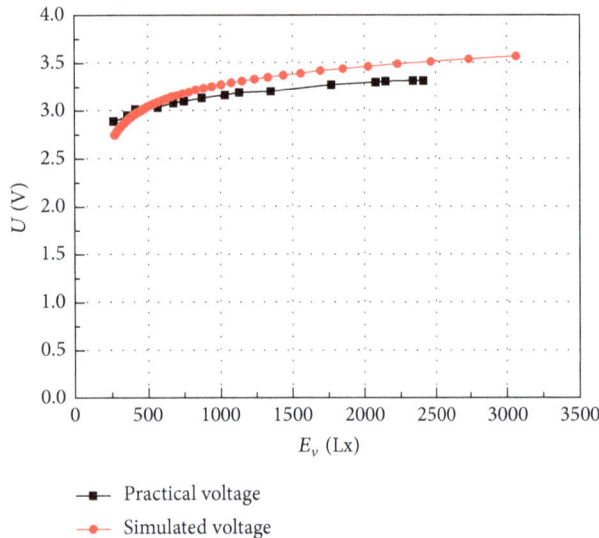

FIGURE 6: Comparison of simulation and experiments result for voltage and illumination.

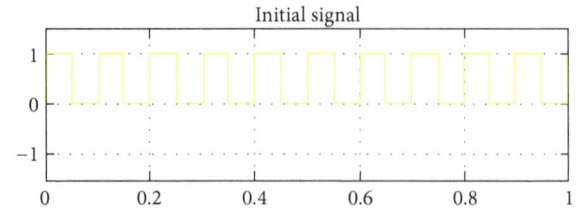

FIGURE 7: Initial signal.

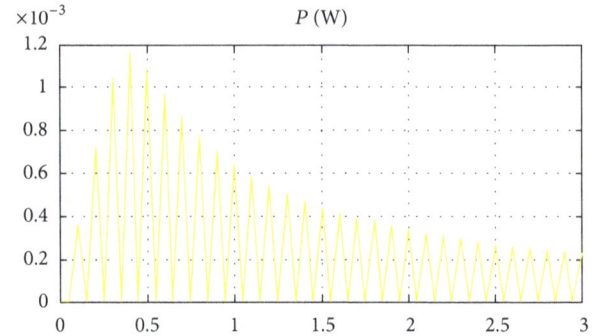

FIGURE 8: Power of solar cell.

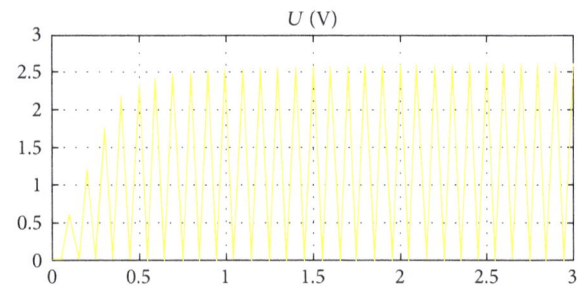

FIGURE 9: Voltage of solar cell.

4. Conclusion

In our works, we set up a model of solar cell VLC system which was simulated in Matlab/Simulink. We had verified the correction of the model and gave reasonable design to optimize the system.

The energy gathering and signals detecting system was demonstrated. The data rate of it is 19200 bps. The DC voltage of photocell was about 2.77 V which is enough for low voltage power supply circuits. The AC voltage of photocell was about 410 mV and could be optimized by one stage amplifier circuit. It was proved that solar cell can act as energy converting and detecting device simultaneously in VLC system.

The channel influences [6], response of solar cell to frequency, room lighting conditions, and other factors were ignored in our model. Further studies can take these factors into consideration. At the same time, we will optimize the design for the actual application.

Competing Interests

The authors declare that they have no competing interests.

different indoor lighting conditions. The distance between the 2×8 photocell array and the 15 w LED is 1.8 m. The illumination value on photocell was 690 lx, when the LED was not modulated. The illumination value on photocell was 637.5 lx, when the LED was modulated. The baud rate of computer's output was 19200. The output data was the repetition of "A5" in HEX form and the polarity was reversed by RS485 converter chip. The yellow line in Figure 11 represents a DC coupled output signal of the silicon photocell which is about 2.77 V. The green line in Figure 11 represents the AC coupled output signal of the silicon photocell, filtered by a $0.1\,\mu$F coupling capacitor. And the AC signal is around 410 mV. The baud rate and AC amplitude could be higher after one stage amplifier circuit [5].

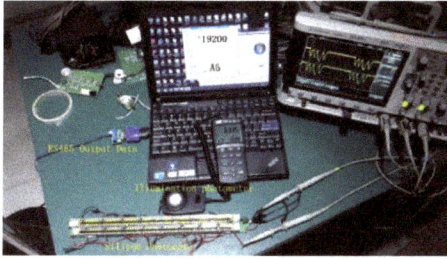

FIGURE 10: Energy gathering and signal detecting demo system.

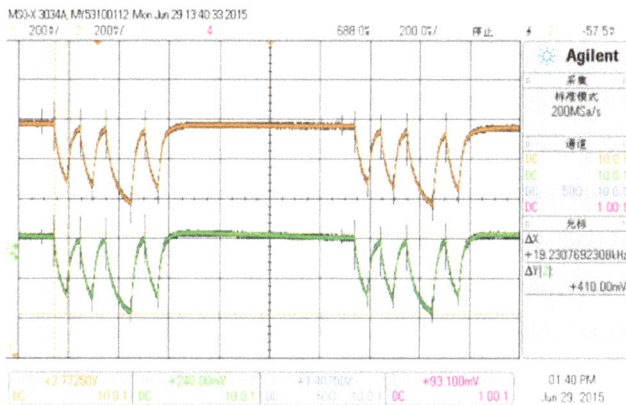

FIGURE 11: Output signals of silicon photocell.

Acknowledgments

This work is supported in part by the National Key Basic Research Program of China (Grant no. 2013CB329204), in part by the National High Technology Research and Development Program of China (Grant no. 2015AA033303), and in part by Science and Technology Planning Project of Guangdong Province, China (Grant no. 2014B010120004).

References

[1] A. Jovicic, J. Li, and T. Richardson, "Visible light communication: opportunities, challenges and the path to market," *IEEE Communications Magazine*, vol. 51, no. 12, pp. 26–32, 2013.

[2] S.-M. Kim, J.-S. Won, and S.-H. Nahm, "Simultaneous reception of solar power and visible light communication using a solar cell," *Optical Engineering*, vol. 53, no. 4, Article ID 046103, 2014.

[3] T.-H. Do and M. Yoo, "Optimization for link quality and power consumption of visible light communication system," *Photonic Network Communications*, vol. 27, no. 3, pp. 99–105, 2014.

[4] E. Koutroulis, K. Kalaitzakis, and N. C. Voulgaris, "Development of a microcontroller-based, photovoltaic maximum power point tracking control system," *IEEE Transactions on Power Electronics*, vol. 16, no. 1, pp. 46–54, 2001.

[5] J. Guo, X. Chen, H. Li, Y. Huang, and H. Chen, "The response properties analysis of silicon photovoltaic cell array in visible light communication system," *Optoelectronics Laser*, vol. 26, no. 3, pp. 475–479, 2015.

[6] K. Lee, H. Park, and J. R. Barry, "Indoor channel characteristics for visible light communications," *IEEE Communications Letters*, vol. 15, no. 2, pp. 217–219, 2011.

3

Impact of Radii Ratios on a Two-Dimensional Cloaking Structure and Corresponding Analysis for Practical Design at Optical Wavelengths

Nadia Anam and Ebad Zahir

American International University-Bangladesh (AIUB), Kemal Ataturk Avenue, Banani, Dhaka 1213, Bangladesh

Correspondence should be addressed to Nadia Anam; neela1303@yahoo.com and Ebad Zahir; ebad.zahir@aiub.edu

Academic Editor: Kin Seng Chiang

This work is an extension to the evaluation and analysis of a two-dimensional cylindrical cloak in the Terahertz or visible range spectrum using Finite Difference Time-Domain (FDTD) method. It was concluded that it is possible to expand the frequency range of a cylindrical cloaking model by careful scaling of the inner and outer radius of the simulation geometry with respect to cell size and/or number of time steps in the simulation grid while maintaining appropriate stability conditions. Analysis in this study is based on a change in the radii ratio, that is, outer radius to inner radius, of the cloaking structure for an array of wavelengths in the visible spectrum. Corresponding outputs show inconsistency in the cloaking pattern with respect to frequency. The inconsistency is further increased as the radii ratio is decreased. The results also help to establish a linear relationship between the transmission coefficient and the real component of refractive index with respect to different radii ratios which may simplify the selection of the material for practical design purposes. Additional performance analysis is carried out such that the dimensions of the cloak are held constant at an average value and the frequency varied to determine how a cloaked object may be perceived by the human eye which considers different wavelengths to be superimposed on each other simultaneously.

1. Introduction

As recent as a decade ago, the idea of making something invisible seemed fitting to the world of fiction only but some revolutionary work in the field of artificially engineered materials called "metamaterials" is gradually bringing this idea into the real world scenario. Metamaterials consist of periodically or randomly structured subunits whose size and separation are much smaller than the wavelength of an electromagnetic field. Consequently, microscopic details of individual structure elements cannot be sensed by the field, but the average of the assembly's collective response matters. The electromagnetic response of this kind of material can be characterized by an effective relative permittivity and permeability. What makes the metamaterials attractive is the fact that the effective permeability can have nonunity and even negative values at the optical wavelengths. In addition, the effective material parameters can be controlled using properly designed structures [1] and suitable materials [2].

The implications of the practical realization of such a system is vast and thus a matter of great interest to researchers. Practical designs [3–5] have already been made and different methods of analysis used for such designs. An increasingly popular numerical method due to its simplicity yet accurate outputs is the Finite Difference Time-Domain method.

2. Materials and Methods

2.1. Metamaterial Cloaking. Cloaking is the ability to make a region of space, and everything in it, invisible to an external observer. A true cloak allows the clear observation of the space behind the cloaked region, and the cloaked region casts no shadow and produces no wavefront changes in the light that has passed through the cloaked region. Cloaking cannot be achieved with materials that exist in nature as they are unable to exhibit negative permittivity and permeability which would lead to a negative index of refraction. Negative

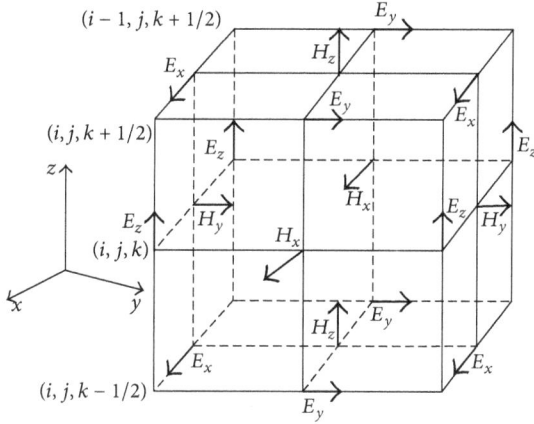

FIGURE 1: Yee's arrangement of field components in a cubic lattice [9].

refractive index is necessary for the way in which light needs to turn around the object that has to be cloaked. A probable solution to this requirement is the use of mematerials. A metamaterial is an artificially structured material which attains its properties from the unit structure rather than the constituent materials. An ordinary material responds to an electric or magnetic field according to the polarization of the atoms and molecules in that material. The structural units of metamaterials can be tailored in shape and size, their composition and morphology can be artificially tuned, and inclusions can be designed and placed in a predetermined manner to achieve prescribed functionalities [3]. In metamaterials, the atoms and molecules are replaced by slightly larger elements which have a physical structure of their own. The response of atoms and molecules is duplicated using tiny circuits [6].

2.2. Finite Difference Time-Domain Cloaking Using a Cylindrical Structure.

The Finite-Difference Time-Domain (FDTD) method achieved discretization of Maxwell's equations in the space and time dimensions. The method is able to solve Maxwell's equations in the time domain for complex structures and geometries [7, 8]. Simulation of electromagnetic waves is made possible by creating a cubic lattice and assigning a staggered arrangement of E and H components to the nodes such as the one shown in Figure 1.

The construction, operation, and characteristics of a 2D cylindrical cloaking structure have been extensively studied in [2, 10–12]. For such a cloaking structure in the Transverse Electric (TE) mode, the customized equations for E and H are [12]

$$E_x^{n+1} = a_e \left(D_x^{n+1} - 2D_x^n + D_x^{n-1} \right) + b_e \left(D_x^{n+1} - D_x^{n-1} \right) \\ + c_e \left(2E_x^n - E_x^{n-1} \right) + d_e \left(2E_x^n + E_x^{n-1} \right) + e_e E_x^{n-1}, \quad (1)$$

where

$$a_e = \frac{4}{g},$$

$$b_e = \frac{\gamma (2\Delta t)}{g},$$

$$c_e = \frac{4\varepsilon_0 \varepsilon_\infty}{g},$$

$$d_e = \frac{\varepsilon_0 \omega_p^2 (\Delta t)^2}{g},$$

$$e_e = \frac{\varepsilon_0 \varepsilon_\infty \gamma_e (2\Delta t)}{g},$$

$$g = 4\varepsilon_0 \varepsilon + \varepsilon_0 \omega_p^2 (\Delta t)^2 + \varepsilon_0 \varepsilon \gamma (2\Delta t); \quad (2)$$

here, D is the electric flux density, ω_p is the plasma frequency, γ is the collision frequency, and Δt is the size of each time step in the FDTD grid.

Similarly, the magnetic field strength, H can be calculated using

$$H_y^{n+1} = a_m \left(B_x^{n+1} - 2B_y^n + B_y^{n-1} \right) + b_m \left(B_y^{n+1} - B_y^{n-1} \right) \\ + c_m \left(2H_y^n - H_y^{n-1} \right) + d_m \left(2H_y^n + H_y^{n-1} \right) \quad (3) \\ + e_m H_y^{n-1}.$$

The FDTD model in [13, 14] uses the TM mode for cloaking. The cloaking parameters are therefore described by

$$\mu_r (r) = \frac{r - r_a}{r},$$

$$\mu_\varphi (r) = \frac{r}{r - r_a}, \quad (4)$$

$$\varepsilon_z (r) = \left(\frac{r_b}{r_b - r_a} \right)^2 \frac{r - r_a}{r},$$

where r_a and r_b are the inner and outer radius, respectively, of the cylindrical cloak and r is an arbitrary radius within these two boundary values.

E_z, H_x, and H_y can be calculated from corresponding flux densities equations:

$$D_z^{n+1} [i, j] = D_z^n [i, j] + \frac{\Delta t}{\Delta} \left(H_y^{n+1/2} \left[i + \frac{1}{2}, j \right] \right. \\ - H_y^{n+1/2} \left[i - \frac{1}{2}, j \right] - H_x^{n+1/2} \left[i, j + \frac{1}{2} \right] \\ \left. + H_x^{n+1/2} \left[i, j - \frac{1}{2} \right] \right),$$

$$B_x^{n+1/2} \left[i, j + \frac{1}{2} \right] = B_x^{n-1/2} \left[i, j + \frac{1}{2} \right] \quad (5) \\ + \frac{\Delta t}{\Delta} \left(-E_z^n [i, j + 1] + E_z^n [i, j] \right),$$

$$B_y^{n+1/2} \left[i + \frac{1}{2}, j \right] = B_x^{n-1/2} \left[i + \frac{1}{2}, j \right] + \frac{\Delta t}{\Delta} \left(E_z^n [i + 1, j] \right. \\ \left. - E_z^n [i, j] \right).$$

The index of refraction was calculated from

$$n_{\text{FDTD}} = \frac{1}{jk_0 (z_1 - z_2)} \log \left| \frac{E_x (\omega, z_2)}{E_x (\omega, z_1)} \right|, \quad (6)$$

TABLE 1: Parametric variations considered for the study of the FDTD cloaking model.

r_a and r_b	Frequency	Δ	Δt	I and J	S_c	PML
Varied	Fixed	Fixed	Fixed	Fixed	Fixed	Fixed
Fixed	Fixed	Fixed	Varied	Fixed	Varied	Fixed
Fixed	Varied	Varied	Varied	Fixed	Fixed	Fixed
Varied	Varied	Varied	Varied	Fixed	Fixed	Fixed
Varied	Varied	Varied	Varied	Varied	Fixed	Fixed

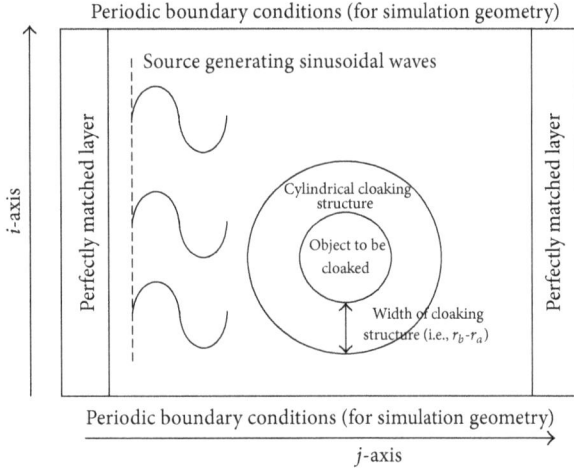

FIGURE 2: Simulation geometry for FDTD method of cloaking [10, 11].

where ω is the angular frequency, k_0 is the wave number set as ω_0/c (ω_0 is the angular frequency of the sinusoidal source wave), and the fields were recorded at locations $z_1 = 1415\Delta z$ and $z_2 = 1424\Delta z$ (Δz is the spatial step).

In order to expand the range of the FDTD cloaking model from a few Gigahertz to hundreds of Terahertz, it was essential to understand the scope of the simulation parameters. The inner and outer radii of the structure r_a and r_b, cell size in the x and y directions Δ, the temporal steps Δt, the spatial steps in the i-j axis I and J, the courant stability S_c, and the perfectly matched layer width PML were some of the prime parameters taken into consideration and several test analyses were performed. A summary is presented in Table 1 and the simulation geometry is shown in Figure 2. Due to reflection restraints from the boundaries, the PML layer(s) needed to be kept constant.

The equation that governs the stability of the FDTD system is

$$S_c = u\frac{\Delta t}{\Delta}, \tag{7}$$

where u is the speed of light in any medium and is taken to be equal to c ($= 2.998 \times 10^8 \text{ m/s}^2$) since light is considered to be propagating through air medium. Altering the stability suggests that the closest to ideal cloaking can be achieved if the stability value is the lowest possible.

However, maintaining a very low value would make the design excessively stringent due to extremely small time step requirements for high frequency values. At the same time, cloaking tends to become more unreliable with higher stability number so it is proposed that the number be not allowed to exceed 0.5 [15]. For this study it is considered to be 0.25. The ratio of the outer radius to inner radius (r_b/r_a) has been varied in the same ratio as the cell size and/or the number of time steps taking into account that the structural units of the metamaterial must be substantially smaller than the wavelength being considered [3]. The cell size Δ can be related to the wavelength and hence frequency by

$$\Delta = \frac{\lambda}{50}. \tag{8}$$

The equation is formed by implying the relevant relationships and test conditions suggested in [10, 12] and modifying it to match the default frequency and cell size values used in [13, 14]. Thus at optical frequencies, the cell size is reduced and the time steps increased accordingly to hold the stability constant at the preconsidered value. Using (8), the stability relation in (7) may be redefined as

$$S_c = 50 \times f \times \Delta t, \tag{9}$$

where f is the frequency of light waves.

It is also suggested in [12] that the ratio of r_b/r_a should be maintained at 2 to obtain reliable cloaking results. To reduce design costs, however, it would be beneficial to set the ratio as close to 1 as possible. The first part of the study here incorporates this idea; three different radii ratios 1.5, 1.75 and 2 are considered for different optical wavelengths in the visible spectrum to observe the level of cloaking for specific colors of light and whether the radii ratio has any impact on the process. For compatibility of the design with our visual capability where the human eye perceives different colors of light simultaneously, the size and hence dimensions of the cloaked object should be constant and not varied with wavelength. This leads to the latter part of the research, where the inner and outer radius of the cloak are fixed and the frequency varied over the entire visible spectrum. To preserve simulation geometry constraints, however, the cell size still needs to be varied with frequency and this in turn requires the number of time steps to be varied so that stability is maintained at the desired value.

The plots in Figures 3–5 depict the changes in the transmitted sinusoidal waveform when no object is present in its path and it has a frequency of 2 GHz, when a cylindrical cloak of $r_a = 0.1$ m and $r_b = 0.2$ m is situated in the centre of the simulation geometry with the wave frequency still 2 GHz and when the propagating wave has a frequency of 545 THz with $r_a = 0.367$ μm and $r_b = 0.733$ μm. It is worth noting in Figure 5 that, due to the increase in frequency, the j-axis limits have been doubled for complete observation of the propagating wave from crest to trough. The cloak is also considered to be "lossless" suggesting that there is no degradation in the intensity of the reflected light compared to the incident one.

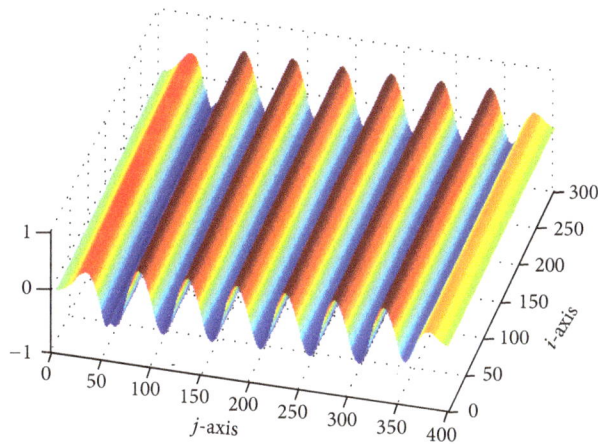

FIGURE 3: Three-dimensional field distribution plot for a 2 GHz sine wave with no object in its path.

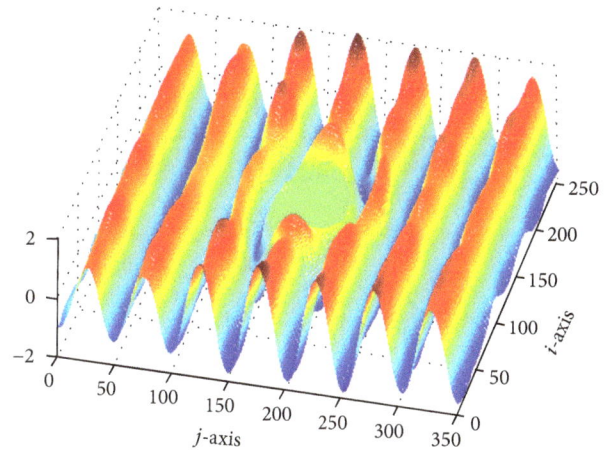

FIGURE 5: Three-dimensional field distribution plot for a 545 THz sine wave with a steady lossless cylindrical cloak in its path and appropriately scaled parameters.

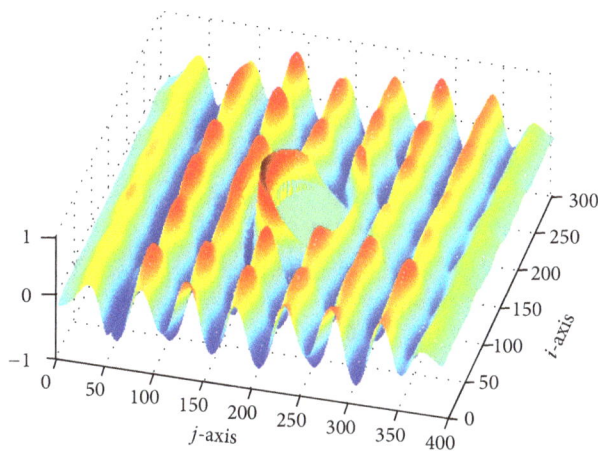

FIGURE 4: Three-dimensional field distribution plot for a 2 GHz sine wave with a steady lossless cylindrical cloak in its path.

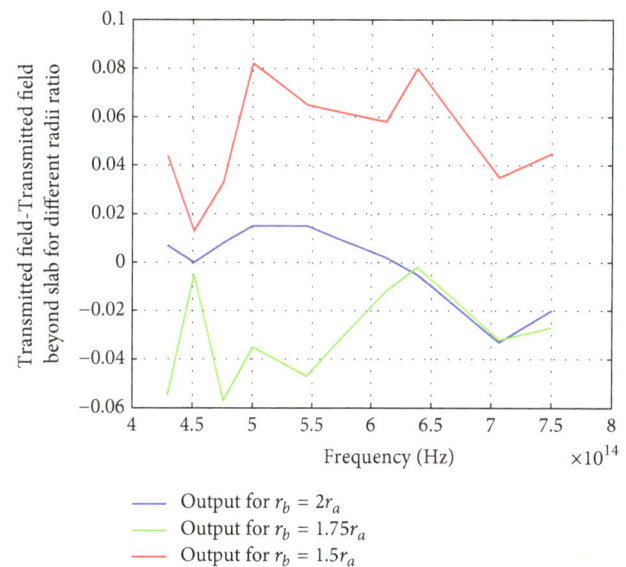

FIGURE 6: Plots for transmitted field-transmitted field beyond slab versus frequency for different radii ratio.

3. Results and Discussion

The model in [13] provides the waveforms of transmitted field, transmitted field beyond the cylindrical slab, that is, the cloaked object, transmission coefficient, reflection coefficient (calculated by subtracting transmission coefficient from unity), and the real and imaginary components of the refractive index. The transmitted fields are determined by averaging the waveforms at various time steps. The transmission and reflection coefficients and refractive indices are calculated at specific optical wavelengths or colors of light as shown in the tables. Tables 2–4 are constructed using three different radii ratios (r_b/r_a) 2, 1.75, and 1.5, respectively. Outputs at a ratio lower than 1.5 are fairly unreliable so they are excluded from this study. Table 5 uses a fixed radius ratio of 2 as well as fixed average radius values ($r_a = 0.365 \, \mu m$ and $r_b = 0.730 \, \mu m$) for various wavelengths.

The outcomes of Tables 2–4 are summarized in Figures 6–10. Figure 6 displays the difference between the transmitted field prior to its entrance into the cylindrical slab and that beyond it with respect to frequency. For ideal cloaking, this

value should be as close to 0 as possible. For all three radii ratio, red color is cloaked most accurately, with the highest accuracy for $r_b = 2r_a$. This high accuracy is obtained again at a wavelength near to aqua but only for a radius ratio of 2. At this same frequency, the accuracy deviates almost 1 percent for $r_b = 1.75r_a$ and almost 6 percent for $r_b = 1.5r_a$. For $r_b = 2r_a$, the maximum deviation occurs around indigo light (3.5 percent approximately); for $r_b = 1.75r_a$, maximum inaccuracy is obtained at orange light (5.8 percent approximately) and for $r_b = 1.5r_a$; this point occurs at yellow light (8.1 percent approximately).

Figure 7 plots transmission coefficients with respect to different optical frequencies. The ideal value is 1 and the behavioral pattern obtained in Figure 6 follows with a few discrepancies, significant of which is the sharp plunge in value to almost 0.68 for blue light at $r_b = 1.5r_a$. One reason may

TABLE 2: Variations in different parameters of an optical cloaking system for a radius ratio of 2.

Color of light	Wavelength (nm)	Frequency (THz)	Transmitted field	Transmitted field beyond slab	Transmission coefficient	Reflection coefficient	Refractive index (real)	Refractive index (imaginary)
Dark red	700	429	0.712	0.705	1.04	−0.04	1.9	0.5
Red	665	451	0.698	0.698	1	0	1.8	0.5
Orange	630	476	0.703	0.695	1.07	−0.07	1.95	0.56
Yellow	600	500	0.713	0.698	1.02	−0.02	1.85	0.5
Green	550	545	0.73	0.715	1.08	−0.08	1.97	0.58
Aqua	490	612	0.705	0.703	1	0	1.8	0.45
Blue	470	638	0.723	0.728	1.04	−0.04	1.9	0.51
Indigo	425	706	0.675	0.708	1.06	−0.06	1.95	0.54
Violet	400	750	0.72	0.74	1.03	−0.03	1.85	0.5

TABLE 3: Variations in different parameters of an optical cloaking system for a radius ratio of 1.75.

Color of light	Wavelength (nm)	Frequency (THz)	Transmitted field	Transmitted field beyond slab	Transmission coefficient	Reflection coefficient	Refractive index (real)	Refractive index (imaginary)
Dark red	700	429	0.725	0.78	1.1	−0.1	2.55	0.36
Red	665	451	0.758	0.763	1.05	−0.05	2.5	0.4
Orange	630	476	0.728	0.785	1.15	−0.15	2.7	0.38
Yellow	600	500	0.735	0.77	1.05	−0.05	2.5	0.38
Green	550	545	0.743	0.79	1.13	−0.13	2.65	0.38
Aqua	490	612	0.738	0.75	1.01	−0.01	2.4	0.38
Blue	470	638	0.743	0.745	1.1	−0.1	2.6	0.39
Indigo	425	706	0.733	0.765	1.14	−0.14	2.64	0.4
Violet	400	750	0.713	0.74	1.05	−0.05	2.5	0.38

TABLE 4: Variations in different parameters of an optical cloaking system for a radius ratio of 1.5.

Color of light	Wavelength (nm)	Frequency (THz)	Transmitted field	Transmitted field beyond slab	Transmission coefficient	Reflection coefficient	Refractive index (real)	Refractive index (imaginary)
Dark red	700	429	0.762	0.718	1.11	−0.11	3.14	0.08
Red	665	451	0.753	0.74	1.04	−0.05	3.1	0.1
Orange	630	476	0.753	0.72	1.15	−0.15	3.15	0
Yellow	600	500	0.765	0.683	1.08	−0.08	3.14	0.03
Green	550	545	0.778	0.713	1.15	−0.15	3.2	0
Aqua	490	612	0.753	0.695	1.04	−0.04	3.1	0.13
Blue	470	638	0.78	0.7	0.67	0.33	2.1	0.58
Indigo	425	706	0.755	0.72	1.14	−0.14	3.15	0
Violet	400	750	0.783	0.738	1.13	−0.13	3	0.15

be the dependency of the metamaterial structure and hence its response on the operational frequency [16]. The reduction in the thickness of the metamaterial cloak impacts its overall response and causes incongruous behavior at blue light. The best results overall remain the same as those depicted by Figure 6; that is, $r_b = 2r_a$ shows minimum deviations.

Figure 8 shows reflection coefficient values that have been directly calculated from the relationship $T = 1 − R$ where T = transmission coefficient and R = reflection coefficient.

The plots in Figures 9 and 10 for refractive index help to determine which materials or metamaterial structures would be suitable for the practical design; the refractive

TABLE 5: Variations in different parameters of an optical cloaking system for a radius ratio of 2 with fixed inner ($r_a = 0.365\,\mu$m) and outer ($r_b = 0.730\,\mu$m) radius.

Color of light	Wavelength (nm)	Frequency (THz)	Transmitted field	Transmitted field beyond slab	Transmission coefficient	Reflection coefficient	Refractive index (real)	Refractive index (imaginary)
Dark red	700	429	0.72	0.77	1.07	-0.07	1.87	-0.25
Red	665	451	0.738	0.784	1.04	-0.04	2.13	-0.33
Orange	630	476	0.736	0.778	1.04	-0.04	2.33	-0.18
Yellow	600	500	0.724	0.744	1.01	-0.01	2.42	0.12
Green	550	545	0.688	0.71	1.02	-0.02	1.9	0.48
Aqua	490	612	0.772	0.826	1.1	-0.1	1.37	-0.13
Blue	470	638	0.77	0.836	1.07	-0.07	1.38	-0.41
Indigo	425	706	0.782	0.86	1.07	-0.07	1.46	1.87
Violet	400	750	0.8	0.69	1.1	-0.1	0	—

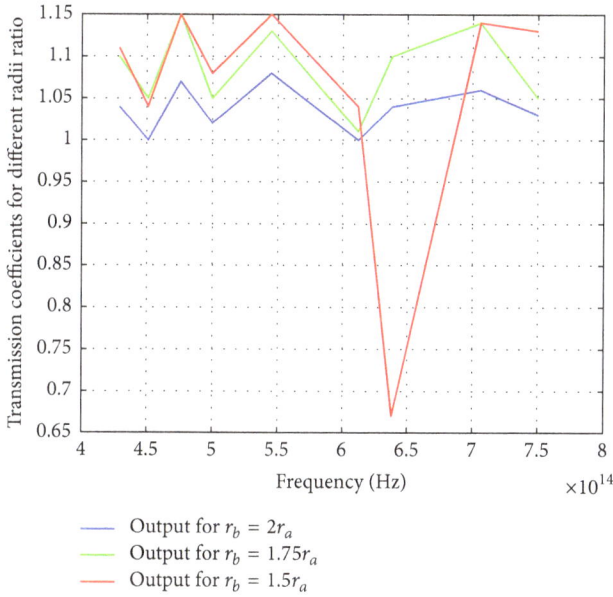

FIGURE 7: Plots for transmission coefficient versus frequency for different radii ratio.

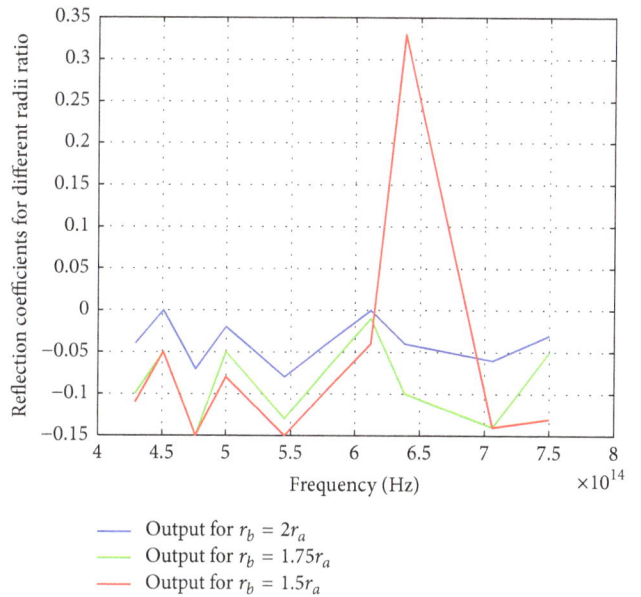

FIGURE 8: Plots for reflection coefficient versus frequency for different radii ratio.

index is a complex value (with real and imaginary components) as the structure requires negative values of permittivity and permeability at certain angles of reflection for cloaking. As expected, the refractive indices for each radius ratio are different because light's path of curvature would vary for individual cloak size. These also suggest that the practical design of an optical cloak for a radius ratio of either 2 or 1.75 would be simpler compared to that for 1.5.

The deviations in refractive index with respect to frequency are lower so easier to incorporate into the metamaterial structure. With a radius ratio of 1.5 however, the necessity to overcome the anomaly existing from aqua to indigo light would make the design quite complex.

Figures 7 and 9 illustrate an additional important aspect, an almost linear relationship between transmission coefficient and real part of refractive index; a clearer presentation of this pattern is demonstrated in Figures 11–13.

This behavior can be mathematically approximated as

$$T \cong 1.8\mathrm{Re}_{\mathrm{real}} \quad \text{(for a radius ratio of 2)},$$

$$T \cong 2.3\mathrm{Re}_{\mathrm{real}} \quad \text{(for a radius ratio of 1.75)} \quad (10)$$

$$T \cong 2.8\mathrm{Re}_{\mathrm{real}} \quad \text{(for a radius ratio of 1.5)}.$$

Otherwise, it is correlated as

$$\frac{T_1}{\mathrm{Re}_1} \times \frac{r_{b1}}{r_{a1}} \times k_1 = \frac{T_2}{\mathrm{Re}_2} \times \frac{r_{b2}}{r_{a2}}, \quad (11)$$

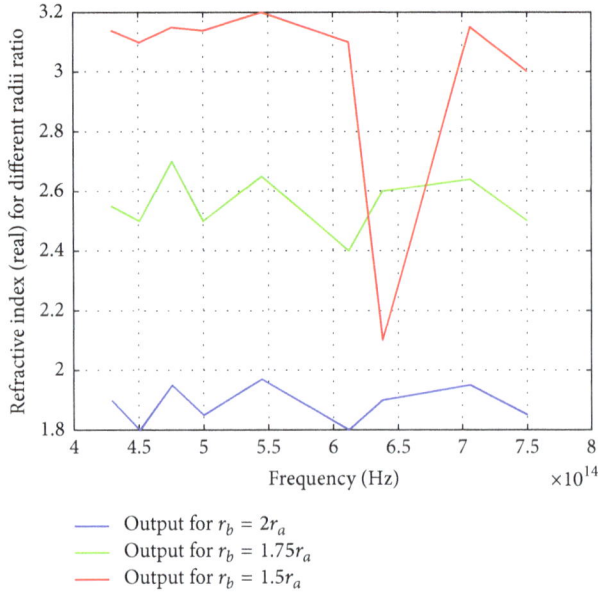

FIGURE 9: Plots for the real component of refractive index versus frequency for different radii ratio.

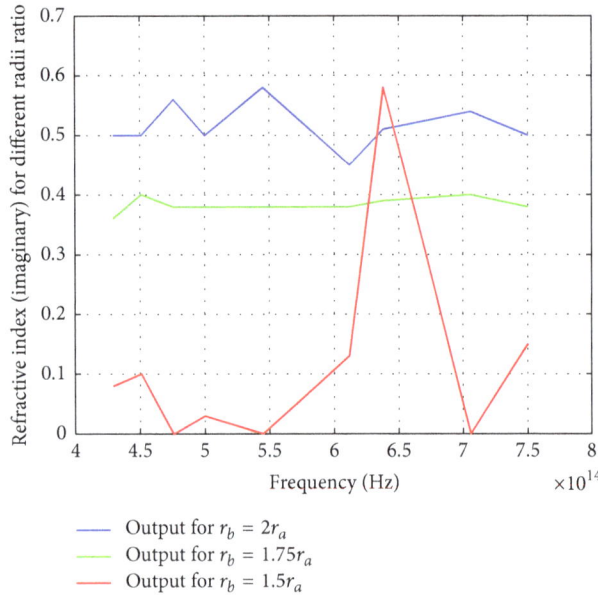

FIGURE 10: Plots for the imaginary component of refractive index versus frequency for different radii ratio.

or

$$\frac{T_2}{\text{Re}_2} \times \frac{r_{b2}}{r_{a2}} \times k_2 = \frac{T_3}{\text{Re}_3} \times \frac{r_{b3}}{r_{a3}}, \qquad (12)$$

where k_1 and k_2 are constants of proportionality. The development of such equations could simplify the choice of the metamaterial structure and composition for the actual design since only the awareness of the radius ratio would be required (and transmission coefficient can be considered to be the ideal value of 1).

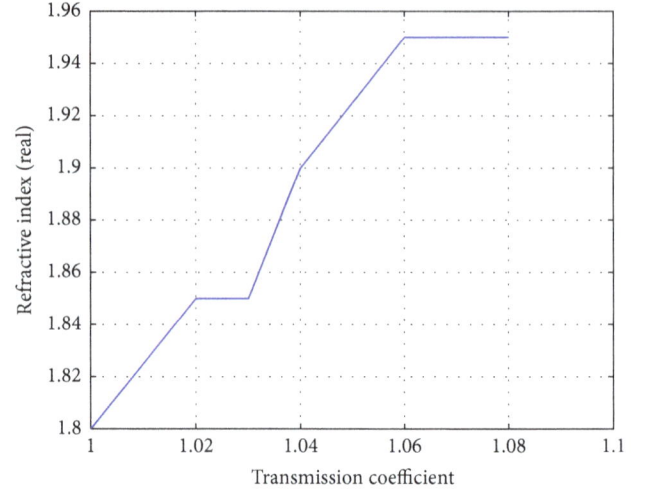

FIGURE 11: Plot depicting an almost linear relationship between the real component of refractive index and transmission coefficient for a radius ratio of 2.

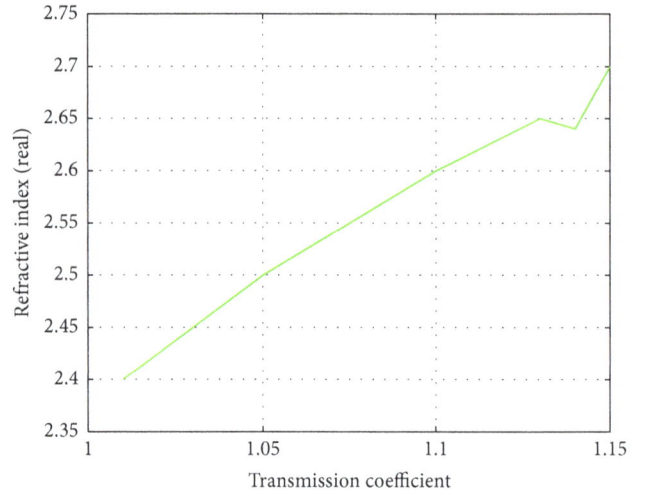

FIGURE 12: Plot depicting an almost linear relationship between the real component of refractive index and transmission coefficient for a radius ratio of 1.75.

If an object has to be completely invisible to the human eye, that is, perfect cloaked, the cloaking structure must not only allow the light to pass through it undisturbed but also ensure that it reflects back to the human eye coherently from what object is present beyond this setup. The incident field, transmission field, transmission field beyond slab, transmission and reflection coefficients, and refractive index versus frequency for red, blue, and green light are plotted simultaneously (Figures 14–18) to obtain an understanding of this matter. The radius needs to be held constant for practical purposes and is chosen as an averaged value of those considered in the preceding part of this study for the entire visible spectrum. The output waveforms for the three basic colors appear quite coherent, particularly for the refractive index values, suggesting easier material selection for designing purposes with an expectation of consistent

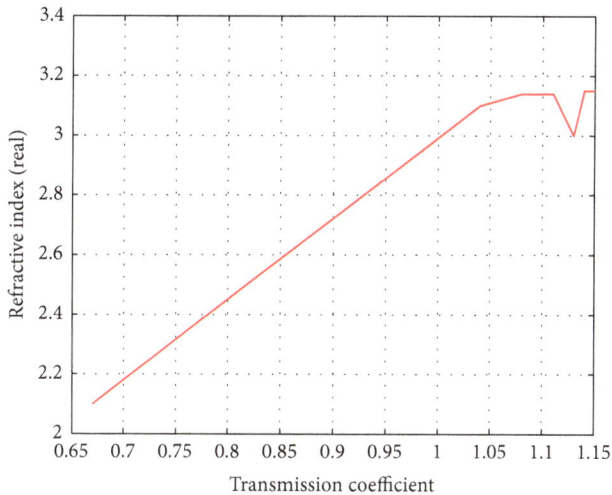

FIGURE 13: Plot depicting an almost linear relationship between the real component of refractive index and transmission coefficient for a radius ratio of 1.5.

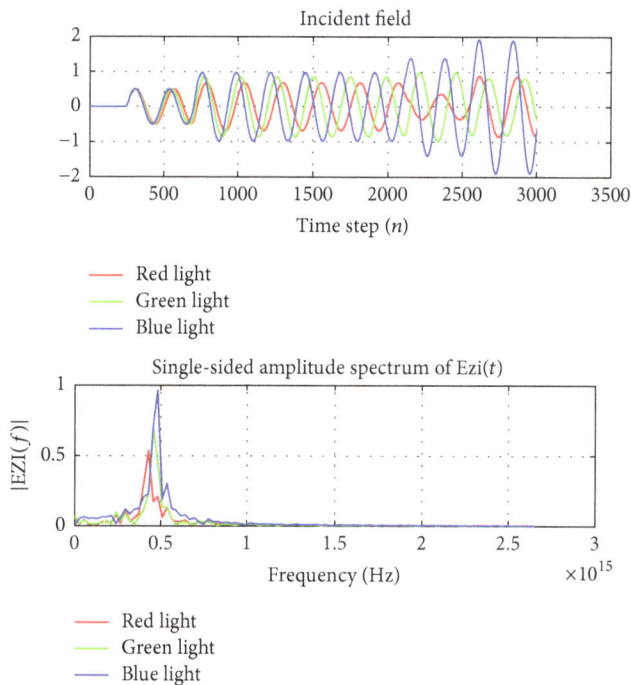

FIGURE 14: Incident field and equivalent amplitude spectrum versus frequency for red, green, and blue light corresponding to conditions in Table 5.

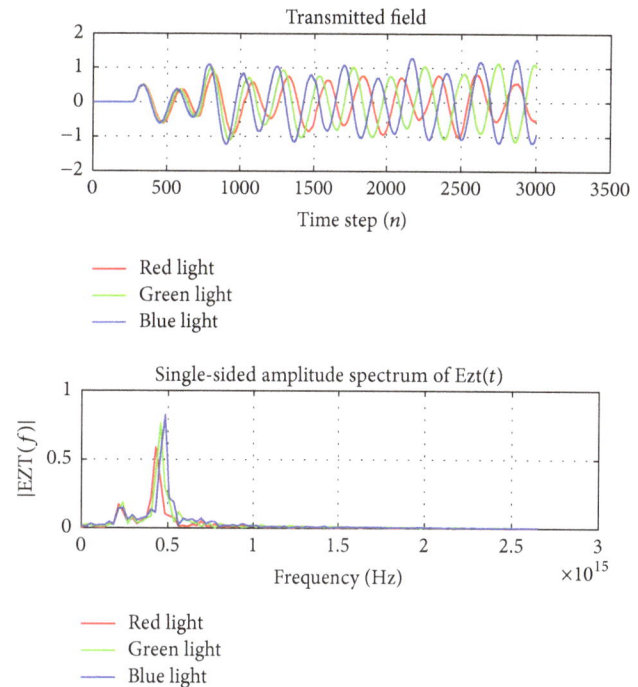

FIGURE 15: Field transmitted into the cloaking slab and equivalent amplitude spectrum versus frequency for red, green, and blue light corresponding to conditions in Table 5.

superposition of all wavelengths. The additional frequency spectrum plots in Figures 14–16 also infer that the structure remains fairly "lossless" and hence the modifications that have been made to the model for optical frequency operations are satisfactory.

Plotting further wavelengths of light on these same axes would make it extremely difficult to differentiate each output. Instead, Table 5 and corresponding Figures 19–21 summarize the overall findings.

Table 5 and Figure 10 suggest that while it may be possible to observe almost every color of light in the visible spectrum with minor inaccuracy (~2 to 7 percent) simultaneously, the overall cloaking quality begins to deteriorate from aqua onwards (inaccuracies rising to 10 percent at maximum). Plots in Figures 12 and 13 present another practical design hurdle; since the real part of refractive index is zero and the imaginary part unobtainable for violet light (shown with arrows), the actual cloaking structure will not be able to reflect violet and other optical wavelengths beyond it unless the radii values are altered.

Due to the frequency independent nature of the coordinate transformation functions used in the FDTD model, the range of the operating frequency can be easily expanded from Gigahertz [12] to Terahertz [10] ranges. Also, altering shapes [17] is quite easily achievable. Such factors are influencing the development of numerous cloaking models based on the FDTD method. However, the performance of such structures can be severely affected if they are not operated at the designed frequency [18]. An ideal cylindrical cloak such as the one in [11] has been proved to work properly for monochromatic incident wave, but when excited with nonmonochromatic radiation, they become nonpractical. Also, losses tend to become significant under such circumstances. The findings in this study are consistent with previous researches and show how the performance of the FDTD model is affected if cloaking parameters are not simultaneously altered with frequencies changes. Since losses are incurred, the object will not be "perfectly" cloaked and

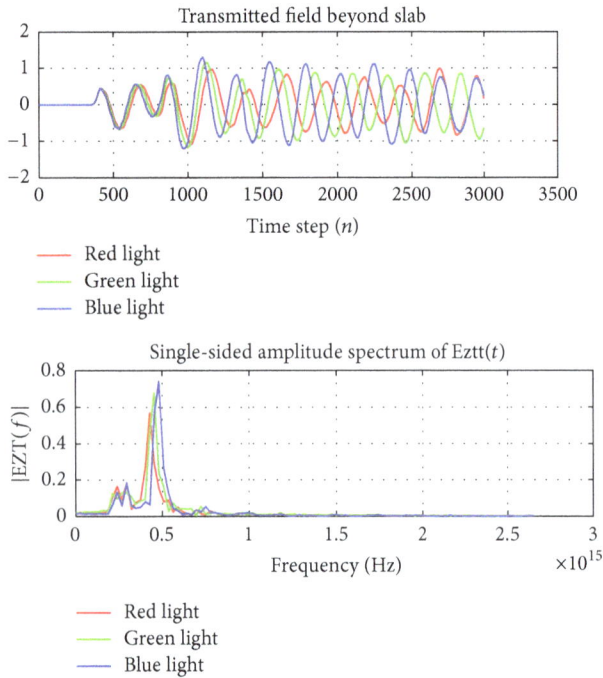

FIGURE 16: Transmitted field beyond the cloaking slab and equivalent amplitude spectrum versus frequency for red, green, and blue light corresponding to conditions in Table 5.

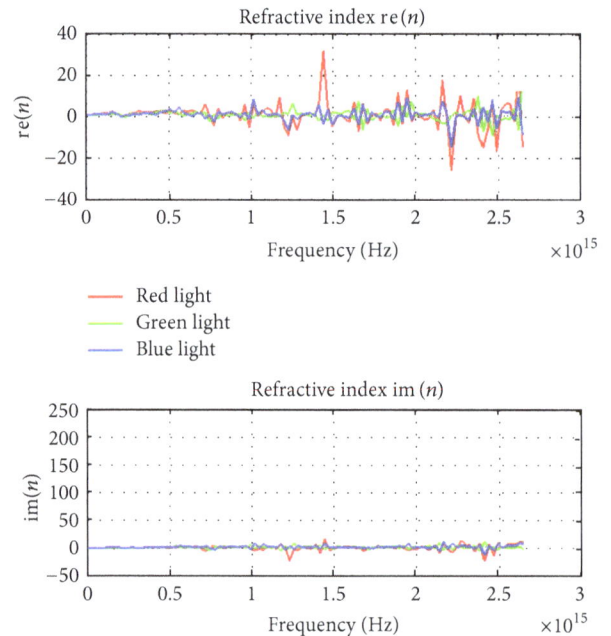

FIGURE 17: Transmission and reflection coefficient versus frequency for red, green, and blue light corresponding to conditions in Table 5.

there will be a shadowing effect; that is, the cloak will appear somewhat gray instead of being fully transparent [11, 19].

4. Conclusion

A previous study of a two-dimensional cylindrical cloak model in the optical frequency or visible spectrum region has been extended here for different radii ratios. The model uses a Finite Difference Time-Domain Method of cloaking and the behavioral transformation for different inner to outer radii ratios in parameters such as transmitted field before and beyond the cloak, transmission and reflection coefficients, and refractive index with respect to different optical wavelengths is studied. Outcomes suggest that the "cloaking" quality is influenced by not just wavelength values but also the radii ratios. If accuracy is most necessary, it is best to use a radius ratio of 2. However, if design costs and thickness are significant, the ratio can be lowered to 1.75 with minor deterioration in output quality. The results also portray an almost linear relationship between the transmission coefficient and the real component of refractive index with respect to radii ratios, implying that the type of material required for cloaking a specific wavelength of light may be easily found if only the thickness of the cloaking structure is known.

In further investigations, the size of the cloak is kept constant and the wavelength varied to obtain an idea of how light would reflect from an entity beyond the cloaked object and appear to our eyes. Results suggest incoherency in cloaking with an acceptable inaccuracy of around 2 to 10 percent. Moreover, the averaging of output values for three different wavelengths (red, blue, and green) of light

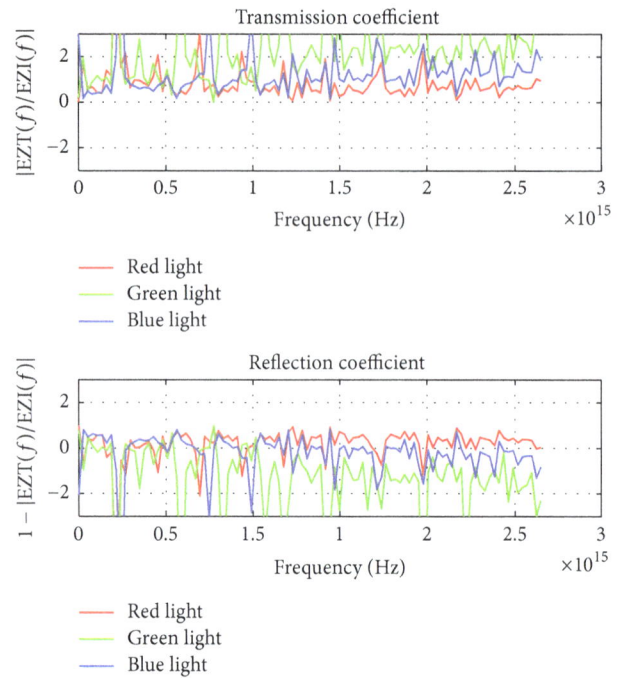

FIGURE 18: Refractive index (real and imaginary) versus frequency for red, green, and blue light corresponding to conditions in Table 5.

suggests that these results may be considered acceptable for the real-life scenario where different wavelengths in the visible spectrum exist in a superimposed form. An important issue, however, is that no outputs can be achieved for violet light and beyond. Thus an entity beyond the cloaked object will not be able to reflect violet light with the average radius

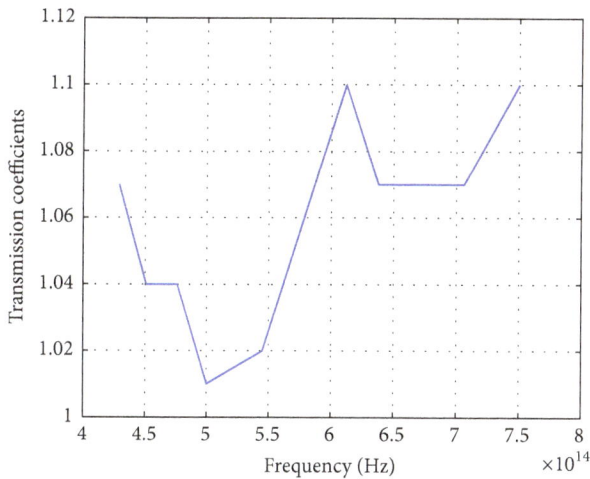

FIGURE 19: Variations in transmission coefficient with frequency for a radius ratio of 2 with fixed inner and outer radius values.

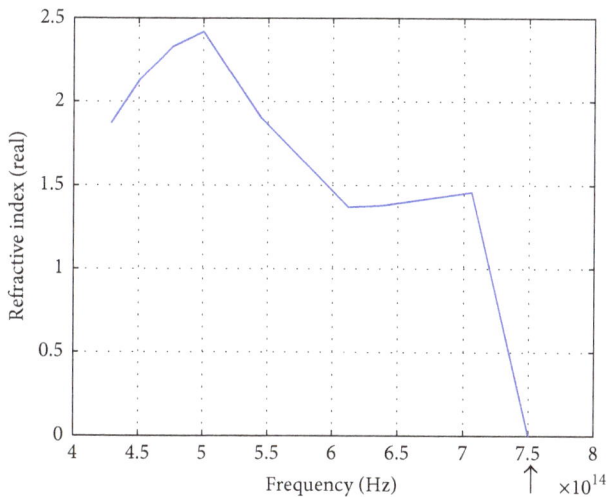

FIGURE 21: Plot showing how the refractive index (imaginary) changes with frequency for a radius ratio of 2 and fixed inner and outer radius value (arrow shows no further values can be achieved beyond a frequency of 7.1 THz).

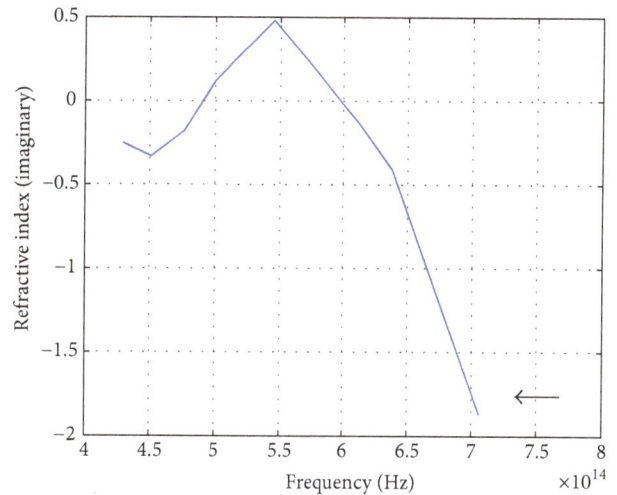

FIGURE 20: Plot showing how the refractive index (real) changes with frequency for a radius ratio of 2 and fixed inner and outer radius value (arrow shows the value reaches zero at a frequency of 7.5 THz).

value used here. Also, the size of this cloak is in the micronano range; for this to increase to a considerable dimension, the parameters, boundary conditions, and mathematical equations that govern this two-dimensional model must be altered in such a way that the output is still stable and reliable while the frequency is sustained in the optical range. These restraints could be overcome if the cloak is constructed from suitable "active" metamaterials.

Competing Interests

The authors declare that there is no conflict of interests regarding the publication of this paper.

References

[1] T. Hakkarainen, *Electromagnetic nanophotonics: superlens imaging of dipolar emitters and cloaking in weak scattering [Ph.D. thesis]*, Department of Applied Physics, Aalto University, Espoo, Finland, 2012.

[2] E. Kallos, C. Argyropoulos, Y. Hao, and A. Alù, "Comparison of frequency responses of cloaking devices under non-monochromatic illumination," *Physical Review B*, vol. 84, no. 4, Article ID 045102, 2011.

[3] W. Cai and V. Shalaev, *Optical Metamaterials Fundamentals and Applications*, Springer, New York, NY, USA, 2010.

[4] T. J. Cui, D. R. Smith, and R. Liu, *Metamaterials Theory, Design, and Applications*, Springer, New York, NY, USA, 2010.

[5] L. Raffensperger, "Ancient Romans' Color-Changing Goblet Was Feat of Nanotechnology," 2013, http://blogs.discovermagazine.com/d-brief/2013/08/29/ancient-romans-color-changing-goblet-was-feat-of-nanotechnology/#.WGzjh0-LXnM.

[6] J. Pendry, *Metamaterials and the Science of Invisibility*, 2014, https://www.youtube.com/watch?v=5ZOV_9Jirp0.

[7] R. A. Shelby, D. R. Smith, and S. Schultz, "Experimental verification of a negative index of refraction," *Science*, vol. 292, no. 5514, pp. 77–79, 2001.

[8] Y. Hao and R. Mittra, *FDTD Modeling of Metamaterials: Theory and Applications*, Artech House, Norwood, Mass, USA, 1st edition, 2008.

[9] M. Sipos, *Optics and cloaking in FDTD [B.Sc. thesis]*, Department of Physics, Ithaca College, 2008.

[10] C. Argyropoulos, E. Kallos, Y. Zhao, and Y. Hao, "Manipulating the loss in electromagnetic cloaks for perfect wave absorption," *Optics Express*, vol. 17, no. 10, pp. 8467–8475, 2009.

[11] C. Argyropoulos, E. Kallos, and Y. Hao, "Dispersive cylindrical cloaks under nonmonochromatic illumination," *Physical Review E*, vol. 81, no. 1, Article ID 016611, 2010.

[12] Y. Zhao, C. Argyropoulos, and Y. Hao, "Full-wave finite-difference time-domain simulation of electromagnetic cloaking structures," *Optics Express*, vol. 16, no. 9, pp. 6717–6730, 2008.

[13] A. Dawood, "Finite difference time-domain modelling of meta-materials: GPU implementation of cylindrical cloak," *Advanced Electromagnetics*, vol. 2, no. 2, pp. 10–17, 2013.

[14] N. Anam and E. Zahir, "Analysis of FDTD cloaking in the visible frequency spectrum," in *Proceedings of the 15th International Conference on Numerical Simulation of Optoelectronic Devices (NUSOD '15)*, pp. 65–66, IEEE, Taipei, Taiwan, September 2015.

[15] N. Anam and E. Zahir, "Courant stability number impact on high frequency EM cloaking using FDTD analysis," *International Journal of Research in Computer Engineering & Electronics (IJRCEE)*, vol. 3, no. 3, 2014.

[16] S. Zhang, W. Fan, K. J. Malloy, S. R. Brueck, N. C. Panoiu, and R. M. Osgood, "Demonstration of metal-dielectric negative-index metamaterials with improved performance at optical frequencies," *Journal of the Optical Society of America B*, vol. 23, no. 3, pp. 434–438, 2006.

[17] N. Okada and J. B. Cole, "FDTD modeling of a cloak with a nondiagonal permittivity tensor," *ISRN Optics*, vol. 2012, Article ID 536209, 7 pages, 2012.

[18] J. A. Silva-Macêdo, M. A. Romero, and B.-H. V. Borges, "An extended FDTD method for the analysis of electromagnetic field rotations and cloaking devices," *Progress in Electromagnetics Research*, vol. 87, pp. 183–196, 2008.

[19] A. A. Maradudin, *Structured Surfaces as Optical Metamaterials*, Cambridge University Press, 2011.

4

Observation of Dissipative Bright Soliton and Dark Soliton in an All-Normal Dispersion Fiber Laser

Chunyang Ma,[1] **Bo Gao,**[2] **Ge Wu,**[1] **Tian Zhang,**[1] **and Xiaojian Tian**[1]

[1]*College of Electronic Science and Engineering, Jilin University, Changchun 130012, China*
[2]*College of Communication Engineering, Jilin University, Changchun 130012, China*

Correspondence should be addressed to Xiaojian Tian; tianxj@jlu.edu.cn

Academic Editor: Xiaohui Li

This paper proposes a novel way for controlling the generation of the dissipative bright soliton and dark soliton operation of lasers. We observe the generation of dissipative bright and dark soliton in an all-normal dispersion fiber laser by employing the nonlinear polarization rotation (NPR) technique. Through adjusting the angle of the polarizer and analyzer, the mode-locked and non-mode-locked regions can be obtained in different polarization directions. Numerical simulation shows that, in an appropriate pump power range, the dissipative bright soliton and dark soliton can be generated simultaneously in the mode-locked and non-mode-locked regions, respectively. If the pump power exceeds the top limit of this range, only dissipative soliton will exist, whereas if it is below the lower bound of this range, only dark soliton will exist.

1. Introduction

With the rapid development of ultrafast optical field, a great many researchers have been focusing on mode-locked fiber laser recently [1–7]. Passive mode-locked fiber lasers have shown many advantages over solid-state systems such as the compact design, low-cost, and stability [8–11]. It has also some potential applications, like laser processing, optical communications, medical equipment, military, and so on [12–14]. Ultrashort pulses are stabilized in an oscillator when the effects of optical nonlinearity are exactly balanced by other processes after one cycle around the cavity. The most widely used method to compensate nonlinearity is group velocity dispersion (GVD). The traditional soliton is formed by the balance between nonlinear and negative dispersion phase changes. The resulting soliton propagates indefinitely without change. However, the traditional soliton is limited to 0.1 nJ of the energy in standard fibers [15]. When the GVD reaches zero, the dispersion managed soliton has been formed, which enables emitting the pulses with the pulse energy reaching the 1 nJ level [16]. Instead of conventional soliton and dispersion managed soliton, self-similar soliton can tolerate strong nonlinearity without wave breaking. But due to the restricted gain bandwidth, self-similar soliton allows the energy to reach the 10 nJ level [17]. Dissipative soliton (DS) has attracted great interest in the development of fiber lasers because it can significantly improve the deliverable energy of pulse, approaching or even exceeding 100 nJ [18, 19], and the DS exists in nonconservative systems whose dynamics are extremely different from those of conventional soliton [20–22]. A fiber laser with pure normal GVD (or large normal GVD together with small anomalous GVD) would presumably have to exploit dissipative process in the mode-locked pulse shaping [23–25].

The research of dark soliton lags behind that of bright soliton due to the difficulty of dark soliton generation. However, dark soliton is more suitable than bright soliton when used in optical communications [26]. Dark soliton is broadened during propagation at nearly half the rate of bright soliton, and dark soliton shows more resistance than bright soliton to perturbation during propagation. Furthermore, as the background noise mainly affects the background of the dark soliton, it is less sensitive to background noise [27–29]. Recently, the dark soliton formation in fiber lasers has been reported in literature [30–33]. Zhang et al. have experimentally observed dark soliton in a fiber ring laser with all-anomalous dispersion fibers [30] and with all-normal dispersion fibers [31]. Tang et al. demonstrated the dark soliton formation in an all-normal dispersion cavity fiber laser without an antisaturable absorber in cavity [33].

In this paper, we propose a novel way for controlling the generation of the dissipative bright soliton and dark soliton operation of lasers. The nonlinear polarization rotation technique is implemented for generating mode-locked and non-mode-locked region. The dissipative soliton and dark soliton are obtained in mode-locked and non-mode-locked regions, respectively.

2. Modeling

To study the feature and dynamic evolution of dissipative and dark soliton in an all-normal dispersion fiber laser, we implement a numerical model that incorporates the most important physical effects like the nonlinear polarization rotation (NPR), spectral filtering (SF), and so forth. The propagation model is shown schematically in Figure 1. It consists of a 4 m erbium-doped fiber (EDF) and 1.8 m dispersion compensating fiber (DCF), and the polarization additive-pulse mode-locking (PAPM) system is made of a polarization-sensitive isolator and two sets of polarization controllers, two quarter waveplates (QWP), and a half waveplate (HWP) made of the polarizer; one QWP and a HWP constitute the analyzer. The PAPM system is used to produce the NPR effect, which relies on the intensity dependent rotation of an elliptical polarization state in a length of optical fiber. By setting the angle of the polarizer and the linear cavity phase delay of the cavity appropriately, we can obtain the mode-locked and the non-mode-locked area in PAPM where mode-locked area means that the light intensity is inversely proportional to the loss and the opposite non-mode-locked area. When the soliton propagates through the PAPM element and the intensity transmission, T is expressed as [11]

$$T = \sin^2(\theta)\sin^2(\varphi) + \cos^2(\theta)\cos^2(\varphi)$$
$$+ 0.5\sin(2\theta)\sin(2\varphi)\cos(\Delta\phi_L + \Delta\phi_{NL}), \quad (1)$$

where θ and φ represent the angle between the fast axis of the birefringent fiber and polarizer (analyzer), $\Delta\phi_L$ is the linear cavity phase delay, $\Delta\phi_L = (2\pi L/\lambda)(n_x - n_y) + \phi_{PC}$, ϕ_{PC} is the phase difference between two polarization directions, and λ and L represent the cavity length of the fiber and the wavelength of light, respectively. $\Delta\phi_{NL}$ is the nonlinear phase delay caused by self-phase modulation (SPM) and cross-phase modulation (XPM). $\Delta\phi_{NL} = -(2\pi L n_2 P/3\lambda A_{eff})\cos 2\alpha$, n_2 is the nonlinear coefficient, A_{eff} is the effective mode area, and P is optical power.

The pulse propagation in the weak birefringent fiber can be modeled well by two coupled nonlinear Schrodinger equations (NLSE). However, the fiber lasers have components which cannot be modeled by several phase modulation terms in NLSEs. For example, a fiber laser has a gain with a limited gain bandwidth (BW) and a saturable absorber (SA). They do not induce the phase modulation, but they definitely cause the amplitude modulation in the frequency and time domain. To count the amplitude modulation effects in a laser cavity properly, more terms needed to be added to

NLSEs. The resulting differential equations are called coupled cubic Ginzburg-Landau equations (CGLE) [34], shown as

$$\frac{\partial u_x}{\partial z} + \frac{\alpha}{2}u_x + \delta\frac{\partial u_x}{\partial T} + i\frac{\beta_2}{2}\frac{\partial^2 u_x}{\partial T^2} - \frac{g}{2}u_x - \frac{g}{2\Omega_g^2}\frac{\partial^2 u_x}{\partial T^2}$$
$$= i\gamma\left(|u_x|^2 + \frac{2}{3}|u_y|^2\right)u_x,$$

$$\frac{\partial u_y}{\partial z} + \frac{\alpha}{2}u_y - \delta\frac{\partial u_y}{\partial T} + i\frac{\beta_2}{2}\frac{\partial^2 u_y}{\partial T^2} - \frac{g}{2}u_y - \frac{g}{2\Omega_g^2}\frac{\partial^2 u_y}{\partial T^2}$$
$$= i\gamma\left(|u_y|^2 + \frac{2}{3}|u_x|^2\right)u_y, \quad (2)$$

where u_x and u_y denote the envelopes of the optical pulses along the two orthogonal polarization axes of the fiber, and α is the loss coefficient of the fiber. $\delta = (\beta_{1x} - \beta_{1y})/2$ is the group velocity difference between the two polarization modes, β_2 represents the fiber dispersion, Ω_g is the bandwidth of the laser gain, $T = t - (\beta_{1x} + \beta_{1y})z/2$ and z indicate the pulse local time and the propagation distance, respectively, and g describes the gain function of EDF which is expressed by $g = g_0 \exp(-E_{pulse}/E_{sat})$ [2], where g_0 is the small signal gain coefficient, related to the doping concentration, and E_{sat} is the gain saturation energy, which corresponds to the pumping strength [2, 34]. The pulse energy E_{pulse} is given by $E_{pulse} = \int_{-T_R/2}^{T_R/2}(|u|^2 + |v|^2)d\zeta$, where T_R is the cavity round-trip time.

3. Simulation Results and Discussion

The model is solved with a standard symmetric split-step Fourier algorithm. The initial u_x denotes an arbitrary signal, while the initial u_y denotes a weak continuous wave (CW) light which contains a small segment dip on the top of it; by appropriately adjusting the angle of the polarizer (analyzer) and the linear cavity phase delay of the cavity, both dissipative and dark solitons can be achieved in different directions of the polarization. Since the saturation energy E_{sat} is proportional to the pumping strength [35], this means increasing E_{sat} corresponds to increasing the pump power in the practical system. The following parameters are applied for the simulations possibly matching the experimental conditions: $\alpha = 0.2$ dB/km, $g_0 = 2$ m^{-1}, $\gamma = 4.2$ W^{-1}km^{-1} for EDF and $\gamma = 1.3$ W^{-1}km^{-1} for DCF, $\Omega_g = 30$ nm, EDF = 4.0 m, DCF = 1.8 m, $\beta_2 = +31 \times 10^{-3}$ ps^2/m for EDF and $\beta_2 = +20 \times 10^{-3}$ ps^2/m for DCF, and net cavity GVD $\beta_{net} \approx 0.16$ ps^2, $\theta = \pi/8$, $\varphi = 5\pi/8$, $\phi_{PC} = 0.96\pi$, and when $E_{sat} = 460$ pJ, it is capable of achieving both dissipative and dark solitons in fiber laser.

The formation and evolution of dissipative and dark solitons at $E_{sat} = 460$ pJ are illustrated in Figure 2. We can find that the dissipative soliton pulse duration, pulse energy, and peak power are about 2.96 ps, 229.19 pJ, and 89.16 W in one polarization, respectively. In the other polarization, we can generate the dark soliton with 960 fs. Figures 2(c) and 2(d) show that dissipative and dark solitons remain stable when they circled 500 times in the cavity separately. According to Figure 2(d), we found that the dip exist not only in the pulse central but also in the edge, because in the simulation we used

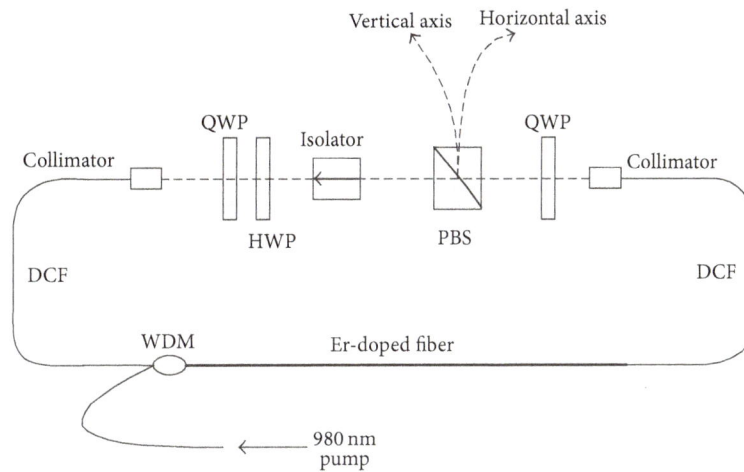

FIGURE 1: A schematic diagram of the all-normal dispersion fiber laser; PBS: polarization beam splitter; HWP: half waveplate; QWP: quarter waveplate; WDM: wavelength division multiplexer.

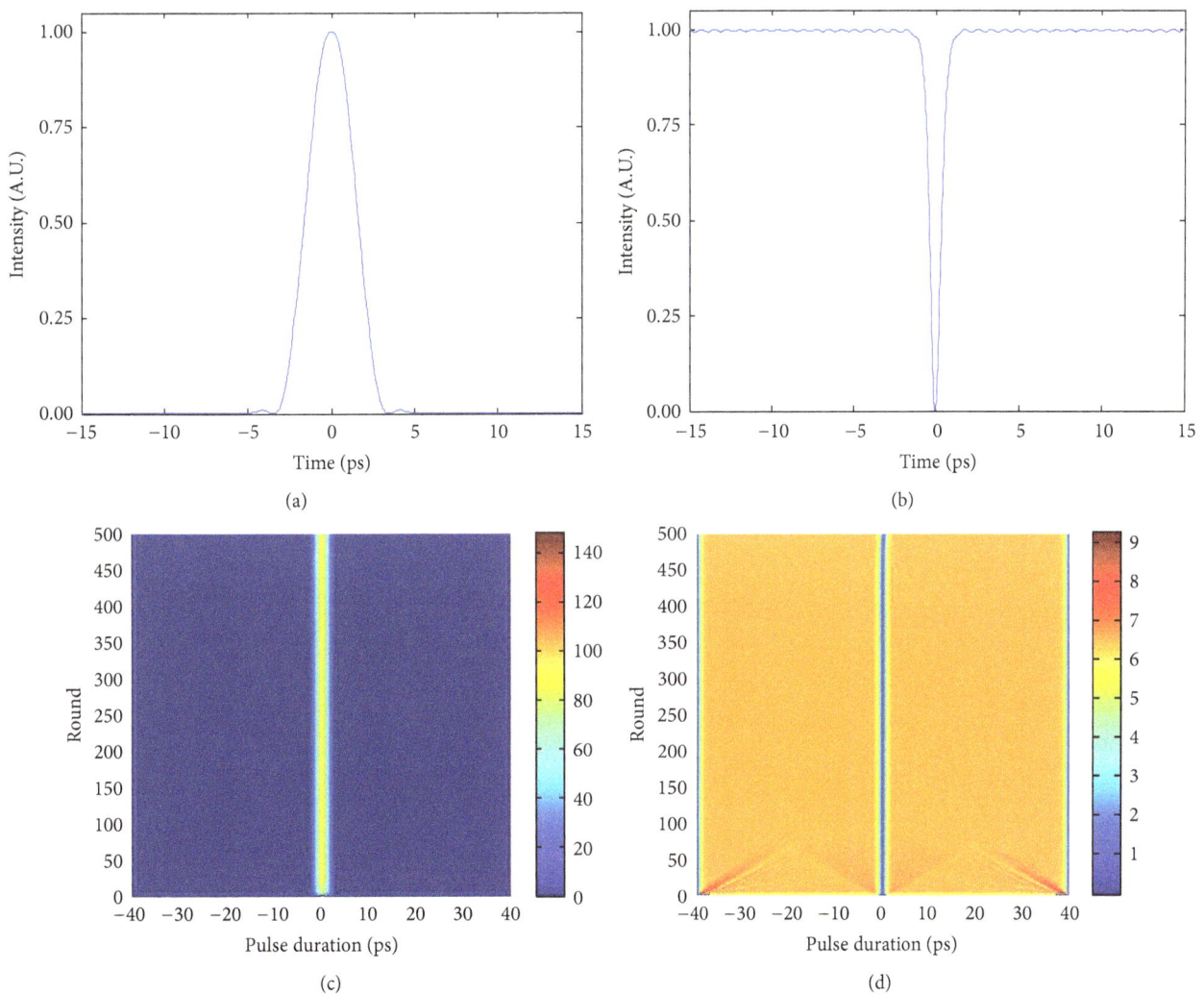

FIGURE 2: (a) Temporal profile of the dissipative soliton at $E_{sat} = 460$ pJ. (b) Temporal profile of the dark soliton at $E_{sat} = 460$ pJ. (c) Dynamic evolution of dissipative soliton. (d) Dynamic evolution of dark soliton.

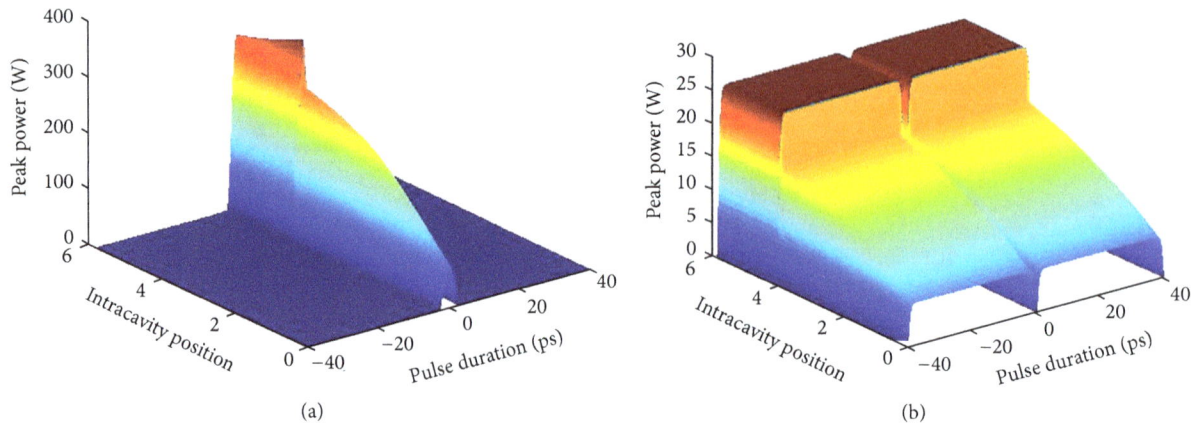

FIGURE 3: (a) Intracavity dissipative soliton evolution (at round = 498) in temporal domain. (b) Intracavity dark soliton evolution (at round = 498) in temporal domain.

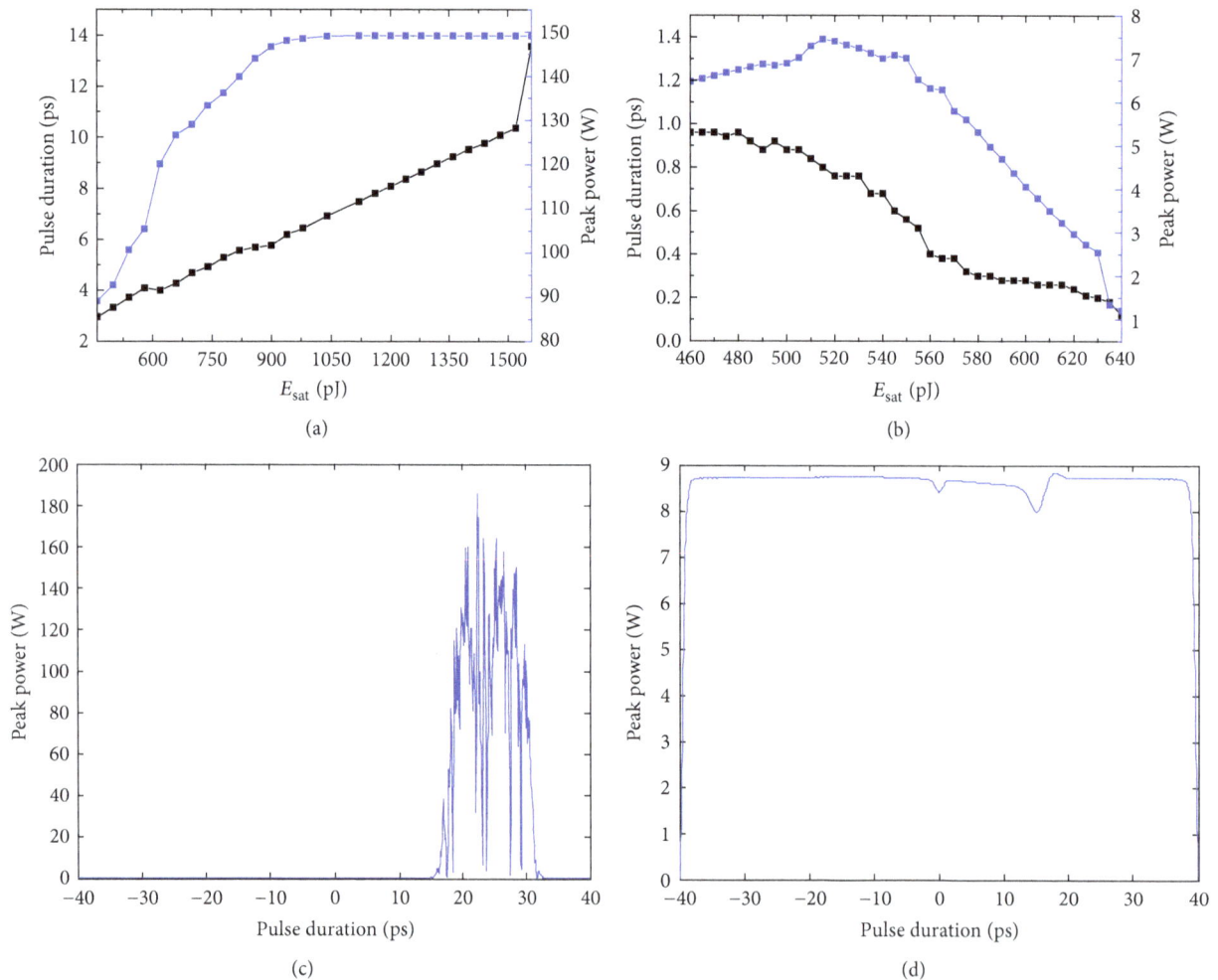

FIGURE 4: (a) With the increase of E_{sat}, dissipative soliton pulse duration and peak power change. (b) With the increase of E_{sat}, dark soliton pulse duration and peak power change. (c) Temporal profile of the dissipative soliton at E_{sat} = 1560 pJ. (d) Temporal profile of the dark soliton at E_{sat} = 640 pJ.

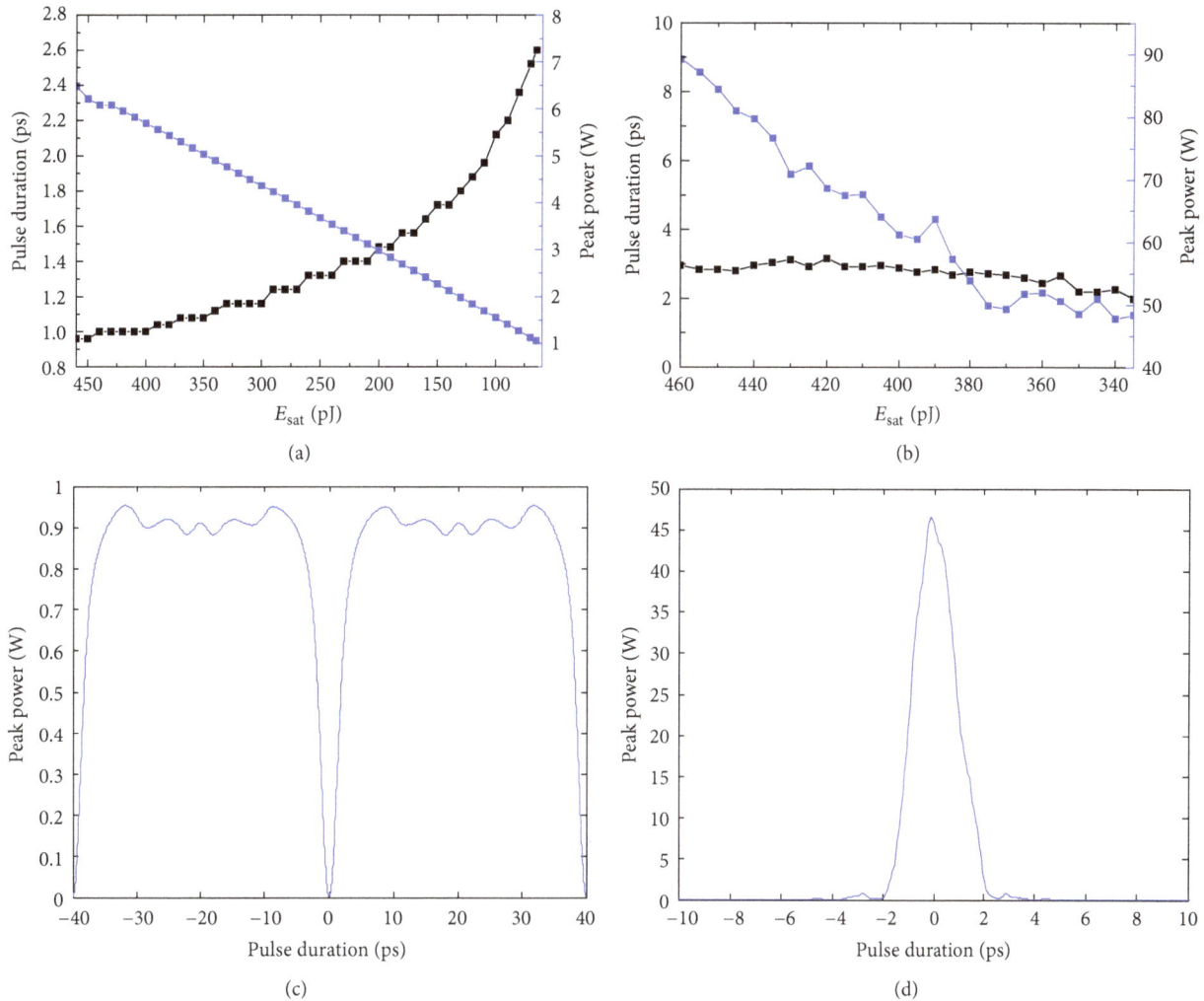

FIGURE 5: (a) With the decrease of E_{sat}, the dark soliton pulse duration and peak power change. (b) With the decrease of E_{sat}, the dissipative soliton pulse duration and peak power change. (c) Temporal profile of the dark soliton at $E_{\text{sat}} = 58$ pJ. (d) Temporal profile of the dissipative soliton at $E_{\text{sat}} = 335$ pJ.

the finite background pulse instead of infinite background pulse to generate dark soliton. Numerical simulation shows that if the background pulse width is 10 times the soliton width, the transmission characteristics of finite and infinite backgrounds of dark solitons are basically the same, so we can ignore the edge dip. From Figures 3(a) and 3(b) we can observe the impact of the various parts of the components for peak power and pulse duration. After propagating the EDF, both the dissipative soliton and the dark soliton peak power show an increase trend. However, as for the pulse duration, the dissipative soliton shows an increase while the dark soliton remains nearly stable. When soliton transmitted to PAPM, both dissipative and dark soliton pulse durations are rapidly compressed while their peak power remains the same. In the DCF, the dissipative soliton peak power related to the previous EDF has been increased rapidly and then decreased slowly, but it has no effect on the pulse duration. As for the dark soliton, the peak power in DCF increases rapidly and then remains unchanged, and the same basic has no effect

on dark soliton pulse width. These results are obtained when $E_{\text{sat}} = 460$ pJ. Next, we analyze the various E_{sat} effects for our system.

Figure 4 shows the change of dissipative and dark solitons with the increase of E_{sat}. Figure 4(a) reflects the notion that dissipative soliton pulse duration and peak power increase with E_{sat}. When E_{sat} increases to about 980 pJ, peak power is nearly unchanged, and when E_{sat} reaches 1560 pJ, dissipative soliton appears split in Figure 4(c). If we continue to increase E_{sat}, the dissipative soliton will generate an unstable pulse which is different from the phenomenon adopted in Liu [2]. However, Figure 4(b) shows that dark soliton varies with different dissipative soliton; firstly the peak power increases with E_{sat}, but the dark soliton pulse duration gradually decreases. When E_{sat} increases to 550 pJ, the peak power reduces gradually and eventually in $E_{\text{sat}} = 640$ pJ, the dark soliton disappears in Figure 4(d).

Figures 5(a) and 5(b) show that with the decrease of E_{sat}, both the dissipative soliton and the dark soliton peak power show a decrease trend. However, as for the pulse duration, the dark soliton shows an increase while the dissipative soliton remains nearly stable. When E_{sat} is reduced to 335 pJ, the dissipative soliton is unstable. When E_{sat} is below 318 pJ, the dissipative soliton cannot be generated. Then we conclude that the generation of dissipative soliton requires a pump power exceeding a threshold. When E_{sat} is reduced to 58 pJ, the dark soliton is unstable, but it is still capable of forming a dark soliton. It can be found that obtaining a dark soliton does not require high E_{sat}, which means the dark soliton does not require high pump power.

4. Conclusion

We propose a novel way for controlling the generation of the dissipative bright soliton and dark soliton operation of lasers. Numerical simulation shows that, in an appropriate pump power range, the dissipative bright soliton and dark soliton can be generated simultaneously in the mode-locked and non-mode-locked regions, respectively. When E_{sat} is up to 640 pJ, only dissipative soliton can be achieved, and the dissipative soliton is unstable when E_{sat} is close to 1560 pJ, whereas it does not generate multipulse phenomenon [2]. Then if we decrease E_{sat} to 335 pJ, only dark soliton can be obtained. Hence, we conclude that the dissipative soliton can tolerate higher pump power but its generation process requires a high threshold, whereas dark soliton can be obtained at a low pump power threshold.

Competing Interests

The authors declare that there are no competing interests regarding the publication of this paper.

References

[1] C. Zhao, H. Zhang, X. Qi et al., "Ultra-short pulse generation by a topological insulator based saturable absorber," *Applied Physics Letters*, vol. 101, no. 21, Article ID 211106, 2012.

[2] X. Liu, "Hysteresis phenomena and multipulse formation of a dissipative system in a passively mode-locked fiber laser," *Physical Review A*, vol. 81, no. 2, Article ID 023811, 2010.

[3] X. Wu, D. Y. Tang, H. Zhang, and L. M. Zhao, "Dissipative soliton resonance in an all-normal-dispersion erbium-doped fiber laser," *Optics Express*, vol. 17, no. 7, pp. 5580–5584, 2009.

[4] Y. Chen, C. Zhao, S. Chen et al., "Large energy, wavelength widely tunable, topological insulator Q-switched erbium-doped fiber laser," *IEEE Journal on Selected Topics in Quantum Electronics*, vol. 20, no. 5, pp. 315–322, 2014.

[5] H. Zhang, D. Y. Tang, L. M. Zhao, and X. Wu, "Observation of polarization domain wall solitons in weakly birefringent cavity fiber lasers," *Physical Review B*, vol. 80, no. 5, Article ID 052302, 2009.

[6] D. Li, D. Tang, L. Zhao, and D. Shen, "Mechanism of dissipative-soliton-resonance generation in passively mode-locked all-normal-dispersion fiber lasers," *Journal of Lightwave Technology*, vol. 33, no. 18, pp. 3781–3787, 2015.

[7] D. D. Han, X. M. Liu, Y. D. Cui, G. X. Wang, C. Zeng, and L. Yun, "Simultaneous picosecond and femtosecond solitons delivered from a nanotube-mode-locked all-fiber laser," *Optics Letters*, vol. 39, no. 6, pp. 1565–1568, 2014.

[8] X. Li, Y. Wang, W. Zhao et al., "Numerical investigation of soliton molecules with variable separation in passively mode-locked fiber lasers," *Optics Communications*, vol. 285, no. 6, pp. 1356–1361, 2012.

[9] X. Li, X. Liu, X. Hu et al., "Experimental observations of separation-changeable soliton pairs in a fiber laser mode-locked by nonlinear polarization rotation technique," *Journal of Optoelectronics and Advanced Materials*, vol. 13, no. 3, pp. 190–195, 2011.

[10] W. H. Renninger, A. Chong, and F. W. Wise, "Self-similar pulse evolution in an all-normal-dispersion laser," *Physical Review A*, vol. 82, no. 2, Article ID 021805, 2010.

[11] X. Liu, "Mechanism of high-energy pulse generation without wave breaking in mode-locked fiber lasers," *Physical Review A*, vol. 82, no. 5, Article ID 053808, 2010.

[12] B. Guo, Y. Yao, P. G. Yan et al., "Dual-wavelength soliton mode-locked fiber laser with a WS2-based fiber taper," *IEEE Photonics Technology Letters*, vol. 28, no. 3, pp. 323–326, 2016.

[13] X. Li, Y. Wang, W. Zhang, and W. Zhao, "Experimental observation of soliton molecule evolution in Yb-doped passively mode-locked fiber lasers," *Laser Physics Letters*, vol. 11, no. 7, Article ID 075103, 2014.

[14] Z.-C. Luo, W.-J. Cao, Z.-B. Lin, Z.-R. Cai, A.-P. Luo, and W.-C. Xu, "Pulse dynamics of dissipative soliton resonance with large duration-tuning range in a fiber ring laser," *Optics Letters*, vol. 37, no. 22, pp. 4777–4779, 2012.

[15] D. Mao, X. M. Liu, L. R. Wang, X. H. Hu, and H. Lu, "Partially polarized wave-breaking-free dissipative soliton with super-broad spectrum in a mode-locked fiber laser," *Laser Physics Letters*, vol. 8, no. 2, pp. 134–138, 2011.

[16] X. Liu, "Dissipative soliton evolution in ultra-large normal-cavity-dispersion fiber lasers," *Optics Express*, vol. 17, no. 12, pp. 9549–9557, 2009.

[17] F. Ö. Ilday, J. R. Buckley, W. G. Clark, and F. W. Wise, "Self-similar evolution of parabolic pulses in a laser," *Physical Review Letters*, vol. 92, no. 21, Article ID 213902, 2004.

[18] W. H. Renninger, A. Chong, and F. W. Wise, "Dissipative solitons in normal-dispersion fiber lasers: exact pulse solutions of the complex ginzburg-landau equation," in *Proceedings of the Bragg Gratings, Photosensitivity, and Poling in Glass Waveguides*, Quebec City, Canada, September 2007.

[19] H. Zhang, D. Y. Tang, L. M. Zhao, Q. L. Bao, and K. P. Loh, "Large energy mode locking of an erbium-doped fiber laser with atomic layer graphene," *Optics Express*, vol. 17, no. 20, pp. 17630–17635, 2009.

[20] X. Li, Y. Wang, W. Zhao et al., "All-fiber dissipative solitons evolution in a compact passively Yb-doped mode-locked fiber laser," *Journal of Lightwave Technology*, vol. 30, no. 15, Article ID 6204196, pp. 2502–2507, 2012.

[21] B. Gao, J. Huo, G. Wu, and X. Tian, "Soliton molecules in a fiber laser mode-locked by a graphene-based saturable absorber," *Laser Physics*, vol. 25, no. 7, Article ID 075103, 2015.

[22] Z. Wang, S.-E. Zhu, Y. Chen et al., "Multilayer graphene for Q-switched mode-locking operation in an erbium-doped fiber laser," *Optics Communications*, vol. 300, pp. 17–21, 2013.

[23] X. Li, X. Liu, X. Hu et al., "Long-cavity passively mode-locked fiber ring laser with high-energy rectangular-shape pulses in anomalous dispersion regime," *Optics Letters*, vol. 35, no. 19, pp. 3249–3251, 2010.

[24] H. Zhang, D. Y. Tang, L. M. Zhao, X. Wu, and H. Y. Tam, "Dissipative vector solitons in a dispersion-managed cavity fiber

laser with net positive cavity dispersion," *Optics Express*, vol. 17, no. 2, pp. 455–460, 2009.

[25] X. He, L. Hou, M. Li et al., "Bound States of dissipative solitons in the single-mode Yb-doped fiber laser," *IEEE Photonics Journal*, vol. 8, no. 2, pp. 1–7, 2016.

[26] Y. Q. Ge, J. L. Luo, L. Li et al., "Initial conditions for dark soliton generation in normal-dispersion fiber lasers," *Applied Optics*, vol. 54, no. 1, pp. 71–75, 2015.

[27] L. Li, F. Y. Song, H. Zhang et al., "Dark soliton operation fiber lasers," in *Proceedings of the 2013 Conference on Lasers and Electro-Optics Pacific Rim (CLEO-PR '13)*, WPB_21, Optical Society of America, Kyoto, Japan, 2013.

[28] D. Y. Tang, L. Li, Y. F. Song, L. M. Zhao, H. Zhang, and D. Y. Shen, "Evidence of dark solitons in all-normal-dispersion-fiber lasers," *Physical Review A*, vol. 88, no. 1, Article ID 013849, 2013.

[29] T. P. Horikis and M. J. Ablowitz, "Constructive and destructive perturbations of dark solitons in mode-locked lasers," *Journal of Optics*, vol. 17, no. 4, Article ID 042001, 2015.

[30] H. Zhang, D. Y. Tang, L. M. Zhao, and X. Wu, "Dark pulse emission of a fiber laser," *Physical Review A—Atomic, Molecular, and Optical Physics*, vol. 80, no. 4, Article ID 045803, 2009.

[31] H. Zhang, D. Y. Tang, L. M. Zhao, and R. J. Knize, "Vector dark domain wall solitons in a fiber ring laser," *Optics Express*, vol. 18, no. 5, pp. 4428–4433, 2010.

[32] Y. Meng, S. Zhang, H. Li, J. Du, Y. Hao, and X. Li, "Bright-dark soliton pairs in a self-mode locking fiber laser," *Optical Engineering*, vol. 51, no. 6, Article ID 064302, 2012.

[33] D. Tang, J. Guo, Y. Song, H. Zhang, L. Zhao, and D. Shen, "Dark soliton fiber lasers," *Optics Express*, vol. 22, no. 16, pp. 19831–19837, 2014.

[34] K.-I. Maruno, A. Ankiewicz, and N. Akhmediev, "Dissipative solitons of the discrete complex cubic-quintic Ginzburg-Landau equation," *Physics Letters A*, vol. 347, no. 4–6, pp. 231–240, 2005.

[35] A. Cabasse, B. Ortaç, G. Martel, A. Hideur, and J. Limpert, "Dissipative solitons in a passively mode-locked Er-doped fiber with strong normal dispersion," *Optics Express*, vol. 16, no. 23, pp. 19322–19329, 2008.

5

Differences in Nanosecond Laser Ablation and Deposition of Tungsten, Boron, and WB$_2$/B Composite due to Optical Properties

Tomasz Moscicki

Institute of Fundamental Technological Research, PAS, Pawinskiego 5B, 02-106 Warsaw, Poland

Correspondence should be addressed to Tomasz Moscicki; tmosc@ippt.pan.pl

Academic Editor: Giulio Cerullo

The first attempt to the deposition of WB$_3$ films using nanosecond Nd:YAG laser demonstrated that deposited coatings are superhard. However, they have very high roughness. The deposited films consisted mainly of droplets. Therefore, in the present work, the explanation of this phenomenon is conducted. The interaction of Nd:YAG nanosecond laser pulse with tungsten, boron, and WB$_2$/B target during ablation is investigated. The studies show the fundamental differences in ablation of those materials. The ablation of tungsten is thermal and occurs due to only evaporation. In the same conditions, during ablation of boron, the phase explosion and/or fragmentation due to recoil pressure is observed. The deposited films have a significant contribution of big debris with irregular shape. In the case of WB$_2$/B composite, ablation is significantly different. The ablation seems to be the detonation in the liquid phase. The deposition mechanism is related mainly to the mechanical transport of the target material in the form of droplets, while the gaseous phase plays marginal role. The main origin of differences is optical properties of studied materials. A method estimating phase explosion occurrence based on material data such as critical temperature, thermal diffusivity, and optical properties is shown. Moreover, the effect of laser wavelength on the ablation process and the quality of the deposited films is discussed.

1. Introduction

In recent years, growing interest in ultraincompressible and superhard materials has been observed. Tungsten triboride WB$_3$ is one of the most promising inexpensive candidates for ultraincompressible, superhard materials [1, 2]. Even in the form of thin films, it has superhard properties [3, 4] and, in the future, may be an alternative to other hard coatings, such as diamond-like DLC or cubic boron nitride (c-BN). One of the most promising methods of obtaining WB$_3$ films is pulsed laser deposition (PLD), because it highly suits deposition of hardly meltable metals, such as tungsten [5]. Pulsed laser deposition is a technique where a pulsed laser beam is focused inside a vacuum chamber to strike a target of the material that is to be deposited. This material is ablated from the target, forms a plasma plume, and subsequently deposits as a thin film on a substrate. Deposited films may have a thickness from several nanometres to several micrometres. The deposition of high-quality films requires knowledge of the first step

of the pulsed laser deposition process, which is laser ablation. The course of target ablation affects plasma plume composition, for example, vapour to nanoparticle and microparticle ratio.

The exact structure of tungsten borides is still a matter of vigorous debate in the literature. Recently, it has been found that the previously experimentally attributed WB$_4$ is in fact the defect-containing WB$_3$ [2, 5]. The first attempt to the deposition of WB$_3$ films using nanosecond Nd:YAG laser demonstrated that deposited coatings are superhard; however, they have very high roughness [5]. Additionally, the deposited debris has regular circular shape like droplets. The mechanism of deposition of droplets on the substrate surface may be explained by the condensation in plasma plume [6, 7]. Ablation plumes induced by nanosecond laser irradiation provide conditions which are well suited to the formation of nanoclusters. The high saturation ratios and presence of ionization lead to extraordinarily high nucleation rates and small critical radii. In the case of condensation, only clusters in the range of 5–50 nm can be produced [6–8], which is above one hundred times less than dimension of droplets deposited in WB_3 films [5]. Moreover, the SEM-EDS study has shown that the composition of deposited film is close to the target one. It is impossible to get such results by the condensation only. All these phenomena indicate that film is deposited mainly from droplets erupted from target. This is a characteristic process of the ablation of very porous targets [9]. However, in this case, the porosity of target was negligible [5].

The simple mechanism of ablation consists of three stages. During the interaction of the laser beam with a material, the target is heated to a temperature exceeding its boiling point and sometimes also its critical temperature. In the second stage, material evaporated from the target forms a thin layer of dense plume, consisting of electrons, ions, and neutrals. This plasma plume absorbs energy from the laser beam (by means of photoionization and inverse Bremsstrahlung) and its temperature and pressure grow. The resulting pressure gradient accelerates the plume to high velocity perpendicular to the target. At the next time steps, the laser pulse terminates and plasma plume expands adiabatically [10, 11]. However, a considerable fraction of the plasma energy is reemitted and coupled back to the target and also affect the dynamics of plasma plume [12]. This ablation pattern can be used up to some limit of fluence in which there is no phase explosion yet [11, 13]. It is assumed that the explosive boiling begins when the temperature exceeds 0.9 of the critical temperature [11]. This type of ablation results in the appearance of nano- and microparticles in the plasma plume. The size and shape of deposited debris depend on material properties and time step during ablation process. The subsurface overheating during laser pulse and subsequent eruption [13] causes deposition of a mixture of vapours, irregular debris, and droplets. After the cessation of laser impulse, the thermal diffusion and subsequent explosive boiling are a potential mechanism. The superheated liquid will undergo a transition into a mixture of vapour and liquid droplets, followed by delayed explosive boiling of the liquid-vapour mix [14, 15]. Lu et al. demonstrated that in the case of ablation of single-crystal silicon and power density 2×10^{10} W/cm^2, large (micron-sized) droplets are ejected from the target after 300–400 ns of impulse cessation [14]. During deposition WB_3 from WB_2/B, the power density was about 10^9 W/cm^2 [5]. According to the presented theory [14], such power density is not sufficient to obtain droplets having a diameter of several micrometres. Therefore, the reasons for, as presented in [5], PLD process must be in the course of ablation of individual components of the SPS sintered WB_2/B target.

Besides the deposition of functional films containing boron [3, 16] or tungsten [4], the laser ablation is also used in applications such as micromachining [17], cleaning [18], and also fabrication of micro/nanostructures [19]. Tungsten is one of very few possible materials for in-vessel components in forward looking thermonuclear reactors. Therefore, any investigation of its properties and behaviour is very important, and in particular the characterisation of its evaporation processes from the inner wall of a tokamak chamber is needed [20].

Despite of numerous applications, the physics of laser ablation process is not yet thoroughly understood. It depends on not only the properties of the material but also laser parameters such as pulse duration, frequency [17], fluence [13], and wavelength [21].

Therefore, detailed knowledge on laser-induced flow dynamics of plasma and understanding of the mechanisms of ablation is necessary for optimizing technological processes of deposition of new materials and promoting powerful lasers with nanosecond pulse durations for cost-effective use in industries.

The main goal of this paper is the comparative analysis of boron, tungsten, and WB_2/B composite ablation processes induced by nanosecond laser radiation in vacuum. The unusual mechanism of ablation of WB_2/B target and deposition of WB_3 films are explained. Moreover, the effect of laser wavelength on the ablation process and the quality of the deposited boron and tungsten films is discussed.

2. Material and Experimental Methods

Irradiation of a tungsten and boron targets was performed using a Nd:YAG laser (Quantel, 981 E) in a chamber evacuated to a residual pressure of 1×10^{-5} Pa. The laser beam (10 ns FWHM) was focused to a spot size of 0.055 mm^2 with a fluence of 10 J/cm^2. The distance from target to substrate was 4 cm. Thin films were deposited on a silicon (100) polished substrates (SPI Supplies) in ambient temperature. During deposition, the target was rotated to avoid crater formation. The deposition time was 30 minutes (18000 pulses) for each experiment. All ablation parameters were the same as in theoretical model. Both harmonics were polarized horizontally. The spatial and time profiles of the laser beam were Gaussian. The surface area of a laser spot was determined by registration of the spot size by ICCD camera after attenuation of the laser beam and was 0.075 mm^2 in the case of 1064 nm and 0.053 mm^2 in the case of 355 nm wavelength. The incident angle of the laser beam was 45° to the surface normal. The diameter of the laser spot used for calculations is equal to the average value of the ellipse diameters. The high-quality targets, boron from Kurt J. Lesker (2.35 g/cm^3 mass density, 99.5% purity) and tungsten from Kurt J. Lesker (19.35 g/cm^3, 99.95% purity), were used. WB_2/B targets were fabricated from boron (~625 mesh, 99.7% purity, Sigma Aldrich) and tungsten (12 μm, 99.9% purity, Sigma Aldrich). Both elements were milled and mixed in the molar ratio 4.5:1 [3, 5]. The WB_2/B target was made in the spark plasma sintering process (SPS). Mersen (Carbone Lorraine) tools were placed in the sintering chamber of the furnace HP D 25-3. Sintering process was conducted in vacuum at a sintering temperature of 1600°C, and under a compaction pressure of 50 MPa. The heating rate was 50°C/min and the sintering time was 2.5 min. The duration of a single current impulse was equal to 15 ms, and the interval between the impulses lasted 3 ms. Cylindrical samples with diameter of 20 mm were produced.

The laser deposited films surface was subject to inspection under a Scanning Electron Microscope: JEOL, JSM-6010PLUS/LV InTouchScope™. In addition, the EDS microanalysis was used to study the elemental distribution of boron and tungsten in the targets and deposited films. The crater shape after 50 impulses was measured with Confocal Microscope: Keyence Vk-×100. The images of the plasma plume were registered with the use of an ICCD camera. The plasma was imaged on the camera using a 180 cm focal length camera lens. The image intensifier was gated for an exposure time of 5 ns, while the delay time between the laser pulse and the pulse triggering the image intensifier was changed gradually.

3. Theory/Calculation

The theoretical model, that describes the target heating, formation of the plasma, and its expansion, was previously presented in [22, 23]. The main goal of the present research is a comparative analysis of ablation mechanism of WB_2/B, boron, and tungsten target and the impact of this phenomenon on the composition of the plasma for different laser wavelengths. It should be emphasized that the absorption of the laser beam depends mainly on the optical properties of the material such as depth of penetration and reflectivity. Calculations were made for two wavelengths of an Nd:YAG laser: 355 nm and 1064 nm. The laser beam with a Gaussian profile (10 ns FWHM) was focused to a spot size of 0.055 mm^2 with a fluence of 10 J/cm^2. It was assumed that the boron or tungsten plume expands to ambient air at a pressure of 10^{-3} Pa.

The intensity of laser beam reaching the target surface I_L was used in the form which fits the shape of the laser pulse used in our experiments:

$$I_L(t,r) = \frac{CF}{\tau} \exp\left(-\left(\frac{t-t_0}{s_2}\right)^2\right) \exp\left(-\left(\frac{r}{s_1}\right)^2\right) \cdot \left(\exp\left(-\int \kappa\, dz\right)\right), \quad (1)$$

where $s_1 = 0.098 \times 10^{-3}$ m and $s_2 = 6.0056 \times 10^{-9}$ s are Gauss parameters, F is the laser fluence, τ is the laser pulse duration, κ is the plasma absorption coefficient, r and z are radial and axial coordinate, respectively, and the integration is over the distance travelled by the beam. The numerical factor C results from normalization; t_0 is the time offset of the beam maximum intensity. The first part of (1) describes temporal evolution of the laser intensity and the second energy distribution in the laser beam and the last exponential component takes into account the attenuation of the laser beam on its way to the point (r,z) in plasma. The 45° incidence angle was taken into account by the multiplying of laser beam way by factor $\sqrt{2}$.

The laser radiation reflected from the surface of target R was included. Due to reflectance, the source component in energy equation is $\kappa I_L(1+R)$ in the case of plasma and $\alpha I_L(1-R)\exp(-\int \alpha\, dz_T)$ for the target. The exponential component takes into account the attenuation of the laser beam on its way to the point (r,z_T) of target. Because, during pulse termination, plasma plume has small dimension and is connected with target, the change of laser beam incidence angle from normal to 45 degrees does not reduce absorption

of energy from laser beam. Oppositely, the extended way (by factor $\sqrt{2}$) increases absorbed energy.

The used ablation model consists of two parts. The first part, which is settled with conduction equation with energy source due to laser irradiation [22, 24], was responsible for the determination of the temperature distribution and mass removal in the target. In the second part, the Eulerian system of equations of continuity and the diffusion equation [22, 24] were solved for plasma.

The system of equations was solved iteratively. The target temperature was calculated and then the stream of ablated particles was determined at the end of the Knudsen layer. These conditions were taken as inlet conditions for plume expansion. Next, the absorption of the laser radiation in developing the plasma plume was determined and the target temperature was recalculated according to actual laser intensity at the target surface. The new target temperature was used to determine the conditions at the end of the Knudsen layer, which were subsequently used as inlet conditions for plume expansion. The other boundary conditions for plasma system of equations were as follows. The stream of boron or tungsten vapour was directed perpendicularly to the target surface. At the wall, the noslip boundary and a fixed temperature condition were applied. At the outflow boundary, the pressure outlet boundary conditions [24] were used, which required the specification of a static pressure at the outlet boundary. This static pressure value is relative to the operating pressure. The axis boundary conditions were used at the centerline of the axisymmetric geometry [24].

The boundary condition at the place where the laser beam strikes the surface of target is

$$-k\frac{\partial T_s}{\partial \vec{n}} = -\rho \vec{u}(t) L_v + \frac{\varepsilon}{2}, \quad (2)$$

where $\vec{u}$ is the recession velocity given by the Hertz-Knudsen equation [13, 25], L_v is the latent heat of vaporization, and $\vec{n}$ is the unit vector perpendicular to the surface. The last part of (2) describes target heating due to plasma radiation ε. Energy losses due to thermal radiation from the surface of target are small compared to other terms and were neglected.

All the material functions of tungsten and boron were described in [22] and depend on the temperature, pressure, and mass fraction. WB_2 target properties also depend on temperature and were as follows: heat capacity at room temperature $c_p = 308$ J/kgK and at melting point $c_p = 364$ J/kgK [26], thermal conductivity at room temperature $k = 43$ W/mK and at melting point $k = 32$ W/mK [27], density $\rho = 12890$ kg/m^3 [26], and the melting temperature $T_m = 2833$ K [26]. Because there is lack of data about boiling point of WB_2, ZrB_2 data was adopted; that is, boiling temperature $T_b = 2400$ K under background pressure $p_b = 2 \times 10^{-3}$ Pa and latent heat of vaporization $L_v = 466.9 \pm 6.5$ kcal/mol [28]. Bolgar et al. [29] determined the vapour pressures and heats of vaporization of fourteen refractory compounds among which were TiB_2, ZrB_2, CrB_2, and AlB_{12}. Latent heat of vaporization all of these borides does not differ more than 20%. Because WB_2 also belongs to the group of refractory borides, the assumption about its boiling point parameters appears to be correct.

WB$_3$ target absorption coefficient α and reflectivity R were 1.7×10^5 m^{-1} and 0.55 for 355 nm laser wavelength and 1.0×10^5 m^{-1} and 0.73 for 1064 nm laser wavelength, respectively [30]. The absorption and emission coefficients of plasma were described in [25].

For the plasma, the calculation domain was $r = 0.01$ m and $z = 0.025$ m with nonuniform grid with 60×200 nodes. The smallest computational cells had dimensions of $50 \times 0.1 \mu$m at the vapour inlet. In the case of the target, the calculation domain was $r = 0.005$ m and $z = 2 \times 10^{-6}$ m with 130×500 nodes, respectively. While the smallest computational cells had dimensions of $12 \times 0.004 \mu$m at the target surface, the cell dimensions were fit to appearing gradients after preliminary calculations. Next, it was checked that further decreasing of cell dimensions did not change the results. The time step was adjusted to the smallest cells. Both cases were time-dependent and were solved in axisymmetric geometry. In the case of the plasma, the system of equations was solved by density-based (coupled) solver [24] with second-order spatial discretization for flow. The default settings [24] were applied for the target.

4. Results and Discussion

4.1. Mechanisms of Ablation. During the interaction of the laser beam with a material, the target is heated to a temperature exceeding its boiling point and sometimes its critical temperature [11, 13]. Assuming a constant fluence, the different mechanisms of ablation can occur depending on the critical temperature. In the case where the target temperature is lower than $0.9T_c$, mainly evaporation occurs. Explosive boiling begins above this limit. This phenomenon results in nanoparticles and microparticles with target composition present in the plasma plume, besides the expected electrons, ions, and neutrals. Both models can be observed during the ablation of tungsten and boron, depending on the laser fluence. Due to the high critical temperature (14778 K [31]) and high absorption coefficients 4.4×10^7 (reflectivity $R_{1064} = 60\%$) and 8.9×10^7 1/m (reflectivity $R_{355} = 47\%$) [32], respectively, for 1064 nm and 355 nm laser wavelengths, tungsten is a material for which evaporation takes place primarily for the laser intensity below 6×10^{10} W/cm^2 [33]. Figure 1 shows evolution of surface temperature at beam centre $r = 0$ for laser wavelengths 1064 nm and 355 nm for tungsten and boron at the laser fluence 10 J/cm^2. For tungsten and the first harmonic of an Nd:YAG laser, the maximum surface temperature is 11700 K and for the third 13700 K (Figure 1(a)). In both cases, the maximum surface temperature is reached in 16 ns. At the next time steps, the temperature suddenly decreases due to absorption of the laser beam in the plasma.

The laser ablation mechanism is different in the case of boron. For both laser harmonics, the maximum temperatures exceed the critical temperature (Figure 1(b)), which is about 10000 K [31]. In this case, the explosive boiling occurs, which results in a film on which there are different sized irregularly shaped contaminants (Figures 2(a) and 2(b)).

The number and size of debris depend mainly on the optical properties of boron. In the case of $\lambda = 1064$ nm,

the absorption coefficient is 1.3×10^6 m^{-1} (reflectivity $R_{1064} = 28.7\%$) [34] and is 23 times lower than for 355 nm (reflectivity $R_{355} = 28.6\%$). The low absorption coefficient results in a much thicker layer of heated material and the critical temperature is exceeded much further (Figure 3).

The result is an increase in the amount of larger fragments of the target in the deposited film. The maximum temperature for 1064 nm is obtained only in 22 ns. Due to the small amount of vaporized material, there is practically no plasma absorption. The high maximum temperature ~19000 K is the result of failure of the model neglecting phenomenon of explosive boiling. The rate of temperature equalization with increasing of distance over time is controlled by the coefficient of thermal diffusivity $a = k/(\rho c_p)$. Above the melting point, where it is assumed that the thermal properties and the density do not change, the value of diffusion coefficient is 2.58×10^{-5} and 1.48×10^{-6} m^2/s for tungsten and boron, respectively. The magnitude of this parameter indicates that subsurface overheating of a boron target is equalized much more slowly than in the case of tungsten. Moreover, in the case of tungsten, in connection with larger absorption coefficients, the laser beam penetrates the target to a lower depth. Both of these phenomena result in the maximum temperature located at the surface and the ablation process mainly by the evaporation of the surface. As a result, the surface of deposited tungsten films is smooth (Figures 2(c) and 2(d)).

This phenomenon is quite different in the case of boron when evaporation from the surface results in maximum of temperature placed at a certain depth. The critical temperature is achieved earlier below surface. The target surface erupts as a result of high stress. In the case of 1064 nm, a large amount of energy is accumulated at a depth of about 1μm. Lower thermal diffusivity of boron causes the increase of target cooling time (Figure 1(b)). The studies of ablation of nonmetallic materials showed that the process should be carried out with a sufficiently low fluence in order to avoid the phase explosion [11, 13]. However, in the case of boron ablation with the first harmonic, such a procedure could result in temperature decrease below the critical but simultaneously could cause the lack of ablation. The comparison of calculated and experimental tungsten crater (made with 355 nm radiation) is presented in Figure 4. The sample crater was formed by 50 laser shots in tungsten target.

The calculated depth of crater for one pulse was determined on the base of Hertz-Knudsen equation [13] described the rate of evaporation $\rho \cdot u$, which was integrated over time and radius. The calculated depth of the crater is 75.8 nm per pulse and is in good agreement with the experimentally measured values in our laboratory and by Spiro et al. [35] (63± 9 nm/pulse). Both experiments were conducted in similar conditions as were implemented in theoretical model. The cross section of crater in boron target ablated with 1064 nm wavelength after 50 laser pulses is presented in Figure 5. The depth of the crater is about 3μm per pulse and is much deeper than the distance from surface where critical temperature is achieved (Figure 3). Formation of deep craters during ablation of boron could not be explained by the evaporation and explosive boiling alone.

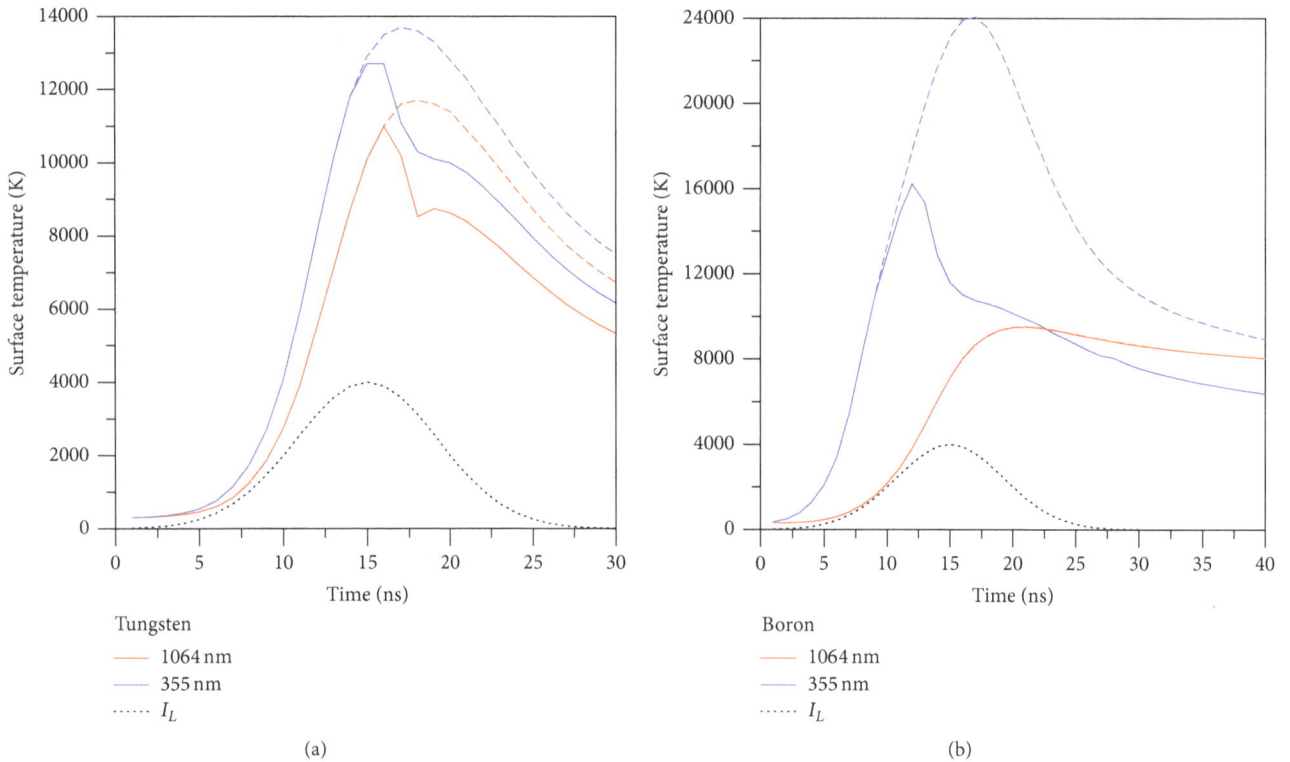

FIGURE 1: Target surface temperature T_s and laser intensity I_L during first 40 ns ($r = 0$). (a) Tungsten target and (b) boron target. Broken line denotes case without plasma absorption.

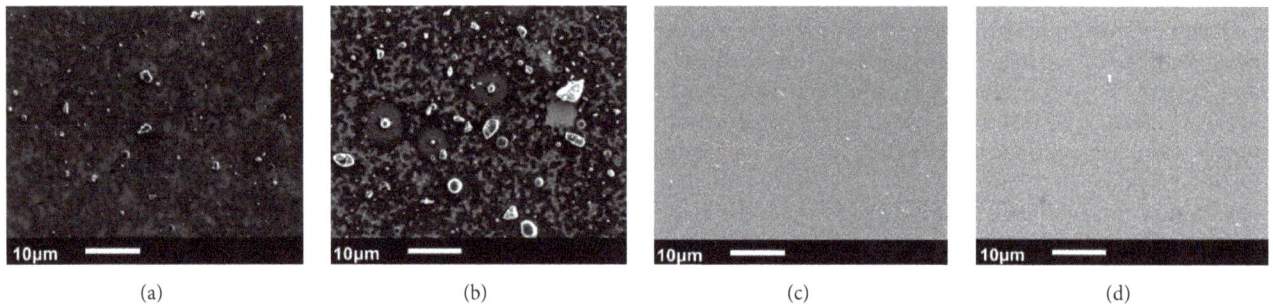

FIGURE 2: SEM micrograph (×1000) of deposited films: (a) boron 355 nm, (b) boron 1064 nm (c) tungsten 1064 nm, and (d) tungsten 355 nm. $F = 10 \, J/cm^2$.

Existing models should be complemented with additional mass removal process. Experimental evidence suggests that this may be the mechanical forces arising as a result of the recoil pressure [36]. As suggested in [36], the recoil pressure at the surface of the target generates compression wave propagating deep into material. It has been shown that the longitudinal compression wave can cause fragmentation. Taking into account the depth and irregular profile of crater and shape and dimension of debris deposited on the target surface, it is very possible mechanism of ablation. The presented model in [36] is novel and should be checked for different materials in the future. In the case of boron, it is not possible currently because of lack of material parameters necessary for calculations.

4.2. Effect of the Ablation of Boron on the Deposition of WB$_3$ Films. As it is shown in Section 4.1, after exceeding of some threshold fluence, ablation of boron has an explosive character. This has an impact on the ablation process of composites containing boron such as WB$_2$/B. Figure 6(a) shows the WB$_3$ film deposited by PLD method from SPS sintered WB$_2$/B target [5]. Experiments were made for 355 nm wavelengths of an Nd:YAG laser and fluence ~6 J × cm^{-2}. The WB$_3$ thin films were deposited on a silicon substrate (SPI Supplies) at temperature of 600°C in vacuum (2×10^{-4} Pa) [5].

Deposited layer mainly consists of droplets of different sizes. The abundances of elements (from three independent spots) in the deposited film are 77.6 ± 0.3% tungsten, 21.8 ± 0.3% boron, and 0.49±0.04% oxygen (percentages by weight),

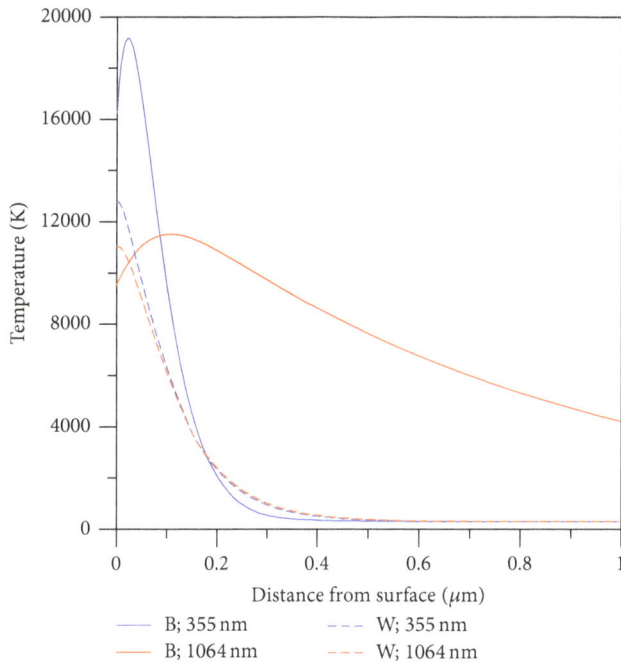

FIGURE 3: Target temperature T and laser intensity I_L along target axis at the time moments when the surface temperature reaches its maximum value.

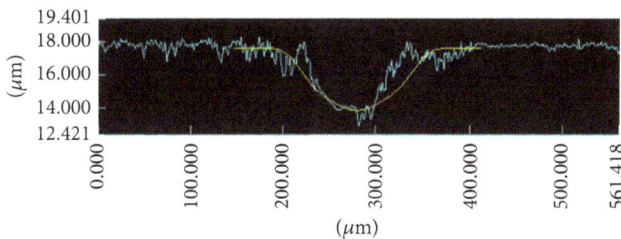

FIGURE 4: The cross section of crater from experiment (blue line) and theoretical calculation (yellow line). Target: tungsten; laser wavelength 355 nm; fluence 10 J/cm^2; 50 laser pulses.

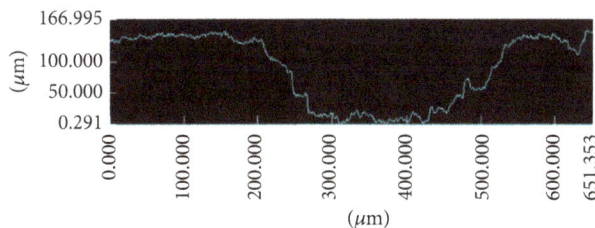

FIGURE 5: The cross section of crater in boron target ablated with 1064 nm wavelength after 50 laser pulses.

Figure 6(c). The B/W ratio found in deposited films is about 4.8, which is very close to the starting material ratio of 4.5 in target. The apparent differences may result from the peak of carbon which overlaps the peak of boron in the vicinity of 0.25 keV (Figure 6(b)). If the contribution of the advantageous carbon peak will be subtracted from the boron peak, the ratio arrived at may actually even be exactly 4.5.

The composition of deposited film is close to the target one. It indicates that film may be deposited mainly from melted debris of target but not due to condensation in plasma plume. Moreover, the high saturation ratios and presence of ionization lead to extraordinarily high nucleation rates and small critical radii. In the case of condensation, only clusters in the range of 5–50 nm can be produced [6, 7], which is above one hundred times less than dimension of droplets deposited in WB$_3$ films [5].

The creating of layers as a result of the deposition of droplets only is not a standard feature of the PLD and should be discussed. In Figure 7(a), SEM image of the WB$_2$/B target surface after ablation is presented. It shows the melted surface with a lot of holes.

The origin of holes can be correlated with the structure of the target presented in Figure 7(b). The dark areas mainly consist of pure boron [5]. Boron surface heats up much faster than the WB$_2$ surface. This is due to the higher reflectivity of laser radiation and a higher rate of temperature equalization for WB$_2$. In the case of tungsten borides, these values are R_{355} = 55% and a = 6.82 × 10^{-6} m^2/s, respectively, while, for boron, 28.5% and 1.48 × 10^{-6} m^2/s, respectively. Due to faster heating of boron, when it evaporates and then achieves critical temperature, the rest of material is melted only. This thesis is confirmed by temperature distributions on target axis for WB$_2$ and B, respectively (Figure 8). In the case of ablation of boron using 355 nm already at 9th nanosecond of pulse duration, the critical temperature (~10000 K) is achieved. At the same time, in the case of the target of pure WB$_2$, the maximum temperature does not exceed 4000 K. Taking into account melting temperature of WB$_2$ (2833 K [26]), the maximum depth of the liquid phase reaches about 0.07 μm at 9th ns (Figures 8 and 9) and increases during next time steps to 0.6 μm. After the cessation of laser impulse, the thermal diffusion and subsequent explosive boiling are a potential mechanism of droplets formation [14, 15]. The superheated liquid will undergo a transition into a mixture of vapour and liquid droplets, followed by delayed explosive boiling of the liquid-vapour mix.

Because of lack of appropriate WB$_2$ material properties, it is impossible to calculate the critical radius and the time for a spherically symmetric bubble in a superheated liquid volume [15]. However, as can be seen in Figure 9, after 100 ns, depth of melted material does not exceed 0.6 μm. At the same time, the maximum temperature falls below the boiling point of the both components of WB$_2$/B. This excludes the possibility of delayed explosive boiling and ejecting droplets with size of a few microns. It should be noted that the calculated depth of melting is in a satisfactory accord with the experimental results [5].

Therefore, the explosive release of the boron and related "detonation in liquid phase" can explain the large droplets found on the substrate. During laser heating, the pressure in boron is growing. Due to Hertz-Knudsen equation [13], the achieved vapour pressure is about 3 × 10^8 Pa when temperature is close to the critical. Abrupt grow of temperature and pressure causes creation of the detonation waves [37, 38]. Next, propagating shocks cause compression

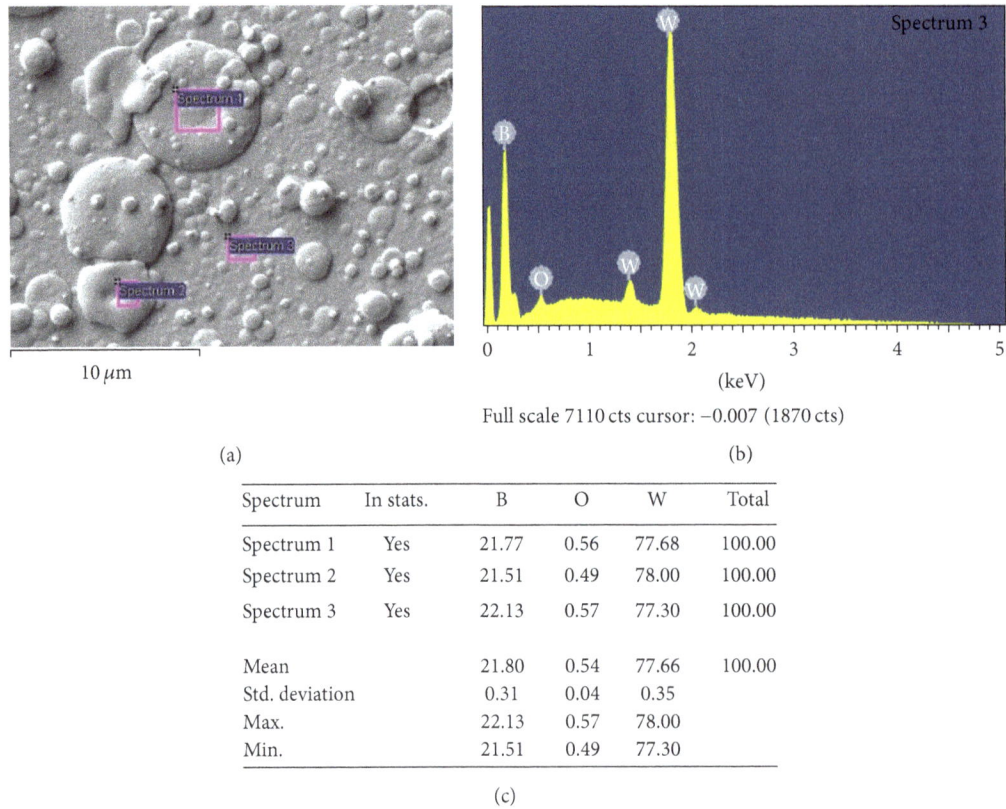

Full scale 7110 cts cursor: −0.007 (1870 cts)

(a)   (b)

Spectrum	In stats.	B	O	W	Total
Spectrum 1	Yes	21.77	0.56	77.68	100.00
Spectrum 2	Yes	21.51	0.49	78.00	100.00
Spectrum 3	Yes	22.13	0.57	77.30	100.00
Mean		21.80	0.54	77.66	100.00
Std. deviation		0.31	0.04	0.35	
Max.		22.13	0.57	78.00	
Min.		21.51	0.49	77.30	

(c)

FIGURE 6: SEM micrograph of deposited WB_3 film (magnification ×5000) with EDS measurement points (a) and the abundances of elements in the deposited film in the spot spectrum 3 (b) and table reporting the composition of the film in examined areas (c).

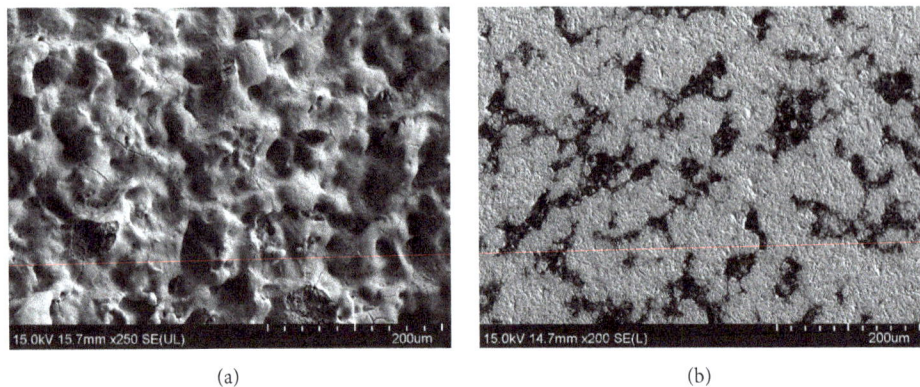

(a)   (b)

FIGURE 7: SEM microphotographs: (a) ablated target surface and (b) cross section of not yet ablated target [5].

of liquid WB_2. When the pressure between two or more waves is large enough to balance the forces associated with surface tension of WB_2, the ejection of droplets occurs. At the places where the distance between boron islands is too high (Figure 7(b)), the evaporation [22], fragmentation [36], and boiling explosion [15, 36] are possible mechanisms of ablation. The comparison of ablation mechanisms is schematically depicted in Figure 10. It should be noted that in all cases ablation parameters are the same changing only material of target.

During the interaction of the laser beam with a material, the target is heated to a temperature exceeding its boiling point. Evaporation takes place when the pressure ratio p_{KL}/p_{sat} across the Knudsen layer KL is lower than unity, where p_{KL} denotes pressure on the end of KL and p_{sat}— saturation pressure. Material evaporated from the target forms a thin layer of dense plume, consisting of electrons, ions, and neutrals (Figure 10(a)). It is assumed that the explosive boiling begins when the target temperature exceeds 0.9 of the critical temperature (Figure 10(b)). This type of

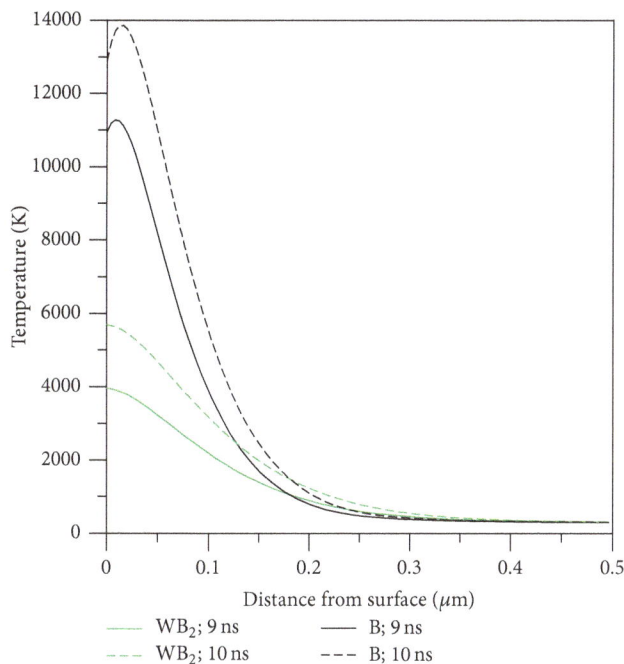

FIGURE 8: Distribution of target temperature along the axis at 9th ns and 10th ns from the beginning of laser pulse. Targets material: WB_2 and B; laser radiation 355 nm.

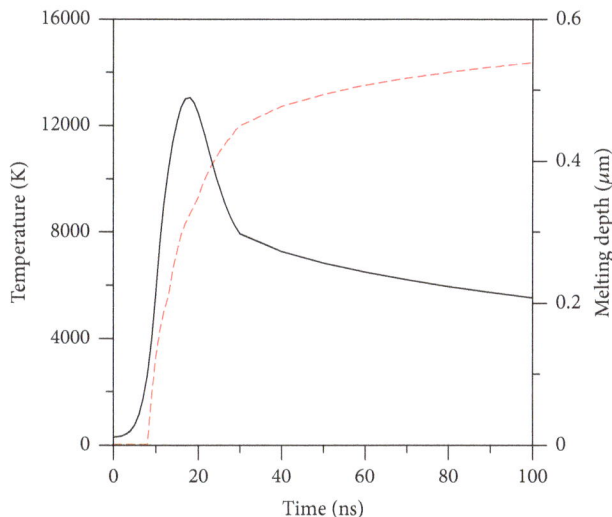

FIGURE 9: Target maximum temperature and melting depth during first 100 nanoseconds ($r = 0$). Broken line denotes melting depth of WB_2 target.

ablation results in the appearance of superheated particles in the plasma plume. In Figure 10(c) described above, "detonation in liquid phase" is shown. Due to such mechanisms of ablation, it is evident that the main constituent of the deposited film is large number of droplets from detonation. However, the smaller droplets can also be formed by gas phase condensation [39]. The overall morphology does not change with laser fluence (2.5–10 J/cm^2) and studied laser wavelength.

It should be noted that described mechanism of ablation is a special case related to the specific target and cannot be generalized, but it is possible in targets with boron excess.

XRD studies and nanoindentation of deposited layer show the additional effect occurring during laser ablation target WB_2/B. The effect is a phase change from WB_2/B to WB_3 [5]. It occurs most likely due to a very large gradients of temperature and pressure in the target and subsequently in the plasma.

4.3. Plasma. The temperature and velocity of plasma are responsible for the deposition rate of the film and its adhesion. Also, the shielding effects due to the absorption and radiation of plasma may have an impact on the course of ablation. All these parameters are dependent on laser wavelengths. The calculated distributions of plasma temperature and density after 100 ns of expansion (Figure 11) show that plasma temperatures are higher in the case of 1064 nm, but the densities and pressures are higher in the case of 355 nm, which is in agreement with experimental findings [40]. Smaller penetration depth and reflection coefficient R in the case of 355 nm causes a higher surface temperature of the target and thus a greater rate of ablation. The greater ablation rate results in larger mass density of the ablated plume and, hence, in higher pressures. An additional consequence of a higher ablation rate is slower expansion and smaller dimensions of the plasma plume. Higher plasma temperature in the case of 1064 nm is the result of lower density and stronger plasma absorption. At 100 ns after the beginning of the laser pulse, the maximum temperature for the third harmonic is 2 times lower than in the case of the first harmonic of laser irradiation and is about 45000 K. The decrease in density due to the expansion and intensive absorption of energy from the laser beam causes a rapid increase of temperature and velocity. After 100 ns, the velocity of the plume ablated by 355 nm is 14850 m/s, while the velocity of the plume ablated by 1064 nm reaches 20750 m/s.

For comparison, the plasma plume propagation was studied by optical imaging at 80–400 ns time delay after the laser shot. The images of tungsten plasma at delay time 100, 200, 300, 400 ns for 355 nm and 1064 nm are presented in Figures 12(a) and 12(b), respectively.

For every single image of the plasma plume, the half-Lorentzian plot was fitted to an axis intensity. For comparison, Figures 12(c) and 12(d) show ICCD images of boron and WB_2/B plasma formed during ablation with a wavelength of 355 nm. The boron plasma plume is much larger than tungsten plume and moves at the speed of 44×10^3 m/s, which is consistent with the results presented in [22]. In the case of WB_2/B plasma, the total velocity is similar to tungsten plasma (18500 m/s). It is worth noting that, contrary to the plumes of pure boron and pure tungsten which move with different velocities (Figures 12(a) and 12(c)) in WB_2/B plasma, both components are mixed and move with approximately the same velocity.

The Lorentzian plot position is taken as a maximum intensity of the plasma plume, and FWHM is a level of plasma expansion.

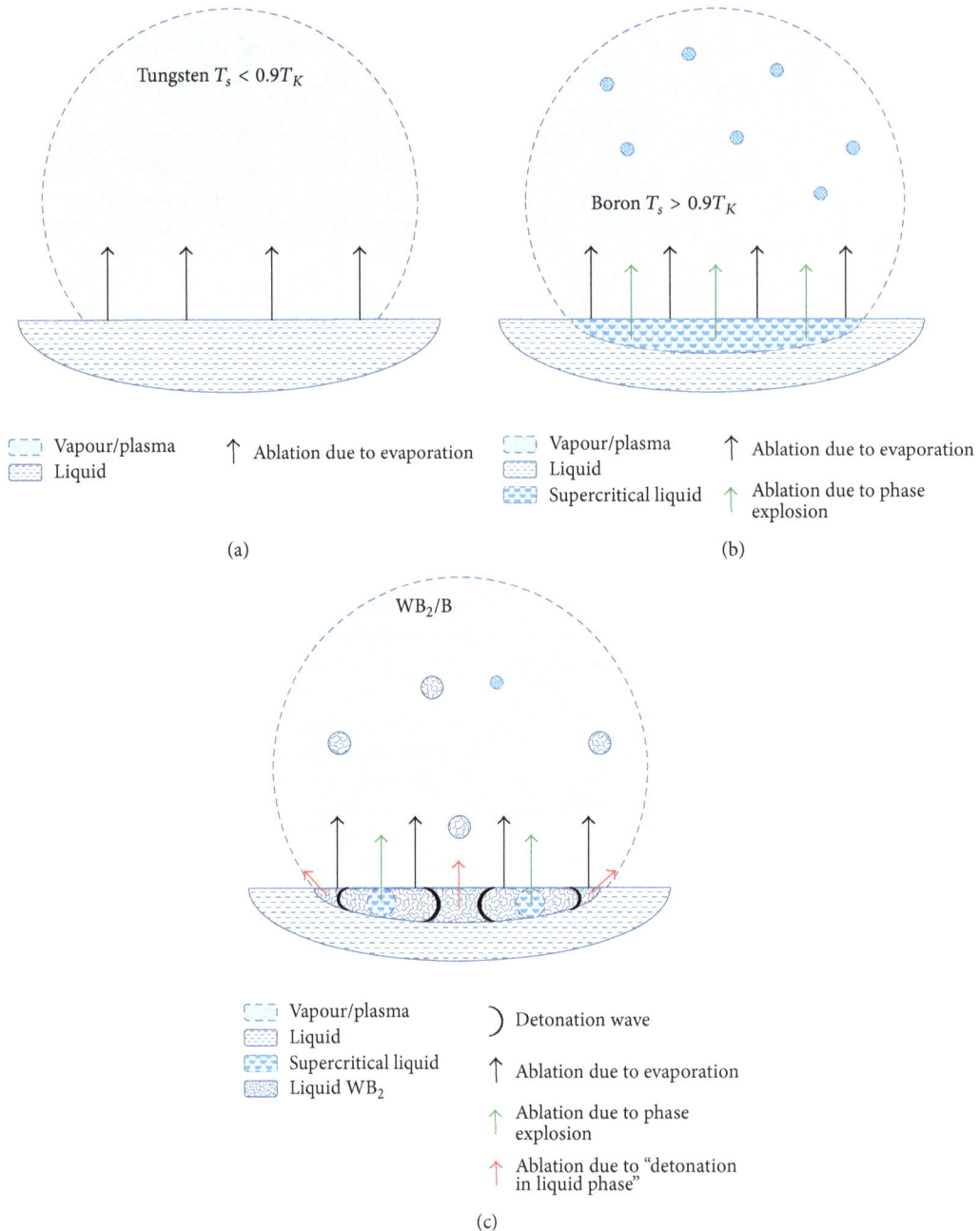

FIGURE 10: The comparison of ablation mechanisms: (a) evaporation, (b) phase explosion, and (c) "detonation in liquid phase."

Figure 13 presents tungsten plasma position and FWHM of Lorentzian plot fitted to plasma intensity distribution. On the basis of plasma propagation and expansion in time, the plasma displacement and expansion velocity were calculated, respectively. Total velocity of plasma front is defined as a sum of both velocities mentioned above. Results are presented in Table 1.

The quantitative comparison of the experimental plasma shape with the theoretical results is difficult because the images show plasma radiation, which depends both on plasma density and temperature. However, plasma dimensions are similar; for example, after 100 ns, the tungsten plasma diameter is about 2 mm. The front velocities obtained from the experimental plasma images are only 10% lower than those calculated in the model. For comparison, Figure 12(c) shows the sequence of ICCD images of boron plasma formed during ablation with a wavelength of 355 nm. The boron plasma plume is much larger than tungsten plume and moves at the speed of 44×10^3 m/s, which is consistent with the results presented in [22].

For the fluence 10 J/cm^2, the total absorption of the laser beam in the plasma plume is 6% in the case of 355 nm

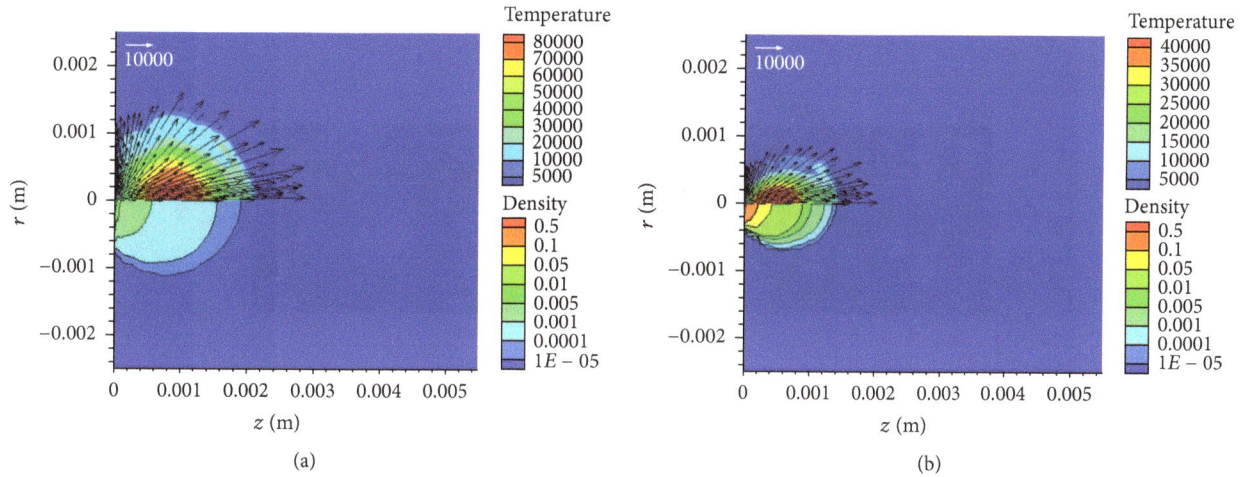

FIGURE 11: Distribution of density, temperature, and velocity in plasma induced during laser ablation of tungsten 100 ns after the beginning of the laser pulse for (a) 1064 nm and (b) 355 nm laser wavelength.

TABLE 1: Comparison of tungsten plasma velocity of W, B, and WB$_2$/B.

	Tungsten 355 nm	Tungsten 1064 nm	Boron 355 nm	WB$_2$/B 355 nm
Plasma displacement [m/s]	3580	5800	9305	8300
Plasma expansion [m/s]	10545	13000	32715	10200
Total velocity [m/s]	14125	18800	42017	18500
Model	14850	20750	~44000 [22]	—

radiation and about 3.5% in the case of 1064 nm. In this case, due to the small amount of absorbed energy, the plasma radiation does not affect the target heating.

5. Summary

In this paper, the interaction of Nd:YAG nanoseconds laser beam with WB$_2$/B, tungsten, and boron target induced during ablation plasma was studied. The investigations for two wavelengths of an Nd:YAG laser—355 nm and 1064 nm—and fluence of 10 J/cm^2 were made. On the base of results from the theoretical model and experimental investigations, the conclusions are as follows:

(1) The studies show the fundamental differences in ablation of WB$_2$/B, tungsten, and boron targets used for deposition of WB$_3$ superhard films. In case of tungsten, the evaporation of material is controlled by the plasma formation and consequently the absorption coefficient. The dense plasma plume can block laser radiation and limit energy transfer from the laser beam to the material. For boron, the explosive ablation is observed. Such behaviour is affected by

subsurface heating and transition to supercritical state. In the case of 1064 nm wavelength, the effect is magnified by the high penetration depth of the laser beam. Explosive nature of the ablation of boron has a crucial influence on the ablation of WB$_2$/B composite. The faster heating of boron in comparison with WB$_2$ case that, at the moment, when the boron exceeds a critical temperature, the rest of material is melted only. The laser beam causes the faster heating of boron than tungsten diboride. As a result, boron exceeds a critical temperature, when the rest of material is melted only. It results in the explosion in liquid phase and deposition of WB$_3$ film mainly from droplets embedded in continuous matrix. The creating of layers as a result of the deposition mainly from droplets is not a standard feature of the PLD. Therefore, in ceramics target with a boron excess [1, 5, 41], there may be adverse phenomena which have impact on the quality of the deposited film.

(2) In the case of boron ablation with the laser wavelength 1064 nm, the main reason of the higher amount of debris on the surface of the deposited films is the low absorption coefficient (1.3×10^6 m^{-1}). It causes the heating of much more material and the critical temperature is exceeded much later and at a greater depth than for 355 nm. It should be noted that, in so far described cases [21], the use of 1064 nm wavelength results in the deposition of films significantly smoother than 355 nm at the same fluence.

(3) Using of shorter wavelength (355 nm here) with the higher absorption coefficient causes faster heating of the target and thus higher ablation rate.

(4) A method estimating phase explosion occurrence based on material data such as critical temperature, thermal diffusivity, and optical properties is shown.

Time (ns)

(a)

Time (ns)

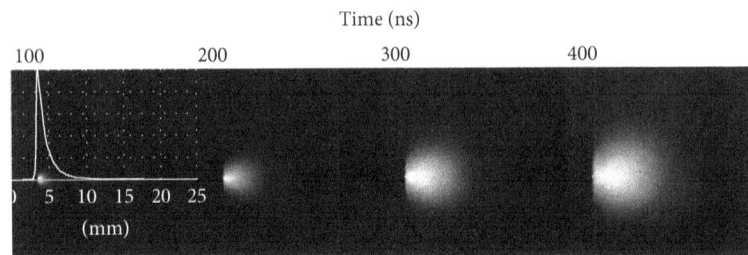

(b)

Time (ns)

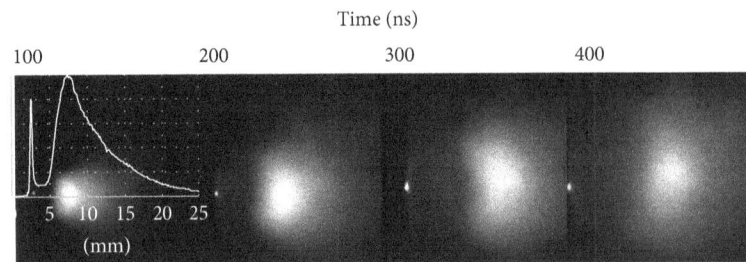

(c)

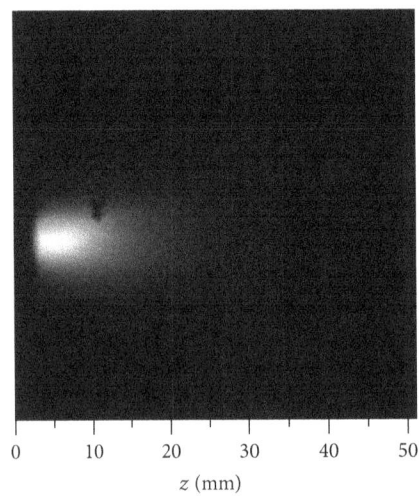

(d)

FIGURE 12: ICCD images of tungsten plasma for (a) tungsten 355 nm, (b) tungsten 1064 nm, (c) boron 355 nm, and (d) WB_2/B 355 nm. Visible light, exposure time 20 ns, and delay time 100, 200, 300, 400 ns, respectively (WB_2/B after 400 ns). Solid line denotes the intensity distribution of plasma radiation on the axis of plasma $r = 0$.

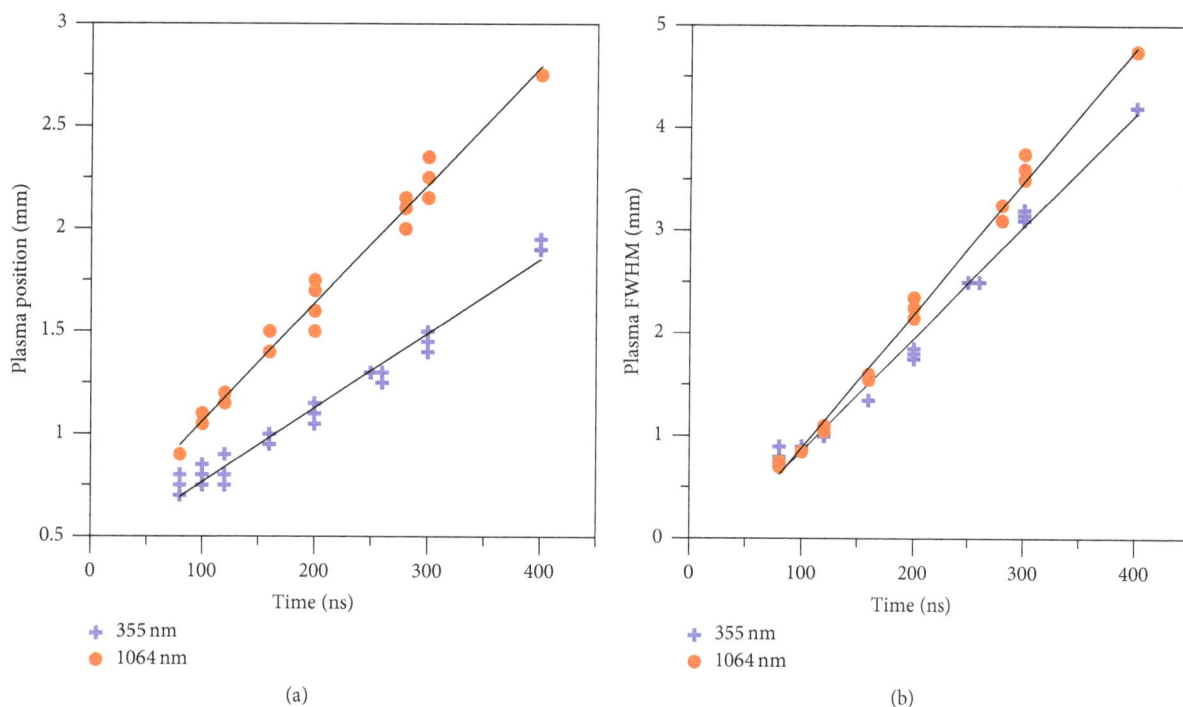

FIGURE 13: Experimental tungsten plasma plume parameters for 1064 and 355 nm laser wavelengths. (a) Velocity of plasma propagation and (b) velocity of plasma expansion.

Additional Points

(1) The genesis of droplets in novel, superhard WB_3 films on the base of optical and material properties is explained.

(2) The influence of laser wavelength on PLA/PLD of tungsten and boron is studied.

(3) The new mechanism of ablation "detonation in liquid phase" is proposed.

Competing Interests

The author declares that there are no competing interests regarding the publication of this paper.

Acknowledgments

This work was supported by the National Science Centre (Poland) and Research Project: UMO-2012/05/D/ST8/03052. The author wishes to thank Dr. Jacek Hoffman and M.S. Justyna Chrzanowska for substantive discussion, suggestions, and help in experiment.

References

[1] R. Mohammadi, A. T. Lech, M. Xie et al., "Tungsten tetraboride, an inexpensive superhard material," *Proceedings of the National Academy of Sciences of the United States of America*, vol. 108, no. 27, pp. 10958–10962, 2011.

[2] Y. Liang, Z. Fu, X. Yuan, S. Wang, Z. Zhong, and W. Zhang, "An unexpected softening from WB_3 to WB_4," *Europhysics Letters*, vol. 98, no. 6, Article ID 66004, 2012.

[3] J. V. Rau, A. Latini, R. Teghil et al., "Superhard tungsten tetraboride films prepared by pulsed laser deposition method," *ACS Applied Materials and Interfaces*, vol. 3, no. 9, pp. 3738–3743, 2011.

[4] D. Dellasega, G. Merlo, C. Conti, C. E. Bottani, and M. Passoni, "Nanostructured and amorphous-like tungsten films grown by pulsed laser deposition," *Journal of Applied Physics*, vol. 112, no. 8, Article ID 084328, 2012.

[5] T. Moscicki, J. Radziejewska, J. Hoffman et al., "WB_2 to WB_3 phase change during reactive spark plasma sintering and pulsed laser ablation/deposition processes," *Ceramics International*, vol. 41, no. 7, pp. 8273–8281, 2015.

[6] T. Scholza and K. Dickmann, "Investigation on particle formation during laser ablation process with high brilliant radiation," *Physics Procedia*, vol. 5, pp. 311–316, 2010.

[7] M. S. Tillack, D. W. Blair, and S. S. Harilal, "The effect of ionization on cluster formation in laser ablation plumes," *Nanotechnology*, vol. 15, no. 3, pp. 390–403, 2004.

[8] Y. Cheng, M. Shigeta, S. Choi, and T. Watanabe, "Formation mechanism of titanium boride nanoparticles by RF induction thermal plasma," *Chemical Engineering Journal*, vol. 183, pp. 483–491, 2012.

[9] L. D'Alessio, D. Ferro, V. Marotta, A. Santagata, R. Teghil, and M. Zaccagnino, "Laser ablation and deposition of Bioglass® 45S5 thin films," *Applied Surface Science*, vol. 183, no. 1-2, pp. 10–17, 2001.

[10] R. K. Singh and J. Narayan, "Pulsed-laser evaporation technique for deposition of thin films: physics and theoretical model," *Physical Review B*, vol. 41, no. 13, pp. 8843–8859, 1990.

[11] R. Kelly and A. Miotello, "Comments on explosive mechanisms of laser sputtering," *Applied Surface Science*, vol. 96–98, pp. 205–215, 1996.

[12] N. M. Bulgakova, A. B. Evtushenko, Y. G. Shukhov, S. I. Kudryashov, and A. V. Bulgakov, "Role of laser-induced plasma in ultradeep drilling of materials by nanosecond laser pulses," *Applied Surface Science*, vol. 257, no. 24, pp. 10876–10882, 2011.

[13] N. M. Bulgakova and A. V. Bulgakov, "Pulsed laser ablation of solids: transition from normal vaporization to phase explosion," *Applied Physics A: Materials Science and Processing*, vol. 73, no. 2, pp. 199–208, 2001.

[14] Q. Lu, S. S. Mao, X. Mao, and R. E. Russo, "Delayed phase explosion during high-power nanosecond laser ablation of silicon," *Applied Physics Letters*, vol. 80, no. 17, pp. 3072–3074, 2002.

[15] J. H. Yoo, S. H. Jeong, X. L. Mao, R. Greif, and R. E. Russo, "Evidence for phase-explosion and generation of large particles during high power nanosecond laser ablation of silicon," *Applied Physics Letters*, vol. 76, no. 6, pp. 783–785, 2000.

[16] T. A. Friedmann, K. F. McCarty, E. J. Klaus et al., "Pulsed laser deposition of BN onto silicon (100) substrates at 600°C," *Thin Solid Films*, vol. 237, pp. 48–56, 1994.

[17] N. M. Bulgakova, V. P. Zhukov, A. R. Collins, D. Rostohar, T. J.-Y. Derrien, and T. Mocek, "How to optimize ultrashort pulse laser interaction with glass surfaces in cutting regimes?" *Applied Surface Science*, vol. 336, pp. 364–374, 2015.

[18] M. Afif, J. P. Girardeau-Montaut, C. Tomas et al., "In situ surface cleaning of pure and implanted tungsten photocathodes by pulsed laser irradiation," *Applied Surface Science*, vol. 96-98, pp. 469–473, 1996.

[19] J.-W. Yoon and K. B. Shim, "Growth of crystalline boron nanowires by pulsed laser ablation," *Journal of Ceramic Processing Research*, vol. 12, no. 2, pp. 199–201, 2011.

[20] A. Suslova, O. El-Atwani, S. S. Harilal, and A. Hassanein, "Material ejection and surface morphology changes during transient heat loading of tungsten as plasma-facing component in fusion devices," *Nuclear Fusion*, vol. 55, no. 3, Article ID 033007, 2015.

[21] Q. Lu, S. S. Mao, X. Mao, and R. E. Russo, "Theory analysis of wavelength dependence of laser-induced phase explosion of silicon," *Journal of Applied Physics*, vol. 104, no. 8, Article ID 083301, 2008.

[22] T. Moscicki, "Expansion of laser-ablated two-component plume with disparate masses," *Physica Scripta T*, vol. 161, Article ID 014024, 2014.

[23] J. Hoffman, T. Moscicki, and Z. Szymanski, "Acceleration and distribution of laser-ablated carbon ions near the target surface," *Journal of Physics D: Applied Physics*, vol. 45, no. 2, Article ID 025201, 2012.

[24] ANSYS, *ANSYS® Academic Research, Release 16.0, Help System, Fluent Documentation*, ANSYS, Canonsburg, Pa, USA, 2015.

[25] T. Moscicki, J. Hoffman, and J. Chrzanowska, "The absorption and radiation of a tungsten plasma plume during nanosecond laser ablation," *Physics of Plasmas*, vol. 22, no. 10, Article ID 103303, 2015.

[26] C. L. Jiang, Z. L. Pei, Y. M. Liu, H. Lei, J. Gong, and C. Sun, "Determination of the thermal properties of AlB_2-type WB_2," *Applied Surface Science*, vol. 288, pp. 324–330, 2014.

[27] M. S. Koval'chenko, L. G. Bodrova, V. F. Nemchenko, and V. F. Kolotun, "Some physical properties of the higher borides of molybdenum and tungsten," *Journal of the Less-Common Metals*, vol. 67, no. 2, pp. 357–362, 1979.

[28] J. M. Leitnaker, M. G. Bowman, and P. W. Gilles, "High-temperature evaporation and thermodynamic properties of zirconium diboride," *The Journal of Chemical Physics*, vol. 36, no. 2, pp. 350–358, 1962.

[29] A. S. Bolgar, T. S. Verkhoglyadova, and G. V. Samsonov, "Vapor pressure and evaporation rate of certain heat-resistant compounds in a vacuum at high temperatures," *Izvestiya Akademii Nauk SSSR, Otdeleniye Tekhnicheskikh Nauk, Metallurgiya i Toplivo*, vol. 1, pp. 142–145, 1961.

[30] Y.-F. Wang, Q.-L. Xia, and Y. Yu, "First principles calculation on electronic structure, chemical bonding, elastic and optical properties of novel tungsten triboride," *Journal of Central South University*, vol. 21, no. 2, pp. 500–505, 2014.

[31] S. Blairs and M. H. Abbasi, "Correlation between surface tension and critical temperatures of liquid metals," *Journal of Colloid and Interface Science*, vol. 304, no. 2, pp. 549–553, 2006.

[32] A. D. Rakić, A. B. Djurišić, J. M. Elazar, and M. L. Majewski, "Optical properties of metallic films for vertical-cavity optoelectronic devices," *Applied Optics*, vol. 37, no. 22, pp. 5271–5283, 1998.

[33] K. Yahiaoui, T. Kerdja, and S. Malek, "Phase explosion in tungsten target under interaction with Nd:YAG laser tripled in frequency," *Surface and Interface Analysis*, vol. 42, no. 6-7, pp. 1299–1302, 2010.

[34] N. Morita and A. Yamamoto, "Optical and electrical properties of boron," *Japanese Journal of Applied Physics*, vol. 14, no. 6, article 825, 1975.

[35] A. Spiro, M. Lowe, and G. Pasmanik, "Drilling rate of five metals with picosecond laser pulses at 355, 532, and 1064 nm," *Applied Physics A*, vol. 107, no. 4, pp. 801–808, 2012.

[36] J. Hoffman, "The effect of recoil pressure in the ablation of polycrystalline graphite by a nanosecond laser pulse," *Journal of Physics D: Applied Physics*, vol. 48, no. 23, Article ID 235201, 2015.

[37] Y. B. Zel'dovich, V. B. Librovich, G. M. Makhviladze, and G. I. Sivashinsky, "On the development of detonation in a non-uniformly preheated gas," *Astronautica Acta*, vol. 15, no. 5-6, pp. 313–321, 1970.

[38] F. Zhang, Ed., *Shock Wave Science and Technology Reference Library, Volume 4: Heterogeneous Detonation*, Springer, 2009.

[39] R. Teghil, L. D'Alessio, M. Zaccagnino, D. Ferro, V. Marotta, and G. De Maria, "TiC and TaC deposition by pulsed laser ablation: a comparative approach," *Applied Surface Science*, vol. 173, no. 3-4, pp. 233–241, 2001.

[40] A. E. Hussein, P. K. Diwakar, S. S. Harilal, and A. Hassanein, "The role of laser wavelength on plasma generation and expansion of ablation plumes in air," *Journal of Applied Physics*, vol. 113, no. 14, Article ID 143305, 2013.

[41] H. H. Itoh, T. Matsudaira, S. Naka, H. Hamamoto, and M. Obayashi, "Formation process of tungsten borides by solid state reaction between tungsten and amorphous boron," *Journal of Materials Science*, vol. 22, no. 8, pp. 2811–2815, 1987.

6

Center Symmetric Local Multilevel Pattern Based Descriptor and Its Application in Image Matching

Hui Zeng,[1] **Xiuqing Wang,**[2] **and Yu Gu**[1]

[1]*School of Automation and Electrical Engineering, University of Science and Technology Beijing, Beijing 100083, China*
[2]*Vocational & Technical Institute, Hebei Normal University, Shijiazhuang 050031, China*

Correspondence should be addressed to Yu Gu; guyu@ustb.edu.cn

Academic Editor: Fortunato Tito Arecchi

This paper presents an effective local image region description method, called CS-LMP (Center Symmetric Local Multilevel Pattern) descriptor, and its application in image matching. The CS-LMP operator has no exponential computations, so the CS-LMP descriptor can encode the differences of the local intensity values using multiply quantization levels without increasing the dimension of the descriptor. Compared with the binary/ternary pattern based descriptors, the CS-LMP descriptor has better descriptive ability and computational efficiency. Extensive image matching experimental results testified the effectiveness of the proposed CS-LMP descriptor compared with other existing state-of-the-art descriptors.

1. Introduction

Image matching is one of the fundamental research areas in the fields of computer vision and it can be used in 3D reconstruction, panoramic image stitching, image registration, robot localization, and so forth. The task of image matching is to search the corresponding points between two images that are projected by the same 3D point. Generally, the image matching has the following three steps. At first, the interest points are detected and their local support regions are determined. Then the descriptors of the feature points are constructed. Finally the corresponding points are determined through matching their descriptors. In the above three steps, the descriptor construction is the key factor that can influence the performance of the image matching [1]. In this paper, we focus on the effective local image descriptor construction method and its application in image matching.

For an ideal local image descriptor, it should have high discriminative power and be robust to many kinds of image transformations, such as illumination changes, image geometric distortion, and partial occlusion [2, 3]. Many research efforts have been made for local image descriptor construction and several comparative studies have shown that the SIFT-like descriptors perform best [4]. The SIFT

(Scale Invariant Feature Transform) descriptor is built by a 3D histogram of gradient locations and orientations where the contribution to bins is weighted by the gradient magnitude and a Gaussian window overlaid over the region [5]. Its dimension is 128 and is invariant to image scale and rotation transforms and robust to affine distortions, changes in 3D viewpoint, addition of noise, and changes in illumination. Because of the good performance of SIFT descriptor, many varieties of SIFT descriptor have been proposed. For example, the PCA-SIFT descriptor is a 36-dimensional vector by applying PCA (Principal Component Analysis) on gradient maps and it can be fast for image matching [6]. The SURF (Speeded Up Robust Features) descriptor uses integral image to compute the gradient histograms and it can speed up the computations effectively while preserving the quality of SIFT [7]. Furthermore, GLOH (Gradient Location-Orientation Histogram) [4], Rank-SIFT [8], and RIFT (Rotation-Invariant Feature Transform) [9] are also proposed based on the construction method of SIFT descriptor.

The LBP (Local Binary Pattern) operator has been proved a powerful image texture feature which has been successfully used in face recognition, image retrieval, texture segmentation, and facial expression recognition [10]. It has several advantages which are suitable for local image descriptor construction, such as computational simplicity and invariance to linear illumination. But the LBP operator tends to produce a rather long histogram, especially when the number of neighboring pixels increases. So it is not suitable for image matching. The CS-LBP (Center Symmetric Local Binary Pattern) descriptor can address the dimension problem while retaining the powerful ability of texture description [11]. It combines the advantages of the SIFT descriptor and LBP operator and performs better than the SIFT descriptor in the field of image matching. To improve the description ability and the robustness to image transformation, several generalized descriptors have been proposed, such as the CS-LTP (Center Symmetric Local Ternary Pattern) descriptor [12], the IWCS-LTP (Improved Weighted Center Symmetric Local Ternary Pattern) descriptor [13], and the WOS-LTP (Weighted Orthogonal Symmetric Local Ternary Pattern) descriptor [14]. Among the above descriptors, the WOS-LTP descriptor has better performance than the SIFT descriptor and IWCS-LTP descriptor. It divides the neighboring pixels into several orthogonal groups to reduce the dimension of the histogram and an adaptive weight is used to adjust the contribution of the code in histogram calculation. For the WOS-LTP descriptor, the quantization level of the intensity values is three. The image intensity variant information of the local neighborhood has not been fully investigated. Furthermore, LDTP (Local Directional Texture Pattern) is another kind of local texture pattern, which includes both directional and intensity information [15]. Although the LDTP descriptor is consistent against noise and illumination changes, its dimension is high. To solve this problem, CLDTP (Compact Local Directional Texture Pattern) is a proposed descriptor [16], which not only reduces the dimension of LDTP descriptor but also retains the advantages of LDTP descriptor.

Vector quantization is an effective method for texture description, and it is robust to noise and illumination variation [17, 18]. The number of quantization levels is an important parameter. The lower the quantization level, the less the discriminative information the descriptor has. However if the quantization level is increased, the descriptor's robustness to noise and illumination variety will degrade. For LBP or LTP operator based descriptor, the differences of the local intensity values are quantized in two or three levels. They have better robustness to illumination variety, but their discriminative abilities are degraded. If we increase the quantization level directly according to the encoding mode of the LBP or LTP operator, the dimension of the descriptor will increase dramatically. LQP (Local Quantized Pattern) was proposed to solve these problems. It uses large local neighborhoods and deeper quantization with domain-adaptive vector quantization. But it uses visual word quantization to separate local patterns and uses a precompiled lookup table to cache the final coding for speed. Its main constraint is the size of the lookup table, and it is not suitable to be used for image matching. So for the multiply quantization level based descriptor, the effective encoding method is the key problem to balance the relationship between the discriminative ability and the dimension of the descriptor.

In this paper, we present a novel encoding method for local image descriptor named as CS-LMP (Center Symmetric Local Multilevel Pattern) operator, which can encode the differences of the local intensity values using multiply quantization levels. The CS-LMP descriptor is constructed based on the CS-LMP operator, and it can describe the local image region more detailedly without increasing the dimension of the descriptor. To make the descriptor containing more spatial structural information, we use a SIFT-like grid to divide the interest region. Compared with binary/ternary pattern based descriptor, it not only has better discriminative ability but also has higher computational efficiency. The performance of the CS-LMP descriptor is evaluated for image matching and the experimental results demonstrate its robustness and distinctiveness.

The rest of the paper is organized as follows. In Section 2, the CS-LBP, CS-LTP, and the WOS-LTP operator and descriptor construction methods are reviewed. Section 3 gives the CS-LMP operator, the CS-LMP histogram, and the construction method of the CS-LMP descriptor. The image matching experiments are conducted and their experimental results are presented in Section 4. Some concluding remarks are listed in Section 5.

2. Related Work

Before presenting in detail the CS-LMP operator and the CS-LMP descriptor, we give a brief review of the CS-LBP, CS-LTP, and WOS-LTP methods that form the basis for our work.

2.1. CS-LBP and CS-LTP. The CS-LBP operator is a modified version of the well-known LBP operator, which compares center symmetric pairs of pixels in the neighborhood [11]. Formally, the CS-LBP operator can be represented as

$$\text{CS-LBP}_{R,N}(u,v) = \sum_{i=0}^{(N/2)-1} s\left(n_i - n_{i+(N/2)}\right) 2^i,$$

$$s(x) = \begin{cases} 1 & x > T \\ 0 & \text{otherwise,} \end{cases}$$

(1)

where n_i and $n_{i+(N/2)}$ correspond to the gray values of center symmetric pairs of pixels of N equally spaced pixels on a circle with radius R. Obviously, the CS-LBP operator can produce $2^{N/2}$ distinct values, resulting in $2^{N/2}$-dimensional histogram. It should be noticed that the CS-LBP descriptor is obtained by the binary codes, which is computed from the differences of the intensity value between pairs of the opposite pixels in a neighborhood. For the CS-LBP descriptor, its dimension is $2^{N/2}$ and the quantization level of the intensity values is two.

The CS-LTP operator is powerful texture operator [12]. It uses the encoding method similar to the CS-LBP operator, and extends the quantization level of the intensity values from two to three. The encoding method of the CS-LTP operator can be formulated as

$$\text{CS-LTP}_{R,N}(u,v) = \sum_{i=0}^{(N/2)-1} s\left(n_i - n_{i+(N/2)}\right) 3^i,$$

$$s(x) = \begin{cases} 2, & x \geq T \\ 1, & -T < x < T \\ 0, & x \leq -T. \end{cases} \tag{2}$$

From (2) we can see that the dimension of the CS-LTP histogram is $3^{N/2}$. Compared with CS-LBP descriptor, the CS-LTP descriptor has better descriptive ability for local textural variants, but its dimension is higher and its computational amount is larger.

2.2. WOS-LTP. The WOS-LTP descriptor is constructed based on the OS-LTP (Orthogonal Symmetric Local Ternary Pattern) operator [14]. The OS-LTP operator is an improved version of the LTP operator to reduce the dimension of the histogram. It takes only orthogonal symmetric four neighboring pixels into account. At first, the neighboring pixels are divided into $N/4$ orthogonal groups. Then the OS-LTP code is computed separately for each group. Given N neighboring pixels equally located in a circle of radius R around a central pixel at (u,v), the encoding method of the OS-LTP operator can be formulated as

$$\text{OS-LTP}_{R,N}^{(i)}(u,v) = s\left(n_{i-1} - n_{(i-1)+2[N/4]}\right) 3^0$$
$$+ s\left(n_{(i-1)+[N/4]} - n_{(i-1)+3[N/4]}\right) 3^1,$$

$$s(x) = \begin{cases} 2, & x \geq T \\ 1, & -T < x < T \\ 0, & x \leq -T, \end{cases} \tag{3}$$

$$i = 1, 2, \ldots, \frac{N}{4}.$$

From (3) we can see that there are $N/4$ different 4-orthogonal-symmetric neighbor operators, each of which consists of turning the four orthogonal neighbors by one position in a clockwise direction. Existing research work has shown that, compared with the LTP, CS-LTP, and ICS-LTP operator, the OS-LTP operator has better discriminative ability for describing local texture structure and could achieve better robustness against noise interference.

The WOS-LTP descriptor is built by concatenating the weighted histograms of the subregions together, which uses the OS-LTP variance of the local region as an adaptive weight to adjust its contribution to the histogram [14]. Suppose the size of the image patch is $W \times H$; the WOS-LTP histogram can be computed as

$$H_i(k) = \sum_{u=1}^{W} \sum_{v=1}^{H} f_i\left(\text{OS-LTP}_{R,N}^{(i)}(u,v), k\right),$$

$$f_i(x,y)$$
$$= \begin{cases} |n_{i-1} - n_{(i-1)+2[N/4]}| + |n_{(i-1)+[N/4]} - n_{(i-1)+3[N/4]}|, & x = y \\ 0, & x \neq y, \end{cases}$$
$$i = 1, 2, \ldots, \left[\frac{N}{4}\right], \tag{4}$$

where $k \in [0, K]$ and K is the maximal value of the OS-LTP operator. Existing experimental results have shown that, compared with SIFT and IWCS-LTP descriptor, the WOS-LTP descriptor can not only better characterize the image texture but also achieve higher computational efficiency. But its quantization level of the intensity values is three, and the intensity variant information has not been fully used.

3. Center Symmetric Local Multilevel Pattern

3.1. CS-LMP Operator. Although the local binary or ternary pattern based descriptors have good performance, they are limited to very coarse quantization and increasing the size of local neighborhood increases the histogram dimensions exponentially. These shortcomings limit the local descriptors' descriptive ability and prevent them from leveraging all the available information. To solve these problems, we proposed a novel encoding method, named CS-LMP operator. It encodes the differences of the local intensity values according to the thresholds, and a pixel has $N/2$ encoding values. The selection method of pixels is the same as the LBP operator. The readers can find the detailed selection steps in [10].

At first, we define the thresholds $T = [-d_m, \ldots, -d_2, -d_1, 0, d_1, d_2, \ldots, d_m]$ to divide the differences of the local intensity values into multiply intervals:

$$g_1 = (-\infty, -d_m],$$
$$g_2 = (-d_m, -d_{m-1}],$$
$$\vdots \tag{5}$$
$$g_{2m} = (d_{m-1}, -d_m],$$
$$g_{2m+1} = (d_m, +\infty).$$

Then the CS-LMP code of the pixel at (u,v) is illustrated as

$$\text{CS-LMP}_{R,N}^{(i)}(u,v) = q\left(n_i - n_{i+(N/2)}\right),$$
$$i = 0, 1, \ldots, \frac{N}{2}, \tag{6}$$
$$q(x) = t, \quad x \in g_t, \ t = 1, 2, \ldots, 2m+1.$$

From (6) we can see that the CS-LMP operator has no exponential computations, and its maximum value is $2m + 1$. Furthermore, the difference of the local intensity value is quantized $2m + 1$ levels. Compared with the local binary/ternary patterns, the CS-LMP can describe the local texture more flexibly and detailedly. Figure 1 shows an example of calculating the CS-LMP operator with 8 neighboring

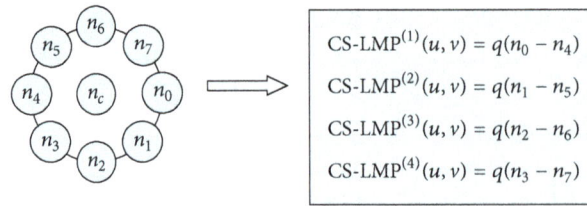

FIGURE 1: Calculation of the CS-LMP operator with 8 neighboring pixels.

FIGURE 2: Examples of four encoding methods ($T = 5$, $N = 8$, $d_1 = 3$, and $d_2 = 26$).

pixels, and the CS-LMP code has 4 values. Figure 2 gives examples of four encoding methods. As shown in Figure 2, for the flat image area and the texture variance image area, the code of the CS-LBP, CS-LTP, and OS-LTP operator remains unchanged. But there exist distinct differences between the CS-LMP code of the flat image area and that of the texture variance image area. So we can conclude that our CS-LMP operator appears to have better discriminative ability for describing local image texture.

3.2. CS-LMP Histogram.
For the local image region, the CS-LMP histogram can be obtained using the CS-LMP code of each pixel. For the CS-LBP, CS-LTP, and WOS-LTP operator, the final code of a pixel has one value by performing binary or ternary computation. Their corresponding histogram can be obtained by computing the number of each kind of code. Different from the above three kinds of operators, the CS-LMP code of a pixel has $N/2$ values; the occurrences of each kind of value should be computed. The CS-LMP histogram can be represented as

$$H_i(k) = \sum_{u=1}^{W}\sum_{v=1}^{H} f\left(\text{CS-LMP}_{R,N}^{(i)}(u,v), k\right),$$

$$i = 0, 1, \ldots, \frac{N}{2}, \quad (7)$$

$$f(x,y) = \begin{cases} 1, & x = y \\ 0, & x \neq y, \end{cases}$$

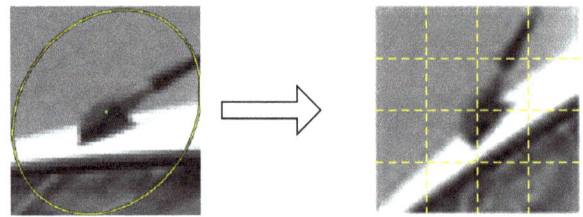

FIGURE 3: The normalization and division of a detected region.

where $k \in [0, 2m+1]$, $2m+1$ is the maximal value of the CS-LMP operator. Based on (7), the CS-LMP descriptor of the local image region can be obtained by concatenating $N/2$ histograms together.

3.3. CS-LMP Descriptor.
To construct the CS-LMP descriptor, the interest regions are firstly detected by the Hessian-Affine detector [19], which are used to compute the descriptors. Then the detected regions are normalized to the circular regions with the same size 41×41. As shown in Figure 3, the detected ellipse region is rotated in order that the long axis of the ellipse is aligned to the positive v-axis of the local u-v image coordinate system, and the rotated elliptical region is geometrically mapped to a canonical circular region by an affine transformation. The normalized regions are invariant to scale, rotation, and affine transformation. In the rest of this paper, the normalized regions are used for local image descriptor construction.

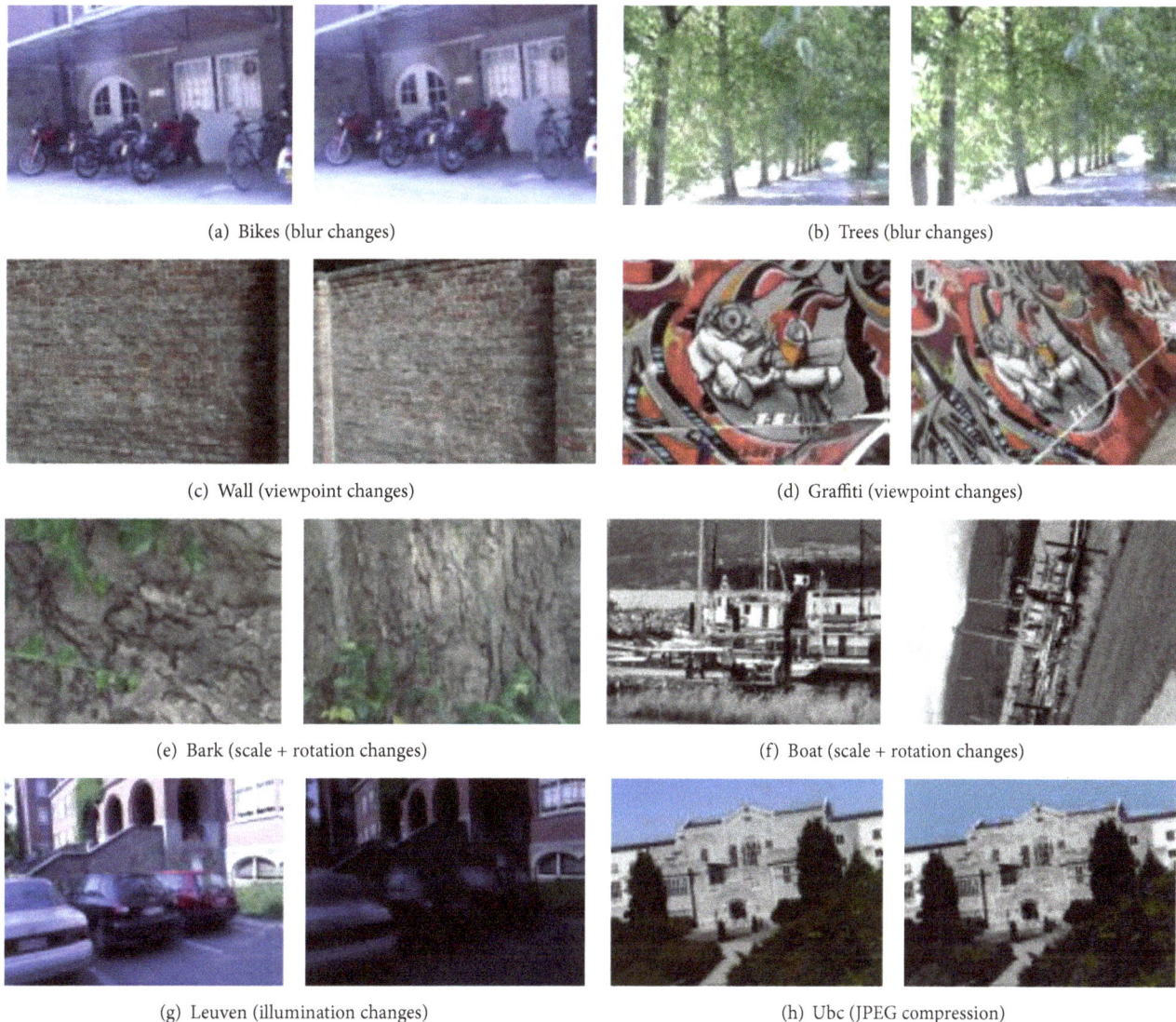

(a) Bikes (blur changes)

(b) Trees (blur changes)

(c) Wall (viewpoint changes)

(d) Graffiti (viewpoint changes)

(e) Bark (scale + rotation changes)

(f) Boat (scale + rotation changes)

(g) Leuven (illumination changes)

(h) Ubc (JPEG compression)

FIGURE 4: Testing image pairs.

In order to integrate the spatial information into the descriptor, we divide the normalized region into 16 (4 × 4) subregions using the grid division method of the SIFT descriptor. For each subregion, we firstly compute the CS-LMP codes of each pixel, respectively. Then the CS-LMP histograms are obtained using (7). For a single subregion, the dimension of the CS-LMP descriptor is $(N/2) \times (2m + 1)$. Finally we connect all the histograms of different subregions together to obtain the final CS-LMP descriptor for the interest region. So the dimension of the CS-LMP descriptor is $16 \times (N/2) \times (2m + 1)$. For example, we compare the dimensions of three descriptors based on the CS-LTP method, WOS-LTP method, and the CS-LMP method, respectively, whose quantization levels are all three. Assume the number of the neighboring pixels is 12; then the variable m is 1 and the dimensions of the CS-LTP, WOS-LTP, and CS-LMP descriptor are 16×729, 16×27, and 16×18, respectively. We can conclude that the dimension of the CS-LMP descriptor is significantly reduced.

Furthermore, two normalization steps are performed on the CS-LMP descriptor to reduce the effects of the illumination. At first, the descriptor vector is normalized to unit length to remove the linear illumination changes. Then the elements of the normalized descriptor vector are truncated by 0.2 in order to reduce the impact of the nonlinear illumination changes. Finally, the descriptor vector is renormalized to unit length and truncated by 0.2 again.

4. Experimental Results

In this paper, we use the Mikolajczyk et al. dataset [20] to evaluate the performance of the SIFT, WOS-LTP, and CS-LMP descriptor by image matching experiments. This dataset includes eight types of scene images with different illumination and geometric distortion transformations and it has the ground-truth matches through estimated homography matrix. As shown in Figure 4, we randomly select one image pair in each category from the dataset. In the image matching

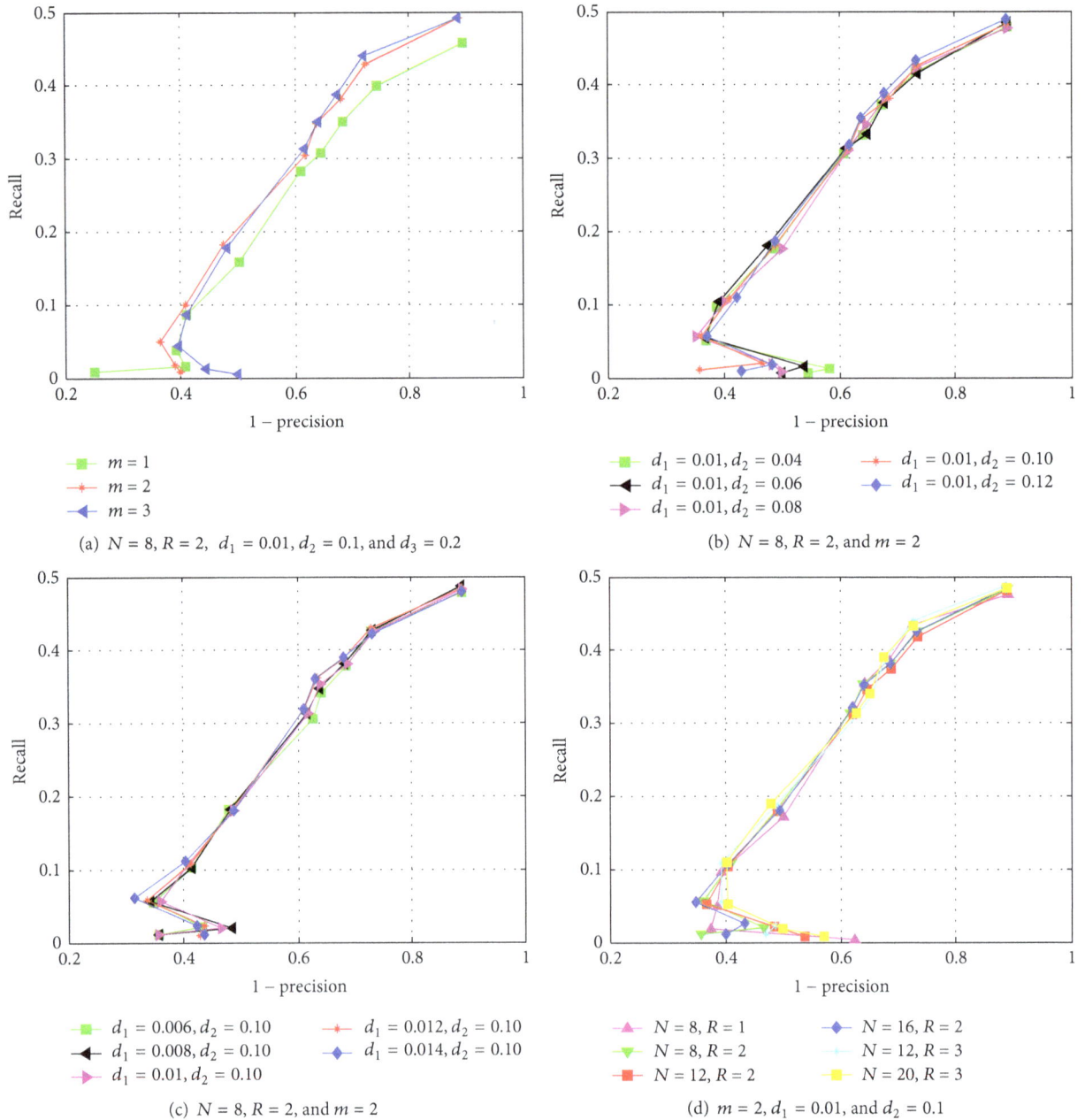

FIGURE 5: The results of the CS-LMP descriptor with different parameter settings.

experiments, we firstly use the Hessian-Affine detector to obtain the interest regions. Then the interest regions are normalized to the circular regions and the gray values of the regions are transformed to lie between 0 and 1. Finally the descriptor of each interest region is constructed and the nearest neighbor distance ratio (NNDR) matching algorithm is performed to obtain the matching points. Here we select the Euclidean distance as similarity measure. The parameter settings of the SIFT descriptor and WOS-LTP descriptor are the same as the original proposed papers [5, 14].

The Recall-Precision criterion is used to evaluate the matching results, which is computed from the number of the correct matches and the number of the false matches between a pair of images. Two interest regions are matched if the distance between their descriptors is below a threshold t, and a match is correct if the overlap error is smaller than 0.5. The Recall-Precision curve can be obtained by changing the distance threshold t. So a perfect descriptor would give a recall equal to 1 for any precision.

4.1. Parameter Evaluation. There are four parameters in the proposed CS-LMP descriptor: the number of neighboring pixels N, the radius of neighboring pixels R, the thresholds $T = [-d_m, \ldots, -d_2, -d_1, 0, d_1, d_2, \ldots, d_m]$, and the variable m. We conducted image matching experiments to investigate the effects of different parameters on the performance of

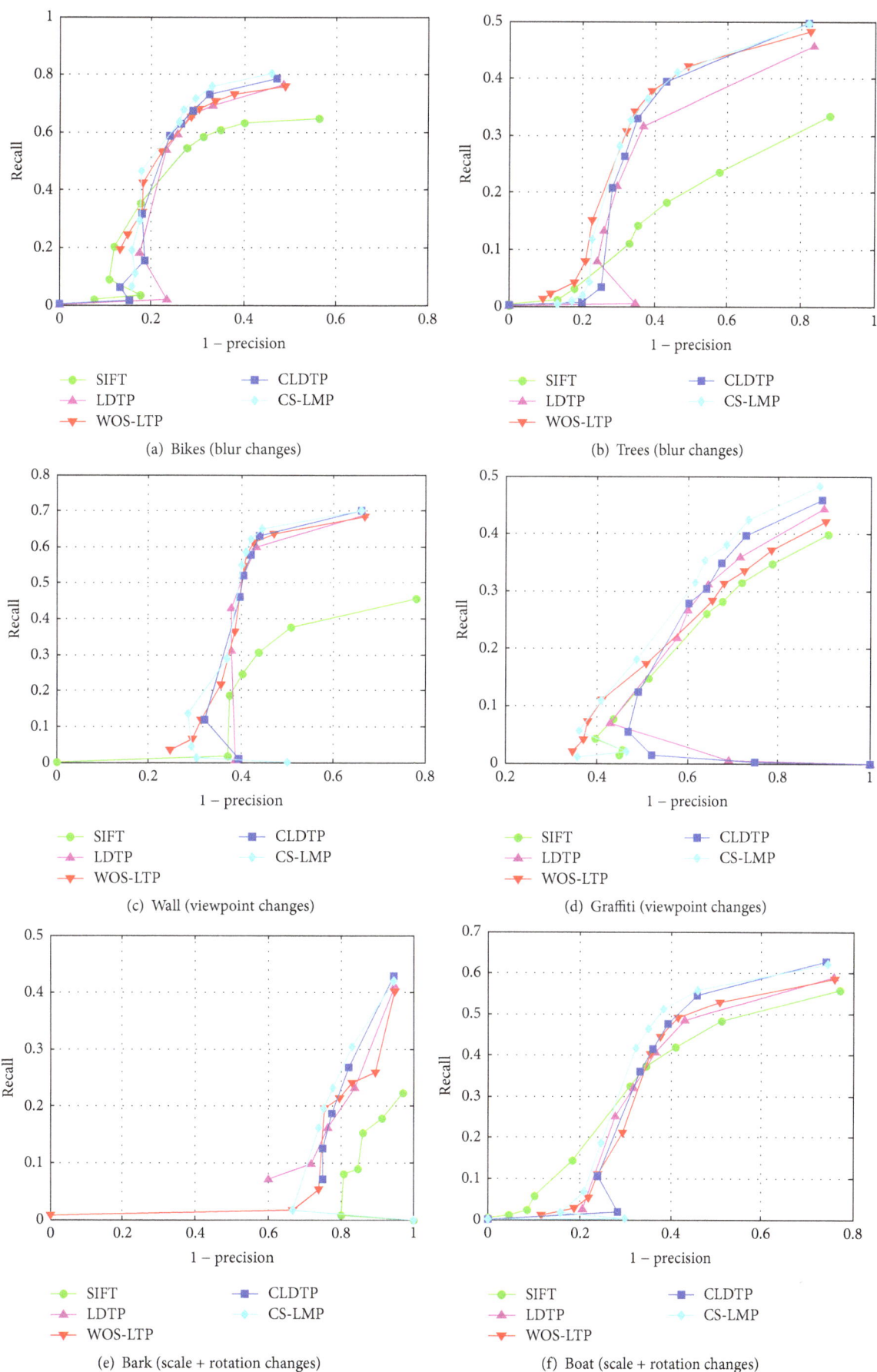

(a) Bikes (blur changes)

(b) Trees (blur changes)

(c) Wall (viewpoint changes)

(d) Graffiti (viewpoint changes)

(e) Bark (scale + rotation changes)

(f) Boat (scale + rotation changes)

FIGURE 6: Continued.

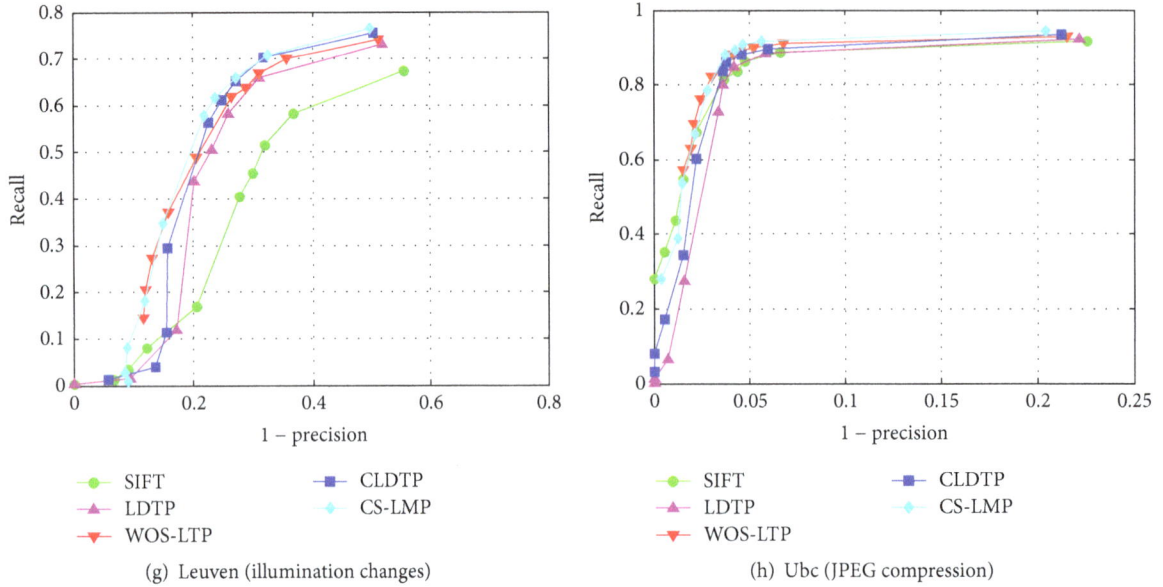

(g) Leuven (illumination changes)

(h) Ubc (JPEG compression)

FIGURE 6: The matching results of the testing image pairs.

the proposed descriptor. The matching results are shown in Figure 5, and only one parameter was varied in one experiment. For simplicity, the parameters N and R were evaluated in pairs, such as $(8,1)$, $(8,2)$, $(12,2)$, $(16,2)$, and $(12,3)$.

Figure 5(a) shows the results with different variable m. From Figure 5(a) we can see that the performances of image matching are similar when $m = 2$ and $m = 3$, and they are better than the performance when $m = 1$. As the dimension of the CS-LMP descriptor with $m = 3$ is much larger than that with $m = 2$, the variable m is fixed to 2 in the following experiments to obtain higher computational efficiency. Figures 5(b) and 5(c) show the results with different thresholds $T = [-d_2, -d_1, 0, d_1, d_2]$. We can see that the CS-LMP descriptor performs similarly under different thresholds, and the best performance is achieved when $T = [-0.1, -0.01, 0, 0.01, 0.1]$. Figure 5(d) shows the results with different (N, R). From the results we can observe that our proposed descriptor is not sensitive to small changes. To achieve the balance between the computation amount and matching performance, the optimal parameter setting of (N, R) is selected as $(8, 2)$. Based on the above analysis, we select the following parameter settings for the following image matching experiments: $N = 8$, $R = 2$, $T = [-0.1, -0.01, 0, 0.01, 0.1]$, and $m = 2$.

4.2. Matching Evaluation. In this section, we compare the performance of the proposed CS-LMP descriptor with the SIFT descriptor, the LDTP descriptor, the WOS-LTP descriptor, and the CLDTP descriptor using the Recall-Precision criterion. The image matching results of the testing images are shown in Figure 6. Figures 6(a) and 6(b) show the results for blur changes. Figure 6(a) is the results for the structured scene and Figure 6(b) for the textured scene. We can see that the SIFT descriptor obtained the lowest

score. The CL-LMP descriptor performs best than other descriptors for the structured scene, and the performance of the WOS-LTP and CS-LMP descriptor is similar for the textured scene. Figures 6(c) and 6(d) show the performance of descriptors for viewpoint changes. Figure 6(c) is the results for the structured scene and Figure 6(d) for the textured scene. Figures 6(e) and 6(f) show the results to evaluate the descriptors for combined image rotation and scale changes. Figure 6(g) shows the results for illumination changes. From Figure 6(c) we can see that the SIFT descriptor obtains worse results and the performances of the other four descriptors are similar. From Figures 6(d)–6(g) we can see that the CS-LMP descriptor obtains the best matching score, and the CLDTP descriptor obtains the second good matching score. Figure 6(h) shows the results to evaluate the influence of JPEG compression. From Figure 6(h) we can see that the five kinds of descriptors perform better than other cases, and the performance of the CS-LMP descriptor is slightly better than the other four descriptors. Based on the above analysis, we can conclude that the CS-LMP descriptor performs better than the well-known state-of-the-art SIFT descriptor, the LDTP descriptor, the WOS-LTP descriptor, and the CLDTP descriptor.

5. Conclusions

This paper presents a novel CS-LMP descriptor and its application in image matching. The CS-LMP descriptor is constructed based on the CS-LMP operator and the CS-LMP histogram, which can describe the local image region using multiply quantization levels. The constructed CS-LMP descriptor not only contains the gradient orientation information, but also contains the spatial structural information of the local image region. Furthermore, the dimension of the CS-LMP descriptor is much lower than the binary/ternary pattern based descriptor when they use the same quantization level. Our experimental results show that the CS-LMP descriptor performs better than the SIFT descriptor, the LDTP descriptor, the WOS-LTP descriptor, and the CLDTP descriptor. So the CS-LMP descriptor is effective for local image description. In the future work, we will further improve its performance and apply it in object recognition.

Competing Interests

The authors declare that there are no competing interests regarding the publication of this paper.

Acknowledgments

This paper is supported by the National Natural Science Foundation of China (Grants no. 61375010, no. 61175059, and no. 61472031) and Beijing Higher Education Young Elite Teacher Project (Grant no. YETP0375).

References

[1] X. Yang and K.-T. T. Cheng, "Local difference binary for ultrafast and distinctive feature description," *IEEE Transactions on Pattern Analysis and Machine Intelligence*, vol. 36, no. 1, pp. 188–194, 2014.

[2] K. Liao, G. Liu, and Y. Hui, "An improvement to the SIFT descriptor for image representation and matching," *Pattern Recognition Letters*, vol. 34, no. 11, pp. 1211–1220, 2013.

[3] C. Zhu, C.-E. Bichot, and L. Chen, "Image region description using orthogonal combination of local binary patterns enhanced with color information," *Pattern Recognition*, vol. 46, no. 7, pp. 1949–1963, 2013.

[4] K. Mikolajczyk and C. Schmid, "A performance evaluation of local descriptors," *IEEE Transactions on Pattern Analysis and Machine Intelligence*, vol. 27, no. 10, pp. 1615–1630, 2005.

[5] D. G. Lowe, "Distinctive image features from scale-invariant keypoints," *International Journal of Computer Vision*, vol. 60, no. 2, pp. 91–110, 2004.

[6] Y. Ke and R. Sukthankar, "PCA-SIFT: a more distinctive representation for local image descriptors," in *Proceedings of the Conference on Computer Vision and Pattern Recognition (CVPR '04)*, pp. 506–513, 2004.

[7] H. Bay, T. Tuytelaars, and L. Van, "SURF: speeded up robust features," in *Computer Vision—ECCV 2006: 9th European Conference on Computer Vision, Graz, Austria, May 7–13, 2006. Proceedings, Part I*, vol. 3951 of *Lecture Notes in Computer Science*, pp. 404–417, Springer, Berlin, Germany, 2006.

[8] B. Li, R. Xiao, Z. Li, R. Cai, B.-L. Lu, and L. Zhang, "Rank-SIFT: learning to rank repeatable local interest points," in *Proceedings of the IEEE Conference on Computer Vision and Pattern Recognition (CVPR '11)*, pp. 1737–1744, Providence, RI, USA, June 2011.

[9] S. Lazebnik, C. Schmid, and J. Ponce, "A sparse texture representation using local affine regions," *IEEE Transactions on Pattern Analysis and Machine Intelligence*, vol. 27, no. 8, pp. 1265–1278, 2005.

[10] T. Ojala, M. Pietikäinen, and T. Mäenpää, "Multiresolution gray-scale and rotation invariant texture classification with local binary patterns," *IEEE Transactions on Pattern Analysis and Machine Intelligence*, vol. 24, no. 7, pp. 971–987, 2002.

[11] M. Heikkilä, M. Pietikäinen, and C. Schmid, "Description of interest regions with local binary patterns," *Pattern Recognition*, vol. 42, no. 3, pp. 425–436, 2009.

[12] R. Gupta, H. Patil, and A. Mittal, "Robust order-based methods for feature description," in *Proceedings of the IEEE Conference on Computer Vision and Pattern Recogntion (CVPR '10)*, pp. 334–341, San Francisco, Calif, USA, June 2010.

[13] H. Zeng, Z.-C. Mu, and X.-Q. Wang, "A robust method for local image feature region description," *Acta Automatica Sinica*, vol. 37, no. 6, pp. 658–664, 2011.

[14] M. Huang, Z. Mu, H. Zeng, and S. Huang, "Local image region description using orthogonal symmetric local ternary pattern," *Pattern Recognition Letters*, vol. 54, pp. 56–62, 2015.

[15] A. R. Rivera, J. R. Castillo, and O. Chae, "Local directional texture pattern image descriptor," *Pattern Recognition Letters*, vol. 51, pp. 94–100, 2015.

[16] H. Zeng, R. Zhang, M. Huang, and X. Wang, "Compact local directional texture pattern for local image description," *Advances in Multimedia*, vol. 2015, Article ID 360186, 10 pages, 2015.

[17] S. Hussain and B. Triggs, "Visual recognition using local quantized patterns," in *Computer Vision—ECCV 2012: 12th European Conference on Computer Vision, Florence, Italy, October 7–13, 2012, Proceedings, Part II*, vol. 7573 of *Lecture Notes in Computer Science*, pp. 716–729, Springer, Berlin, Germany, 2012.

[18] V. Ojansivu and J. Heikkilä, "Blur insensitive texture classification using local phase quantization," in *Proceedings of the 3rd International Conference on Image and Signal Processing (ICISP '08)*, A. Elmoataz, O. Lezoray, F. Nouboud, and D. Mammass, Eds., vol. 5099 of *Lecture Notes in Computer Science*, pp. 236–243, Cherbourg-Octeville, France, July 2008.

[19] K. Mikolajczyk and C. Schmid, "Scale & affine invariant interest point detectors," *International Journal of Computer Vision*, vol. 60, no. 1, pp. 63–86, 2004.

[20] K. Mikolajczyk, T. Tuytelaars, C. Schmid et al., "A comparison of affine region detectors," *International Journal of Computer Vision*, vol. 65, no. 1-2, pp. 43–72, 2005.

7

Nonparaxial Propagation of Vectorial Elliptical Gaussian Beams

Wang Xun, Huang Kelin, Liu Zhirong, and Zhao Kangyi

Department of Applied Physics, East China Jiaotong University, Nanchang, Jiangxi 330013, China

Correspondence should be addressed to Liu Zhirong; liuzhirong_2003@126.com

Academic Editor: Roberto Morandotti

Based on the vectorial Rayleigh-Sommerfeld diffraction integral formulae, analytical expressions for a vectorial elliptical Gaussian beam's nonparaxial propagating in free space are derived and used to investigate target beam's propagation properties. As a special case of nonparaxial propagation, the target beam's paraxial propagation has also been examined. The relationship of vectorial elliptical Gaussian beam's intensity distribution and nonparaxial effect with elliptic coefficient α and waist width related parameter f_ω has been analyzed. Results show that no matter what value of elliptic coefficient α is, when parameter f_ω is large, nonparaxial conclusions of elliptical Gaussian beam should be adopted; while parameter f_ω is small, the paraxial approximation of elliptical Gaussian beam is effective. In addition, the peak intensity value of elliptical Gaussian beam decreases with increasing the propagation distance whether parameter f_ω is large or small, and the larger the elliptic coefficient α is, the faster the peak intensity value decreases. These characteristics of vectorial elliptical Gaussian beam might find applications in modern optics.

1. Introduction

With the development of laser technology, research on semiconductor lasers [1], microoptical technologies [2–4], and highly focusing field [5–7] has become deeper. In practical application, the problem that would be confronted is of a beam with large divergence angle or small spot size that is of the order of light wavelength. In this case, the theory of optical propagation and transformation based on paraxial approximation is no longer valid [8], and it needs strict electromagnetic field theory to solve the problem of beam's nonparaxial propagation. In recent decades, several research methods about solving beam's nonparaxial propagation have been developed, such as vectorial Rayleigh-Sommerfeld diffraction integral method [9], perturbation power series method [10], transition operators [11], angular spectrum representation [12], and virtual source point technique [13]. And vectorial Rayleigh-Sommerfeld diffraction method has been used to treat various beam's nonparaxial propagation problems [14–17].

An elliptical Gaussian beam can be radiated and realized by semiconductor diode laser [18]. In the past few years, some nonparaxial propagation properties of vectorial elliptical Gaussian beams have been reported, such as the far-field beam divergence angle [19], diffracted at a circular and a rectangular aperture [20, 21]. Since the semiconductor laser

beam has a large divergence angle, it would become necessary to consider the target beam's nonparaxial propagation. In this work, we use the vectorial Rayleigh-Sommerfeld diffraction integral formulae to solve the nonparaxial propagation of a vectorial elliptical Gaussian beam. Target beam's nonparaxial propagation analytical expressions are derived and used to investigate its propagation properties, including the evolution of intensity and shape of elliptical Gaussian beam with different elliptic coefficient α and different waist width related parameter f_ω, and the relationships of elliptical Gaussian beam's nonparaxial effect and its intensity distributions with elliptic coefficient α as well as parameter f_ω are analyzed.

2. Nonparaxial Propagation of Vectorial Elliptical Gaussian Beams in Free Space

Let us consider the incident field of elliptical Gaussian beam, which is polarized in the x direction and can be defined by

$$
\begin{aligned}
\begin{array}{l} E_x\left(\mathbf{r}_0, 0\right) \\ E_y\left(\mathbf{r}_0, 0\right) \end{array}
&= \begin{cases} E_x\left(x_0, y_0, 0\right) \\ 0 \end{cases} \\
&= \begin{cases} E_0 \exp\left[-\dfrac{x_0^2 + \left(\alpha y_0\right)^2}{\omega^2}\right] \\ 0, \end{cases}
\end{aligned} \tag{1}
$$

where $\mathbf{r}_0 = x_0\mathbf{i} + y_0\mathbf{j}$ and $\mathbf{i}$ and $\mathbf{j}$ are the unit vectors in x and y directions, respectively. E_0 is a constant, ω is the waist width, and α is elliptic coefficient, which denotes the ratio of elliptical Gaussian beam's waist width in x and y directions.

According to the vectorial Rayleigh-Sommerfeld diffraction integral formulae, the nonparaxial propagation of light beam in the half-space $z > 0$ turns out to be [9]

$$E_x(x, y, z) = -\frac{1}{2\pi} \iint_{-\infty}^{\infty} E_x(x_0, y_0,$$
$$0) \frac{\partial \mathbf{R}(\mathbf{r}, \mathbf{r}_0)}{\partial z} dx_0\, dy_0, \tag{2a}$$

$$E_y(x, y, z) = -\frac{1}{2\pi} \iint_{-\infty}^{\infty} E_y(x_0, y_0,$$
$$0) \frac{\partial \mathbf{R}(\mathbf{r}, \mathbf{r}_0)}{\partial z} dx_0\, dy_0, \tag{2b}$$

$$E_z(x, y, z) = \frac{1}{2\pi} \iint_{-\infty}^{\infty} \left(E_x(x_0, y_0, 0) \frac{\partial \mathbf{R}(\mathbf{r}, \mathbf{r}_0)}{\partial x} \right.$$
$$+ E_y(x_0, y_0, \tag{2c}$$
$$\left. 0) \right) dx_0\, dy_0 \frac{\partial \mathbf{R}(\mathbf{r}, \mathbf{r}_0)}{\partial y},$$

where $\mathbf{r} = x\mathbf{i} + y\mathbf{j} + z\mathbf{k}$ and $\mathbf{k}$ denotes the unit vector in z direction. $E_{x,y,z}(x, y, z)$ are components of the E vector along x, y, and z directions in an arbitrary plane z, respectively,

$$\mathbf{R}(\mathbf{r}, \mathbf{r}_0) = \frac{\exp(ik|\mathbf{r} - \mathbf{r}_0|)}{|\mathbf{r} - \mathbf{r}_0|}, \tag{3}$$

where $k = 2\pi/\lambda$ is the wave number and λ is the incident wavelength. When $|\mathbf{r} - \mathbf{r}_0| \gg \lambda$, $|\mathbf{r} - \mathbf{r}_0|$ can be approximately expanded into [19]

$$|\mathbf{r} - \mathbf{r}_0| \approx r + \frac{x_0^2 + y_0^2 - 2xx_0 - 2yy_0}{2r}. \tag{4}$$

So (3) can be expressed as

$$\mathbf{R}(\mathbf{r}, \mathbf{r}_0) = \frac{1}{r} \exp\left[ik\left(r + \frac{x_0^2 + y_0^2 - 2xx_0 - 2yy_0}{2r} \right) \right], \tag{5}$$

where $r = (x^2 + y^2 + z^2)^{1/2}$.

Substituting (5) into (2a)–(2c), we obtain

$$E_x(x, y, z) = -\frac{iz}{\lambda r^2} \iint_{-\infty}^{\infty} E_x(x_0, y_0, 0)$$
$$\cdot \exp\left[ik\left(r + \frac{r_0^2 - 2xx_0 - 2y_0 y}{2r} \right) \right] dx_0\, dy_0, \tag{6a}$$

$$E_y(x, y, z) = -\frac{iz}{\lambda r^2} \iint_{-\infty}^{\infty} E_y(x_0, y_0, 0)$$
$$\cdot \exp\left[ik\left(r + \frac{r_0^2 - 2xx_0 - 2y_0 y}{2r} \right) \right] dx_0\, dy_0, \tag{6b}$$

$$E_z(x, y, z) = \frac{i}{\lambda r^2} \iint_{-\infty}^{\infty} \left[(x - x_0) E_x(x_0, y_0, 0) \right.$$
$$\left. + (y - y_0) E_y(x_0, y_0, 0) \right] \tag{6c}$$
$$\cdot \exp\left[ik\left(r + \frac{r_0^2 - 2xx_0 - 2y_0 y}{2r} \right) \right] dx_0\, dy_0.$$

Substituting (1) into (6a), we can obtain

$$E_x(x, y, z) = -\frac{izE_0}{\lambda r^2} \exp(ikr)$$
$$\cdot \int_{-\infty}^{\infty} \exp\left(-\frac{x_0^2}{\omega_0^2} + \frac{ikx_0^2}{2r} - \frac{ikxx_0}{r} \right) dx_0 \tag{7}$$
$$\cdot \int_{-\infty}^{\infty} \exp\left(-\frac{\alpha^2 y_0^2}{\omega_0^2} + \frac{iky_0^2}{2r} - \frac{ikyy_0}{r} \right) dy_0.$$

By utilizing the following integral formula [20]

$$\int_{-\infty}^{\infty} x^n \exp\left(-\mu x^2 + 2vx \right) dx$$
$$= n! \sqrt{\frac{\pi}{\mu}} \left(\frac{v}{\mu} \right)^n \exp\left(\frac{v^2}{\mu} \right) \sum_{s=0}^{[n/2]} \frac{1}{(n - 2s)! s!} \left(\frac{\mu}{4v^2} \right)^s, \tag{8}$$
$$[\operatorname{Re} \mu > 0],$$

(7) can be expressed as follows:

$$E_x(x, y, z)$$
$$= \frac{iE_0\pi z}{\lambda r^2 \sqrt{pq}} \exp(ikr) \exp\left[\frac{k^2}{4r^2} \left(\frac{x^2}{p} + \frac{y^2}{q} \right) \right], \tag{9}$$

with p and q being given by

$$p = k^2 f_\omega^2 - \frac{ik}{2r}, \tag{10a}$$

$$q = \alpha^2 k^2 f_\omega^2 - \frac{ik}{2r}, \tag{10b}$$

$$f_\omega = \frac{1}{k\omega}. \tag{10c}$$

Similarly, substituting (1) into (6b) and (6c), and recalling integral formula (8), we can obtain other elements of the elliptical Gaussian beam:

$$E_y(x, y, z) = 0, \tag{11}$$

$$E_z(x, y, z)$$
$$= \frac{iE_0\pi x}{\lambda r^2 \sqrt{pq}} \tag{12}$$
$$\cdot \left(1 + \frac{ik}{2pr} \right) \exp(ikr) \exp\left[-\frac{k^2}{4r^2} \left(\frac{x^2}{p} + \frac{y^2}{q} \right) \right].$$

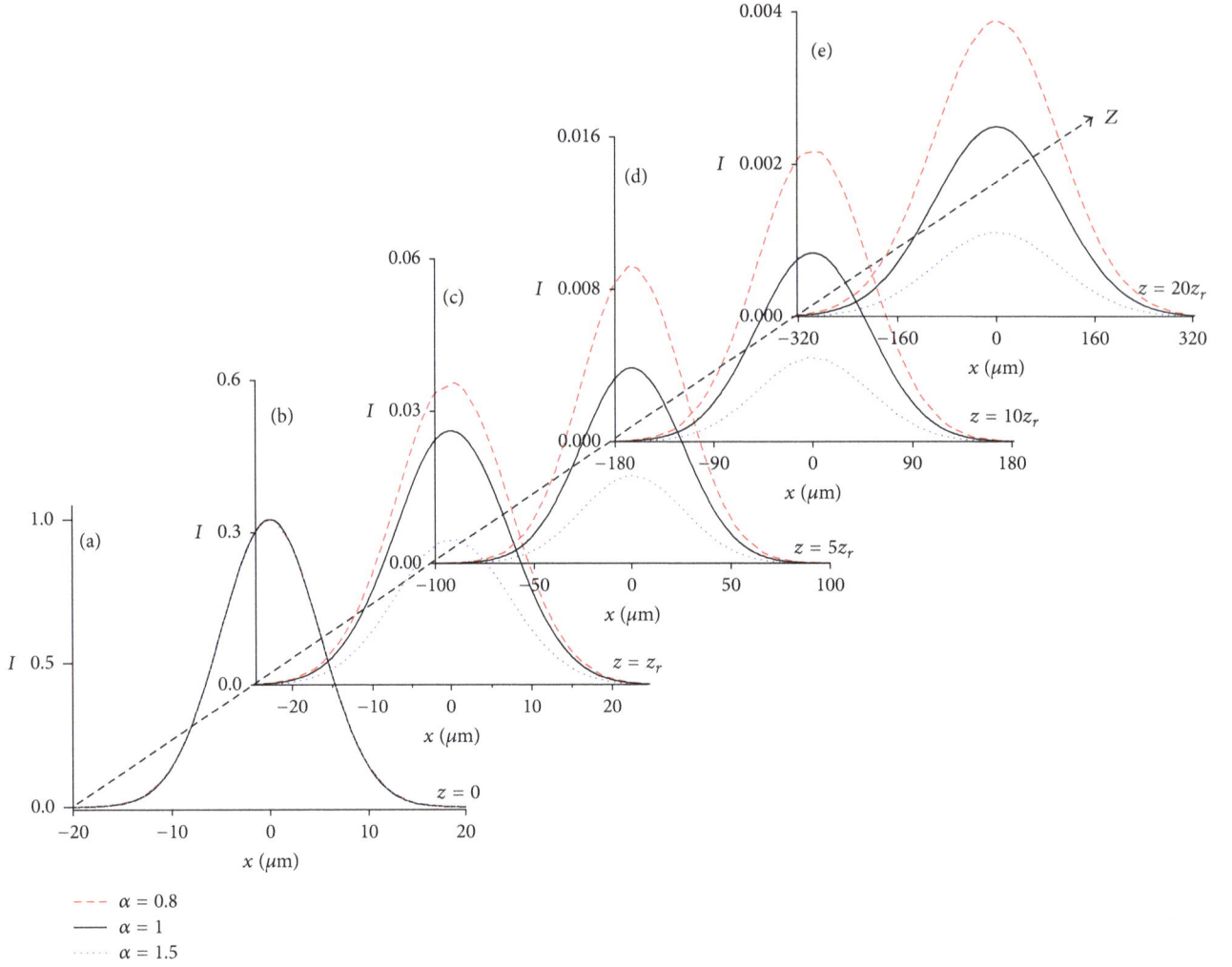

FIGURE 1: Intensity distribution of elliptical Gaussian beam for $f_\omega = 0.01$ as a function of x in free space in different planes: (a) $z = 0$; (b) $z = z_r$; (c) $z = 5z_r$; (d) $z = 10z_r$; (e) $z = 20z_r$. Several curves correspond to different elliptic coefficient α: $\alpha = 0.8$ (dashed curve); $\alpha = 1$ (solid curve); $\alpha = 1.5$ (dotted curve).

The intensity distribution of nonparaxial propagation of the elliptical Gaussian beam at the point (x, y, z) can be expressed as follows:

$$I(x, y, z) = I_x(x, y, z) + I_y(x, y, z) + I_z(x, y, z)$$

$$= \left| E_x(x, y, z) \right|^2 + \left| E_y(x, y, z) \right|^2 \qquad (13)$$

$$+ \left| E_z(x, y, z) \right|^2,$$

where $I_x(x, y, z)$, $I_y(x, y, z)$, and $I_z(x, y, z)$ are the intensity distributions of the x, y, and z components of the field, respectively.

The paraxial propagation of elliptical Gaussian beam can be dealt with as a special case by using the paraxial expansion

$$r \approx z + \frac{x^2 + y^2}{2z}. \qquad (14)$$

Accordingly, (7) can be reduced to

$$E_p(x, y, z) = \frac{iE_0\pi}{\lambda z \sqrt{p'q'}} \exp(ikz) \exp\left(ik\frac{x^2 + y^2}{2z}\right)$$

$$\cdot \exp\left[\frac{k^2}{4z^2}\left(\frac{x^2}{p'} + \frac{y^2}{q'}\right)\right], \qquad (15)$$

where

$$p' = k^2 f_\omega^2 - \frac{ik}{2z}, \qquad (16a)$$

$$q' = \alpha^2 k^2 f_\omega^2 - \frac{ik}{2z}. \qquad (16b)$$

Equations (9) and (12) are the main analytical results for elliptical Gaussian beam's nonparaxial propagating in free space, and (15) is the paraxial analytical formula for elliptical Gaussian beam's paraxial propagating in free space.

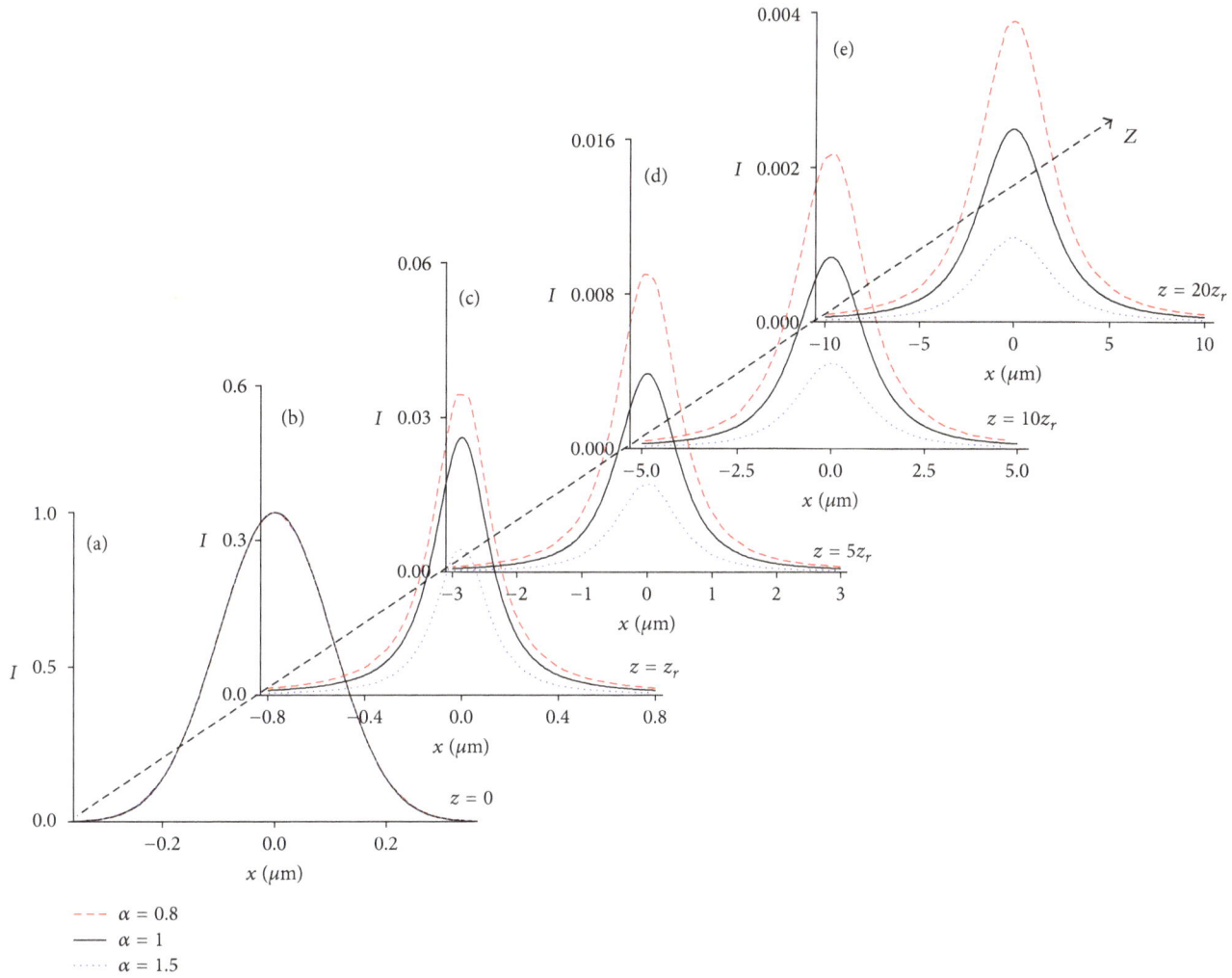

FIGURE 2: Intensity distribution of elliptical Gaussian beam for $f_\omega = 0.5$ as a function of x in free space in different planes: (a) $z = 0$; (b) $z = z_r$; (c) $z = 5z_r$; (d) $z = 10z_r$; (e) $z = 20z_r$. Several curves correspond to different elliptic coefficient α: $\alpha = 0.8$ (dashed curve); $\alpha = 1$ (solid curve); $\alpha = 1.5$ (dotted curve).

3. Numerical Simulations and Analysis

In order to confirm the relationship of elliptical Gaussian beam's intensity distribution and nonparaxial effects with elliptic coefficient α as well as parameter f_ω, according to the analytical expressions obtained above, we have carried out the numerical simulations of intensity distributions of vectorial elliptical Gaussian beam's nonparaxial propagating in free space. For the convenience of comparison, the light peak intensity in the input plane $z = 0$ is set to 1. The propagation distance is normalized to z/z_r, where $z_r = \pi\omega^2/\lambda$ is the Rayleigh distance, and the incident wavelength is 632.8 nm.

The evolution behavior of intensity distributions of nonparaxial elliptical Gaussian beams with several elliptic coefficients α in different observation planes is depicted in Figures 1 and 2, which correspond to two different waist width related parameters f_ω, respectively. From Figure 1, for small value $f_\omega = 0.01$—that is, elliptical Gaussian beam's waist width ω is large—one can see that all the normalized intensity distributions of nonparaxial elliptical Gaussian beams would

preserve Gaussian type when the propagation distance ranges from $z = 0$ to $z = 20z_r$, while for large value $f_\omega = 0.5$—that is, elliptical Gaussian beam's waist width ω is small (see Figure 2)—we can find that, with the increase of propagation distance z, the transverse intensity profiles turn into Gaussian-like shape quickly. Besides, numerical results also show that the peak intensity value decreases when the propagation distance increases, and the larger the value of elliptic coefficient α is, the faster the peak intensity value decreases, no matter whether f_ω is large or small.

Figure 3 gives the intensity distributions of elliptical Gaussian beam in the plane $z = 10z_r$ for different parameter f_ω. The elliptic coefficient α is fixed to 0.8, 1, and 1.5 from the first row to the third row, respectively. The corresponding longitudinal component I_z of nonparaxial elliptical Gaussian beam and paraxial result I_p of elliptical Gaussian beam are also depicted together for comparison. From Figures 3(a1)– 3(a3), one can see that no matter what value of the elliptic coefficient α is, for small value of $f_\omega = 0.1$, I_z is very small and can be neglected; hence, the curves of total intensity

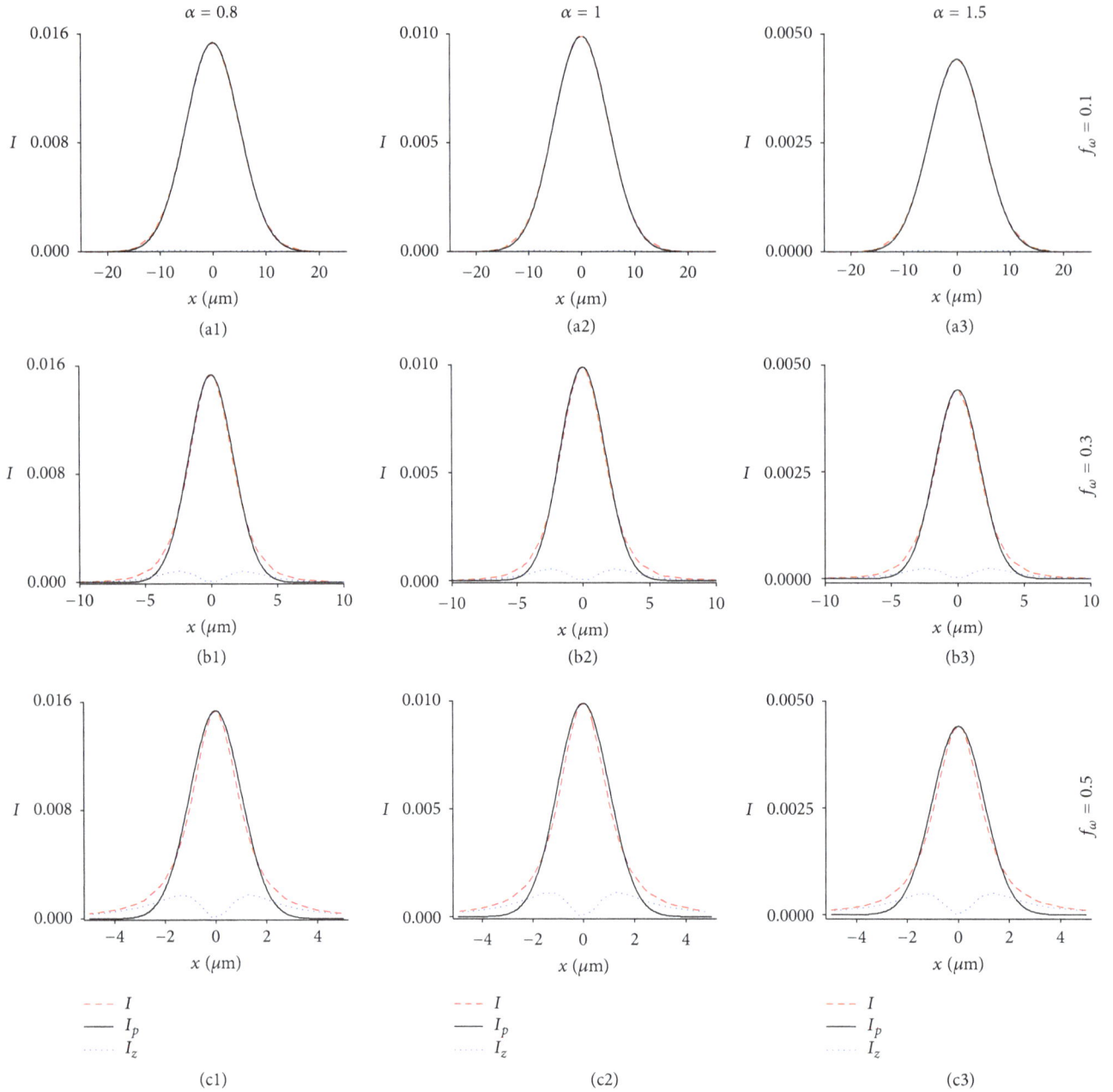

FIGURE 3: Intensity distributions $I(x, y, z)$, $I_z(x, y, z)$, and $I_p(x, y, z)$ of elliptical Gaussian beam in the plane $z = 10z_r$ for different parameters f_ω and elliptic coefficient α: (a1) $f_\omega = 0.1$, $\alpha = 0.8$; (b1) $f_\omega = 0.3$, $\alpha = 0.8$; (c1) $f_\omega = 0.5$, $\alpha = 0.8$; (a2) $f_\omega = 0.1$, $\alpha = 1$; (b2) $f_\omega = 0.3$, $\alpha = 1$; (c2) $f_\omega = 0.5$, $\alpha = 1$; (a3) $f_\omega = 0.1$, $\alpha = 1.5$; (b3) $f_\omega = 0.3$, $\alpha = 1.5$; (c3) $f_\omega = 0.5$, $\alpha = 1.5$.

distribution I and corresponding paraxial result I_p are almost coincident. While the parameters f_ω are increased to 0.3 (see Figures 3(b1)–3(b3)), I_z becomes strong, so I and I_p began to show slight difference. When f_ω is further increased to 0.5, I_z becomes more strong, and the difference between I and I_p increased obviously (see Figures 3(c1)–3(c3)). As a result, no matter what the value of α is, the nonparaxial conclusions of the elliptical Gaussian beam should be considered when f_ω is large. Conversely, the paraxial approximation of elliptical Gaussian beam is valid when f_ω is small. Furthermore, the light peak intensity value will decrease with increasing

the elliptic coefficient α in the same observation plane, no matter what value of f_ω is. However, the larger the value of parameters f_ω is, the smaller the spot size of beam is, no matter what value of elliptic coefficient α is.

Figure 4 shows the contour graphs of intensity distributions I_x, I_z, and I of nonparaxial elliptical Gaussian beams for elliptic coefficient $\alpha = 1.5$ in the plane $z = 10z_r$, and the corresponding paraxial result I_p is also given in Figure 4. The parameter f_ω is chosen as 0.1, 0.3, and 0.5 from the first row to the third row, respectively. As shown in Figure 3, when f_ω is small, the longitudinal component I_z is very

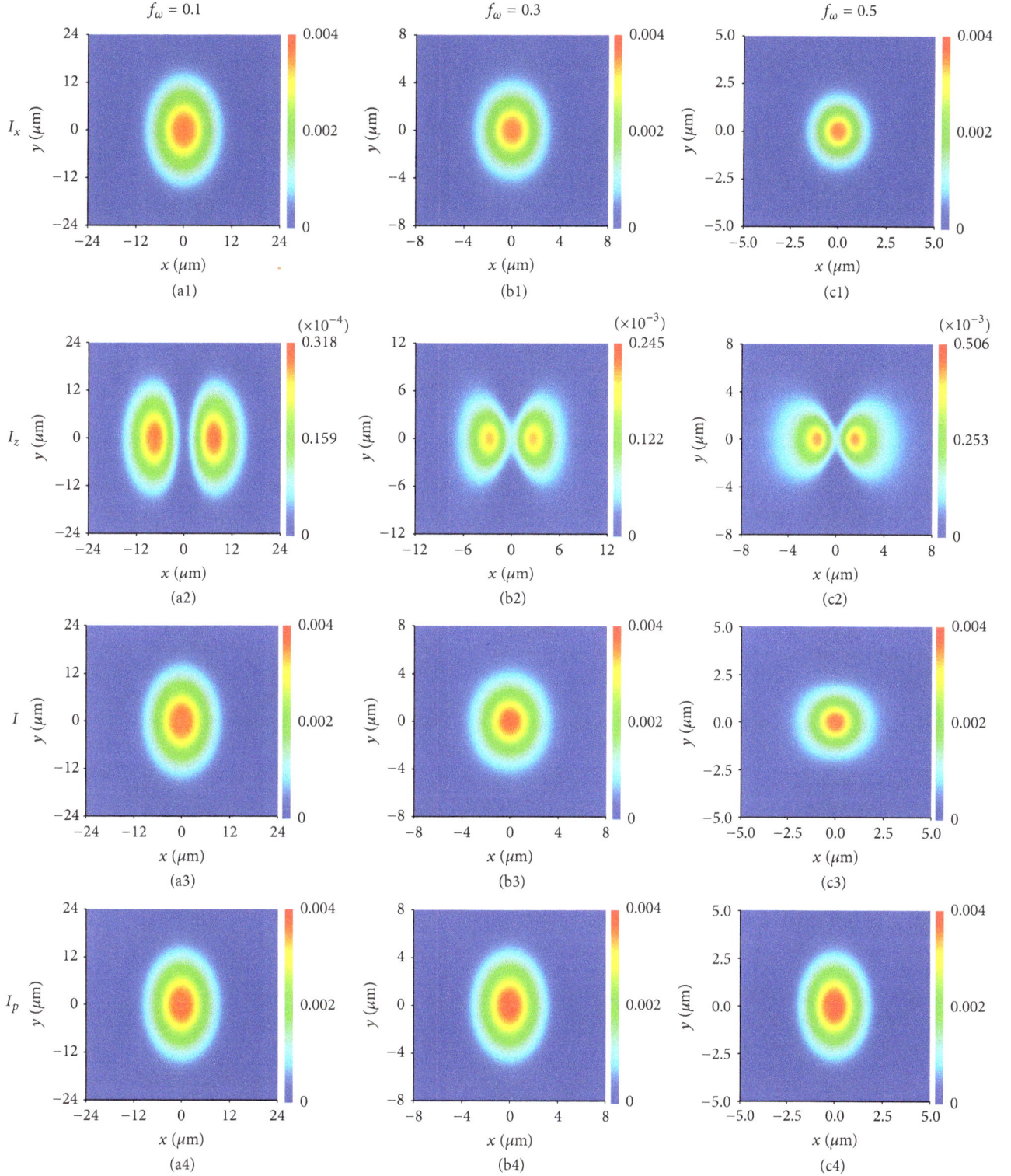

FIGURE 4: Contour graphs of intensity distributions $I_x(x, y, z)$, $I_z(x, y, z)$, $I(x, y, z)$, and $I_p(x, y, z)$ of elliptical Gaussian beam with elliptic coefficient $\alpha = 1.5$ in the plane $z = 10z_r$ for different parameters f_ω: (a1)–(a4) $f_\omega = 0.1$; (b1)–(b4) $f_\omega = 0.3$; (c1)–(c4) $f_\omega = 0.5$.

small and can be neglected; hence the beam profiles of total intensity distribution I and corresponding paraxial result I_p are visibly similar. Figures 4(a1)–4(a4) also show that I_z can be neglected, and the paraxial approximation is valid when f_ω is small. However, when f_ω is chosen as 0.3 (see Figures 4(b1)–4(b4)), I_z becomes strong, and the spots of I and I_p show a little distinction. As f_ω is further increased to 0.5, I and I_p show obvious difference (see Figures 4(c1)–4(c4)). In other words, the contribution of the longitudinal component I_z would become significant, and the nonparaxial conclusions of elliptical Gaussian beam should be adopted when f_ω is large.

4. Conclusions

In this paper, based on the vectorial Rayleigh-Sommerfeld diffraction integral formulae, we have derived the analytical expressions for a vectorial elliptical Gaussian beam's nonparaxial propagating in free space, and the paraxial approximation expression has also been examined as a special case. The evolution of the beam's intensity and shape with different elliptic coefficient α and different waist width related parameter f_ω is illustrated by numerical examples. Results show that, with increasing propagation distance z, all contours of the transverse cross sections of nonparaxial propagation of the elliptical Gaussian beams preserve Gaussian type when f_ω is small, while all contours of the transverse cross sections of nonparaxial propagation of the elliptical Gaussian beams would change to Gaussian-like type when f_ω is large. Meanwhile, whether parameter f_ω is large or small, the peak intensity value decreased with increasing the propagation distance, and the larger the elliptic coefficient α is, the faster the peak intensity value decreases. In addition, numerical results also show that no matter what value of elliptic coefficient α is, when parameter f_ω is small, the paraxial approximation of elliptical Gaussian beam is effective; when parameter f_ω is large, the nonparaxial conclusions of the elliptical Gaussian beam should be adopted. These characteristics of vectorial elliptical Gaussian beam might find applications in modern optics.

Conflict of Interests

The authors declare that there is no conflict of interests regarding the publication of this paper.

Acknowledgments

This work is supported by the National Science Foundation of China (11547002 and 11447235), Jiangxi Provincial Natural Science Foundation of China (20142BAB212003), and China Scholarship Council 201508360027.

References

[1] E. Kapon, J. Katz, and A. Yariv, "Supermode analysis of phase-locked arrays of semiconductor lasers," *Optics Letters*, vol. 9, no. 4, pp. 125–127, 1984.

[2] J. Faist, F. Capasso, D. L. Sivco, C. Sirtori, A. L. Hutchinson, and A. Y. Cho, "Quantum cascade laser," *Science*, vol. 264, no. 5158, pp. 553–556, 1994.

[3] D. A. Fletcher, K. E. Goodson, and G. S. Kino, "Focusing in microlenses close to a wavelength in diameter," *Optics Letters*, vol. 26, no. 7, pp. 399–401, 2001.

[4] K. J. Vahala, "Optical microcavities," *Nature*, vol. 424, no. 6950, pp. 839–846, 2003.

[5] R. Borghi, M. Santarsiero, and M. A. Alonso, "Highly focused spirally polarized beams," *Journal of the Optical Society of America A: Optics and Image Science, and Vision*, vol. 22, no. 7, pp. 1420–1431, 2005.

[6] Z. R. Liu and D. M. Zhao, "Radiation forces acting on a Rayleigh dielectric sphere produced by highly focused elegant Hermite-cosine-Gaussian beams," *Optics Express*, vol. 20, no. 3, pp. 2895–2904, 2012.

[7] Z. R. Liu and D. M. Zhao, "Optical trapping Rayleigh dielectric spheres with focused anomalous hollow beams," *Applied Optics*, vol. 52, no. 6, pp. 1310–1316, 2013.

[8] S. Nemoto, "Nonparaxial Gaussian beams," *Applied Optics*, vol. 29, no. 13, pp. 1940–1946, 1990.

[9] R. K. Luneburg, *Mathematical Theory of Optics*, University of California Press, Berkeley, Calif, USA, 1966.

[10] M. Lax, W. H. Louisell, and W. B. McKnight, "From Maxwell to paraxial wave optics," *Physical Review A*, vol. 11, no. 4, pp. 1365–1370, 1975.

[11] A. Wünsche, "Transition from the paraxial approximation to exact solutions of the wave equation and application to Gaussian beams," *Journal of the Optical Society of America B*, vol. 9, no. 5, pp. 765–774, 1992.

[12] C. G. Chen, P. T. Konkola, J. Ferrera, R. K. Heilmann, and M. L. Schattenburg, "Analyses of vector Gaussian beam propagation and the validity of paraxial and spherical approximations," *Journal of the Optical Society of America A: Optics and Image Science, and Vision*, vol. 19, no. 2, pp. 404–412, 2002.

[13] S. R. Seshadri, "Virtual source for a Hermite-Gauss beam," *Optics Letters*, vol. 28, no. 8, pp. 595–597, 2003.

[14] Z. R. Mei and D. M. Zhao, "Nonparaxial analysis of vectorial Laguerre-Bessel-Gaussian beams," *Optics Express*, vol. 15, no. 19, pp. 11942–11951, 2007.

[15] D. G. Deng, H. Yu, S. Q. Xu, G. L. Tian, and Z. X. Fan, "Nonparaxial propagation of vectorial hollow Gaussian beams," *Journal of the Optical Society of America B: Optical Physics*, vol. 25, no. 1, pp. 83–87, 2008.

[16] G. Q. Zhou, "Nonparaxial propagation of a Lorentz-Gauss beam," *Journal of the Optical Society of America. B*, vol. 26, no. 1, pp. 141–147, 2009.

[17] B. Gu and Y. Cui, "Nonparaxial and paraxial focusing of azimuthal-variant vector beams," *Optics Express*, vol. 20, no. 16, pp. 17684–17694, 2012.

[18] A. Naqwi and F. Durst, "Focusing of diode laser beams: a simple mathematical model," *Applied Optics*, vol. 29, no. 12, pp. 1780–1785, 1990.

[19] K. L. Duan and B. D. Lü, "Propagation properties of vectorial elliptical Gaussian beams beyond the paraxial approximation," *Optics and Laser Technology*, vol. 36, no. 6, pp. 489–496, 2004.

[20] B. D. Lü and K. L. Duan, "Nonparaxial propagation of vectorial Gaussian beams diffracted at a circular aperture," *Optics Letters*, vol. 28, no. 24, pp. 2440–2442, 2003.

[21] K. L. Duan and B. D. Lü, "Vectorial nonparaxial propagation equation of elliptical Gaussian beams in the presence of a rectangular aperture," *Journal of the Optical Society of America A*, vol. 21, no. 9, pp. 1613–1620, 2004.

8

Experimental Research of Reliability of Plant Stress State Detection by Laser-Induced Fluorescence Method

Yury Fedotov, Olga Bullo, Michael Belov, and Viktor Gorodnichev

Faculty of Radioelectronics and Laser Techniques, Department of Laser and Optoelectronics Systems, Bauman Moscow State Technical University, 2-ya Baumanskaya Ulitsa, Moscow 105005, Russia

Correspondence should be addressed to Yury Fedotov; fed@bmstu.ru

Academic Editor: Giulio Cerullo

Experimental laboratory investigations of the laser-induced fluorescence spectra of watercress and lawn grass were conducted. The fluorescence spectra were excited by YAG:Nd laser emitting at 532 nm. It was established that the influence of stress caused by mechanical damage, overwatering, and soil pollution is manifested in changes of the spectra shapes. The mean values and confidence intervals for the ratio of two fluorescence maxima near 685 and 740 nm were estimated. It is presented that the fluorescence ratio could be considered a reliable characteristic of plant stress state.

1. Introduction

Fluorescence analysis is a widely used high-sensitivity method that is applied in many scientific and technical fields. A viable application of the technique is the analysis of plant state [1–14]. External factors can cause plants stress and make their growth abnormal. Stress conditions are difficult to detect by visual observation during the early growth stages of a plant; however, the laser-induced fluorescence method is effective in the remote detection of plant stress state.

Chlorophyll is the basic fluorescent component of green leaf in the red and far-red regions. The fluorescence spectrum of a green leaf at room temperature exhibits two maxima in the red band (680–690 nm) and in the far-red band (730–740 nm) [1, 6, 8, 15]. The fluorescence spectrum of a stressed plant is deformed in comparison with that of a plant in a nonstressed state. This effect is caused by disturbing the photosynthetic process of a plant under stressed conditions. The fluorescence spectrum depends on different factors such as excitation wavelength, type of stress factor, and plant species.

There are wide experimental data on the fluorescence spectra of various plant species, both stressed and non-stressed, excited at wavelength ranges of 266–635 nm [2, 10, 11, 15, 16]. However, a number of points remain to be investigated. One such point is the reliability of the plant state

detection based on the differences in fluorescence spectra of samples of a plant species, grown under identical conditions, except that some samples were stressed and the others were not.

In this paper, experimental results of the analysis of fluorescence spectra variation of different samples of a plant species in both normal and stressed states are presented.

2. Materials and Methods

2.1. Laboratory Setup Description. The fluorescence spectra were excited at a wavelength of 532 nm. It is common to use lasers with wavelengths at 337, 335, and 532 nm for fluorescence excitation in experimental research. The laser source used in this study was selected because of the advantages offered by the solid-state YAG:Nd laser at the wavelength of 532 nm (for remote sensing equipment development), in comparison with both the nitrogen gas laser at the wavelength of 337 nm and the solid-state YAG:Nd laser at the wavelength of 355 nm (the third harmonic of the YAG laser has lower pulse intensity than its second harmonic).

The laboratory configuration used to measure fluorescence spectra is shown in Figure 1.

An EKSPLA NL210 solid-state YAG:Nd laser with diode pumping and frequency doubling was used as the source of

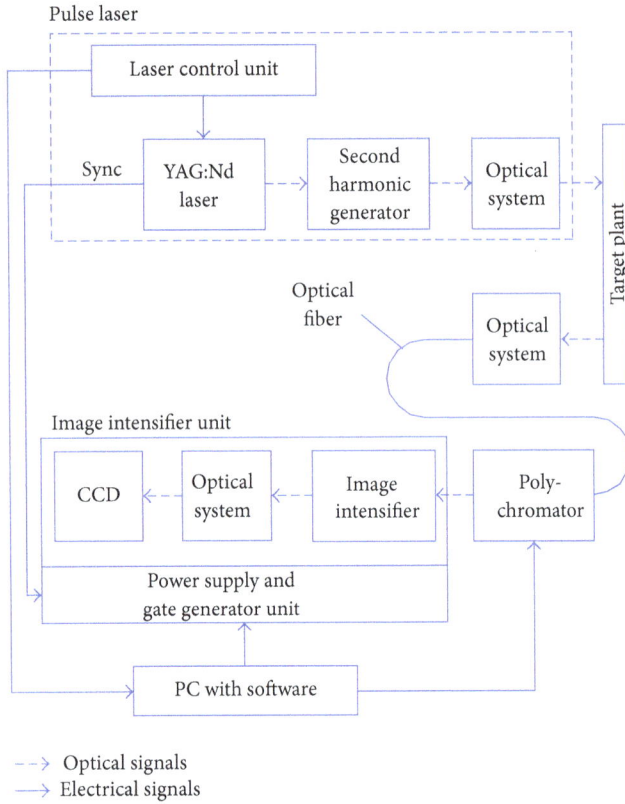

FIGURE 1: Laboratory configuration for laser-induced fluorescence experiments.

TABLE 1: Specifications of laboratory setup.

Specifications	Value
Laser pulse energy, mJ	2.1
Laser wavelength, nm	532
Laser pulse duration, ns	<7
Laser repetition rate, Hz	<500
Laser beam spread, mrad	<3
Spectral band of registration, nm	595–800
Spectral resolution, nm	6
Diameter of optical detection system, mm	15
Distance to sample, m	1

wavelength using a calibration light source based on a mercury-argon lamp (SL2 StellarNet Inc.) with a linear spectrum. The test was performed at the wavelength of 546.07 nm. Calibration of sensitivity of the registration system was performed using a light source based on halogen lamp (DH-2000-CAL Ocean Optics Inc.) with a continuous spectrum. Known spectrum of the lamp was acquired for sensitivity calculation.

2.2. Plant Samples. The experimental research of laser-induced fluorescence spectra was performed using easy to keep fast-growing plant species, that is, salads, watercress, mustard, common borage, cucumbers, and lawn grass. The experimental measurements of fluorescence spectra of watercress (*Lepidium sativum*) and lawn grass (that comprised a mixture of 30% perennial ryegrass (*Lolium perenne*), 65% creeping red fescue (*Festuca rubra*), and 5% sheep's ovina (*Festuca ovina*)) are presented in this paper. The research was conducted on plants in their normal state and under the influence of stress factors, for example, mechanical damage (leaf cutting and laying, root system damage), root system overwatering, and soil pollution (copper sulfate, $CuSO_4$, ferric sulfate, $FeSO_4$, and sodium chloride, NaCl).

2.3. Normal and Stress Conditions. The plants in normal state were grown in favorable condition for their development. The watercress plants have height of approximately 4 cm, and the lawn grass plants 8 cm.

By the leaf cutting of the watercress, the half of one leaf of each plant was dissected. The leaf laying was conducted using 7×7 cm flat plate with 200 g weight during approximately 1 min. For root system damage in seedlings pots was cut a slit at the depth of 2 cm, the root system has been damaged through the slit by means of utility knife, and then the slit was closed.

The overwatering stress condition was implemented by placing the pot of the watercress sample in a watering can. The level of water in the watering can was always slightly below the level of soil in the plant pot; thus, it was not visually obvious that the root system of the plant sample was constantly in overwatered soil.

fluorescence excitation. Laser light was transmitted by means of the optical system to the target plant located at a distance of 1 m from the optical system. The apparent diameter of the laser beam on the plant sample was approximately 25 mm. The laser spot has covered 15–20 plants. The fluorescent radiation of the plants was collected from the same spot size together with the reflected laser light by the optical system and directed into the optical fiber. The optical fiber was used to transmit light to the input of the polychromator. The reflected light from the laser beam was prevented from entering a polychromator by using an NF01-532U Semrock filter. Fluorescent radiation from 595 to 800 nm was detected. An M266 Solar LS polychromator was used as the spectral device and all transitions within the polychromator fully automated (i.e., the swapping of diffraction grids and optical filters and slit width selection).

The fluorescence spectrum was detected using a highly sensitive detector (Matrix-430k-ns Deltatekh) based on CCD array with an image intensifier. The image intensifier (generation II+, diameter 18 mm) has quantum efficiency 15% at the wavelength 550 nm. The image was transferred by the optical system from the image intensifier to the CCD. The image was converted into a digital array and transmitted to the computer. Special software developed with LabVIEW National Instruments was used to control the setup. The major specifications of the setup are presented in Table 1.

The experiment included equipment calibration as a preparatory step. The polychromator was calibrated by

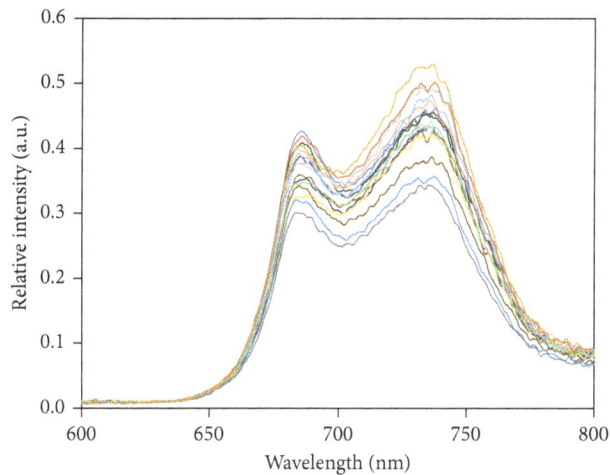

FIGURE 2: Fluorescence spectra of different watercress samples in normal state.

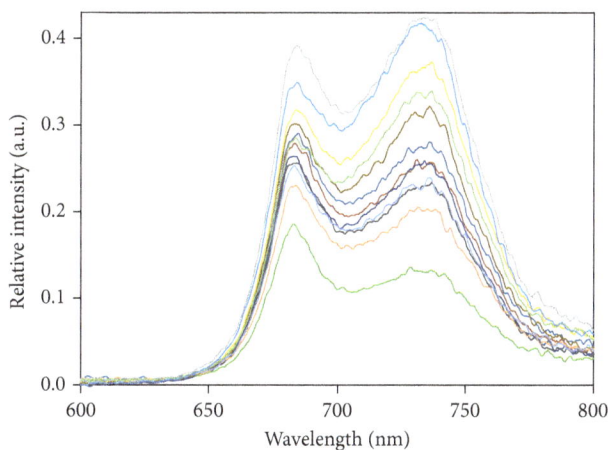

FIGURE 3: Fluorescence spectra of different watercress samples in stressed state caused by leaf laying.

3. Results and Discussion

3.1. Fluorescence Spectra of Different Watercress Samples in Normal State. The fluorescence spectra of different samples of watercress grown under normal conditions are shown in Figure 2. The different plots in Figure 2 correspond to different plant samples that were planted at the same time and grown under the same conditions. The measurements were conducted in 16 days after planting.

As it can be seen in Figure 2, there are insignificant changes in the shapes of the fluorescence spectra from one sample to another, despite the differences in spectra intensity.

3.2. Fluorescence Spectra of Watercress Stressed by Mechanical Damage. The fluorescence spectra of different samples of watercress stressed by leaf laying mechanical damage are shown in Figure 3. There have been several experimental researches on the fluorescence spectra of plants in stressed states caused by different types of mechanical damage [9, 13, 14], but few or none investigated the fluorescence spectra at an excitation wavelength of 532 nm [9].

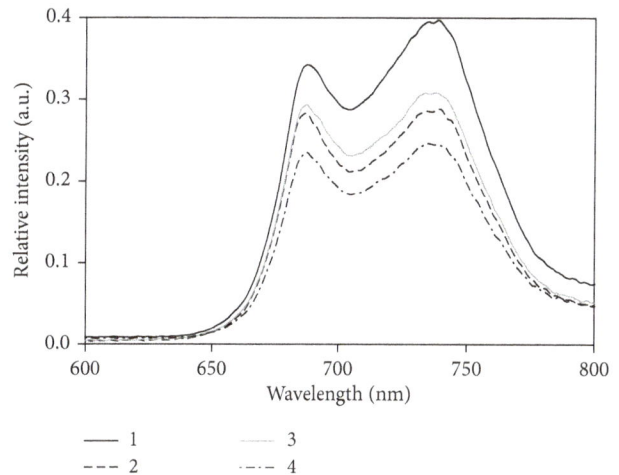

FIGURE 4: Averaged fluorescence spectra of watercress samples in normal and stressed conditions: (1) normal state, (2) leaf laying stress, (3) leaf cutting stress, and (4) root system damage stress.

The fluorescence spectra of watercress in a stressed state caused by leaf laying fluctuate considerably (Figure 3) and differ from those of watercress in a normal state (Figure 2). A similar difference is found between the fluorescence spectra of watercress in normal and stressed states when the stress is caused by mechanical damage of root system.

The differences between the fluorescence spectra of plants under normal and stressed conditions are illustrated clearly by averaging the measurements of the fluorescence spectra. Figure 4 displays the averaged fluorescence spectra of watercress in a normal state (plot 1) and stressed state by the mechanical damage of leaf laying (plot 2), leaf cutting (plot 3), and root system damage (plot 4). Plot 1 in Figure 4 corresponds to the averaged fluorescence spectra over the result of 20 measurements. Plots 2, 3, and 4 in Figure 4 correspond to the averaged fluorescence spectra over 11 measurements for each stress factor; thus, a single measurement corresponds to the measurement of a single fluorescence spectrum of a plant sample in definite time intervals from 20 to 40 min from the start of the stress factor influence.

It is clearly illustrated in Figure 4 that the shapes of the laser-induced fluorescence spectra of watercress in stressed conditions were caused by various types of mechanical damage change significantly. The ratios of fluorescence intensity in the red region (680–690 nm) and far-red region (730–740 nm) increase in stress conditions.

3.3. Fluorescence Spectra of Watercress in Stress State Caused by Overwatering. The laser-induced fluorescence spectra of watercress in stressed state caused by overwatering are comparable with those presented in Figures 3 and 4. Figure 5 shows the fluorescence spectra of watercress in stressed state caused by overwatering during 24 days (different spectra correspond to different measurements and plant samples).

As it is clearly illustrated in Figure 5 the fluorescence spectra of the watercress in a stressed state caused by overwatering during 24 days differ from those of watercress in a normal state. Furthermore, the spectra of the stressed samples

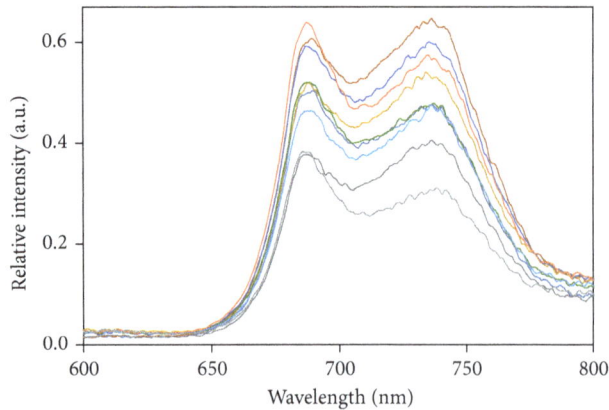

FIGURE 5: Fluorescence spectra of different watercress samples in stressed state caused by overwatering during 24 days.

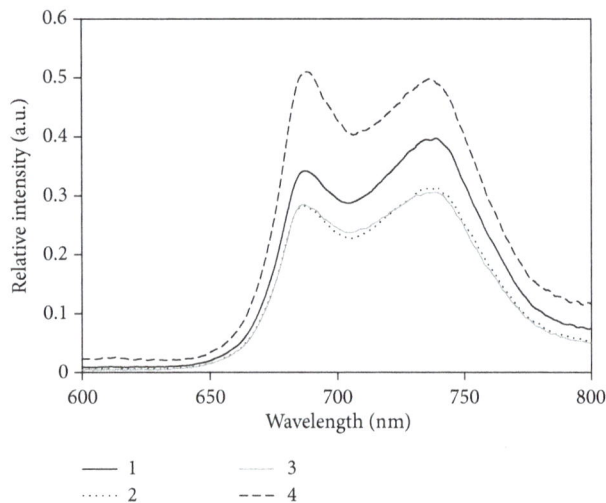

FIGURE 6: Fluorescence spectra of different watercress samples in stressed state caused by overwatering: (1) normal state, (2) 11-day overwatering, (3) 17-day overwatering, (4) 24-day overwatering.

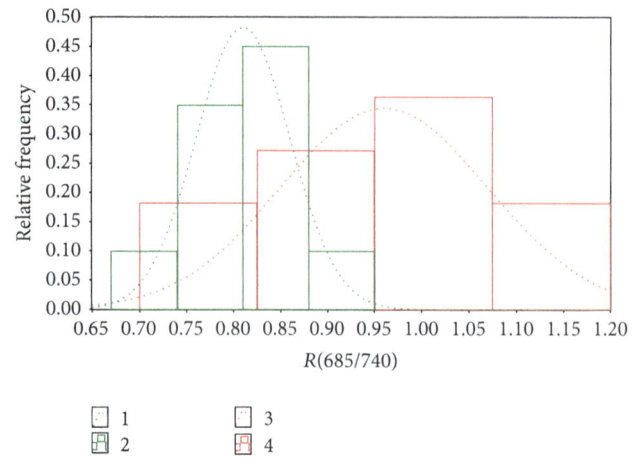

FIGURE 7: Histograms of distribution of fluorescence ratio, for watercress in a normal state and in a stressed state caused by leaf laying: (1) histogram for normal state, (2) histogram approximation for normal state, (3) histogram for stressed state, and (4) histogram approximation for stressed state.

fluctuate considerably, comparable with the fluorescence spectra of the watercress stressed by mechanical damage.

Figure 6 shows the laser-induced fluorescence spectra of watercress averaged over the number of plant samples and measurements (18 measurements for watercress in normal condition and 9 measurements for watercress in stressed condition).

Plot 1 in Figure 6 corresponds to the averaged fluorescence spectrum of watercress in a normal state. Plots 2, 3, and 4 in Figure 6 correspond to the averaged fluorescence spectra of watercress in stressed condition caused by overwatering during 11, 17, and 24 days, respectively. It can be clearly seen that the influence of the stress factor (overwatering in this case) accumulates gradually over the time of abnormal watering, increasing the fluorescence intensity.

The results presented in Figures 2–6 are in agreement with those of other experimental researches [8, 12] on plants under nitrogen stress and soil pollution using a fluorescence excitation source at the wavelength of 532 nm.

3.4. Fluorescence Ratio.

The ratio of fluorescence intensities in the 680–690 and 730–740 nm spectral bands is widely used in experimental research to characterize the fluorescence spectrum shape. Analysis of experimental data indicated that the ratio of fluorescence intensities near 685 and 735 nm can be used to characterize plant stress state.

Histograms of distribution of fluorescence intensities ratio (R) at 685 and 740 nm in narrow spectral bands with bandwidths of 10 nm, for watercress in a normal state in 16 days after planting and in a stressed state caused by leaf laying, are shown in Figure 7. Histograms were approximated by Gaussian function.

The mean value of the fluorescence ratio is 0.81 and the standard deviation is 0.05 for the watercress plants in a normal state. The mean value of the fluorescence ratio is 0.96 and the standard deviation is 0.11 for the watercress plants in a stressed state caused by leaf laying.

Histograms of distribution of the fluorescence intensities ratio at 685 and 740 nm for watercress in normal condition in 16 days after planting and in a stressed condition after 24 days of overwatering are shown in Figure 8.

The mean value of the fluorescence ratio is 0.81 and the standard deviation is 0.05 for watercress plants in a normal state. The mean value of the fluorescence ratio is 0.97 and the standard deviation is 0.07 for watercress in a stressed state caused by overwatering during 24 days.

As it is shown in Figures 7 and 8, it is possible to mistake, using a single measurement of fluorescent ratio R, defining whether a plant is under normal or stressed conditions because the distributions are overlapped. A far reliable method for defining the condition of a plant consists in using mean value of the fluorescence ratio, even in the case of small set of measurements.

The mean values (with 95% confidence intervals) of the experimental laser-induced fluorescence spectra of watercress under different stress conditions (leaf cutting, leaf

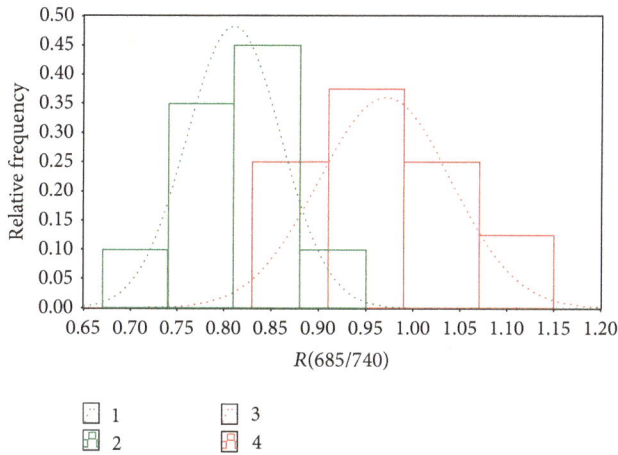

FIGURE 8: Histograms of distribution of fluorescence ratio, for watercress in a normal state and stressed state caused by overwatering for 24 days: (1) histogram for normal state, (2) histogram approximation for normal state, (3) histogram of stressed state, and (4) histogram approximation for stressed state.

laying, root system damage, and root system overwatering during 11, 17, and 24 days) are shown in Figure 9.

Columns 1, 3, 5, 7, 9, and 11 in Figure 9 correspond to the plants in a normal state and columns 2, 4, 6, 8, 10, and 12 correspond to plants in a stressed state (2: leaf laying, 4: leaf cutting, 6: root system damage, 8: overwatering during 11 days, 10: overwatering during 17 days, and 12: overwatering during 24 days).

The changes of fluorescence spectra for plants in stress conditions described above are typical not only for watercress but also for other plants in stress conditions caused by different impact. The effect of soil pollution on lawn grass is considered below.

The aggregated statistical results (mean values and 95% confidence intervals) of the experimental laser-induced fluorescence spectra of lawn grass under different stress conditions caused by soil pollution (copper sulfate, $CuSO_4$, ferric sulfate, $FeSO_4$, and sodium chloride, $NaCl$) are shown in Figure 10.

Columns 1, 3, 5, 7, and 9 in Figure 10 correspond to lawn grass in a normal state (experimental research was conducted six weeks after planting, directly before the soil was polluted). Columns 2, 4, 6, 8, and 10 in Figure 10 correspond to lawn grass in a stressed state; measurements were performed 2 weeks after the initial influence of the stress factor for columns 2 and 6 and 4 weeks after the initial influence of the stress factor for columns 4, 8, and 10. The stress factor was soil pollution by sodium chloride, $NaCl$ (5 g per plant sample, columns 2 and 4), ferric sulfate, $FeSO_4$ (3 g per plant sample, columns 6 and 8), and copper sulfate, $CuSO_4$ (2 g per plant sample, column 10).

It is clearly illustrated in Figures 9 and 10 that the fluorescence ratio (R) is characterized by stable and sufficient difference. The confidence intervals of the fluorescence ratio for plants under normal and stressed states were not large (≤0.1 in the majority of the cases). The sum of confidence intervals of the fluorescence ratio (R) for plants under normal and stressed states is not usually more than the difference

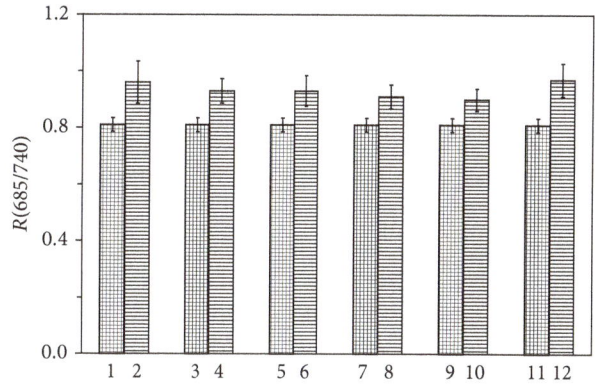

FIGURE 9: Fluorescence ratio (R) mean values and 95% confidence intervals of watercress in normal conditions (columns 1, 3, 5, 7, and 9) and under stressed conditions caused by mechanical damage and overwatering (columns 2, 4, 6, 8, 10, and 12).

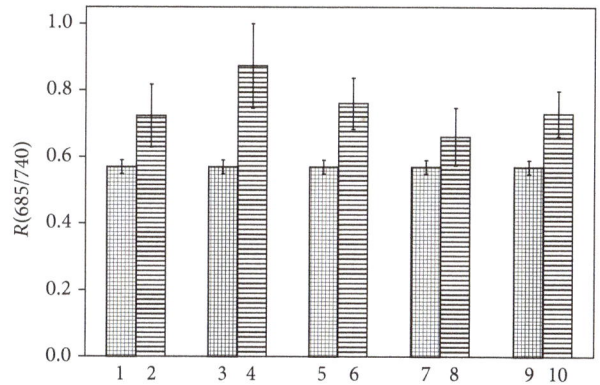

FIGURE 10: Fluorescence ratio (R) mean values and 95% confidence intervals of lawn grass under normal conditions (columns 1, 3, 5, 7, and 9) and stressed conditions caused by different soil pollutants (columns 2, 4, 6, 8, and 10).

between ratio R for plants in a normal state and ratio R for plants in a stressed state caused by various factors (mechanical damage, overwatering, and soil polluting).

This means that fluorescence excitation at 532 nm wavelength and the ratio of fluorescence intensities in the red (685 nm) and far-red (740 nm) bands can be used as signatures of plant stress state caused by various factors.

4. Conclusions

By the processing of the experimental results of fluorescence spectra (induced by a 532 nm wavelength laser) of plants in normal and stressed states caused by mechanical damage, overwatering, and soil pollution the following conclusions can be postulated.

(i) The fluorescence spectra of different samples of a plant species revealed repeatability of the spectra shapes. Ratio R of the fluorescence intensity at 685 and 740 nm demonstrated sufficient stability. However, it is possible to mistake defining the plant stress state (normal or stressed) using only single measurement of ratio R. We proposed more reliable method to define plant condition using mean value of R ratio, which is suitable for small set of measurements.

(ii) The difference between the mean value of ratio R for a plant in a normal state and that in a stressed state, in the majority of cases, is greater than the difference between ratio R values for different samples of one plant species.

The experimental results obtained allow us to develop a remote laser system for detecting plant stress state. However, to ensure the reliability of the measurements, it is necessary to calculate the mean value of ratio R for several measurements for several plants.

Competing Interests

The authors declare that they have no competing interests.

References

[1] A. S. Ndao, A. Konté, M. Biaye, M. E. Faye, N. A. B. Faye, and A. Wagué, "Analysis of chlorophyll fluorescence spectra in some tropical plants," *Journal of Fluorescence*, vol. 15, no. 2, pp. 123–129, 2005.

[2] A. Takeuchi, Y. Saito, T. D. Kawahara, and A. Njmura, "Possibility of disease process monitoring of plants by laser-induced fluorescence method," in *Proceedings of the Development and Evaluation of LIF Measurement Systems*, vol. 4153 of *Proceedings of SPIE*, pp. 22–29, February 2001.

[3] Y. Saito, "Laser-induced fluorescence spectroscopy/technique as a tool for field monitoring of physiological status of living plants," in *14th International School on Quantum Electronics: Laser Physics and Applications*, 66041W, vol. 6604 of *Proceedings of SPIE*, 12 pages, March 2007.

[4] H. A. Hristov, E. G. Borisova, L. A. Avramov, and I. N. Kolev, "Applications of laser-induced fluorescence for remote sensing," in *Proceedings of the 11th International School on Quantum Electronics: Laser Physics and Applications*, vol. 4397 of *Proceedings of SPIE*, pp. 496–500, Varna, Bulgaria, April 2001.

[5] K. J. Lee, Y. Park, A. Bunkin, R. Nunes, S. Pershin, and K. Voliak, "Helicopter-based lidar system for monitoring the upper ocean and terrain surface," *Applied Optics*, vol. 41, no. 3, pp. 401–406, 2002.

[6] L. A. Corp, J. E. McMurtrey, E. M. Middleton, C. L. Mulchi, E. W. Chappelle, and C. S. T. Daughtry, "Fluorescence sensing systems: in vivo detection of biophysical variations in field corn due to nitrogen supply," *Remote Sensing of Environment*, vol. 86, no. 4, pp. 470–479, 2003.

[7] M. V. Grishaev, V. V. Zuev, and O. V. Kharchenko, "Fluorescent channel of the Siberian Lidar Station," in *15th Symposium on High-Resolution Molecular Spectroscopy*, 65800U, vol. 6580 of *Proceedings of SPIE*, 6 pages, December 2006.

[8] G. Matvienko, V. Timofeev, A. Grishin, and N. Fateyeva, "Fluorescence lidar method for remote monitoring of effects on vegetation," in *Lidar Technologies, Techniques, and Measurements for Atmospheric Remote Sensing II*, 63670F, vol. 6367 of *Proceedings of SPIE*, 8 pages, October 2006.

[9] J. Belasque, M. C. G. Gasparoto, and L. G. Marcassa, "Detection of mechanical and disease stresses in citrus plants by fluorescence spectroscopy," *Applied Optics*, vol. 47, no. 11, pp. 1922–1926, 2008.

[10] A. S. Gouveia-Neto, E. A. Silva, R. A. Oliveira et al., "Water deficit and salt stress diagnosis through LED induced chlorophyll fluorescence analysis in *Jatropha curcas* L. oil plants for biodisiel," in *Imaging, Manipulation, and Analysis of Biomolecules, Cells, and Tissues IX*, 79020A, vol. 7902 of *Proceedings of SPIE*, 10 pages, February 2011.

[11] R. Maurya, S. M. Prasad, and R. Gopal, "LIF technique offers the potential for the detection of cadmium-induced alteration in photosynthetic activities of *Zea Mays* L.," *Journal of Photochemistry and Photobiology C: Photochemistry Reviews*, vol. 9, no. 1, pp. 29–35, 2008.

[12] E. Middleton, J. E. McMurtrey, P. K. Entcheva Campbell, L. A. Corp, L. M. Butchera, and E. W. Chappellea, "Optical and fluorescence properties of corn leaves from different nitrogen regimes," in *Remote Sensing for Agriculture, Ecosystems, and Hydrology IV*, vol. 4879 of *Proceedings of SPIE*, pp. 72–83, March 2003.

[13] H. K. Lichtenthaler and U. Rinderle, "The role of chlorophyll fluorescence in the detection of stress conditions in plants," *CRC Critical Reviews in Analytical Chemistry*, vol. 19, no. 1, pp. 29–85, 1988.

[14] G. G. Matvienko, A. I. Grishin, O. V. Kharchenko, and O. A. Romanovskii, "Remote sounding of vegetation characteristics by laser-induced fluorescence," in *Proceedings of the Laser Radar Technology and Applications IV*, vol. 3707 of *Proceedings of SPIE*, pp. 524–532, Orlando, Fla, USA, 1999.

[15] Z. G. Cerovic, G. Samson, F. Morales, N. Tremblay, and I. Moya, "Ultraviolet-induced fluorescence for plant monitoring: present state and prospects," *Agronomie*, vol. 19, no. 7, pp. 543–578, 1999.

[16] A. B. Utkin, R. Felizardo, C. Gameiro, A. R. Matos, and P. Cartaxana, "Laser induced fluorescence technique for environmental applications," in *Proceedings of the 2nd International Conference on Applications of Optics and Photonics*, vol. 9286 of *Proceedings of SPIE*, Aveiro, Portugal, August 2014.

9

Simultaneous Wood Defect and Species Detection with 3D Laser Scanning Scheme

Zhao Peng, Li Yue, and Ning Xiao

Information and Computer Engineering College, Northeast Forestry University, Harbin 150040, China

Correspondence should be addressed to Zhao Peng; bit_zhao@aliyun.com

Academic Editor: Sulaiman Wadi Harun

Wood grading and wood price are mainly connected with the wood defect and wood species. In this paper, a wood defect quantitative detection scheme and a wood species qualitative identification scheme are proposed simultaneously based on 3D laser scanning point cloud. First, an Artec 3D scanner is used to scan the wood surface to get the 3D point cloud. Each 3D point contains its X, Y, and Z coordinate and its RGB color information. After preprocessing, the Z coordinate value of current point is compared with the set threshold to judge whether it is a defect point (i.e., cavity, worm tunnel, and crack). Second, a deep preferred search algorithm is used to segment the retained defect points marked with different colors. The integration algorithm is used to calculate the surface area and volume of every defect. Finally, wood species identification is performed with the wood surface's color information. The color moments of scanned points are used for classification, but the defect points are not used. Experiments indicate that our scheme can accurately measure the surface areas and volumes of cavity, worm tunnel, and crack on wood surface with measurement error less than 5% and it can also reach a wood species recognition accuracy of 95%.

1. Introduction

Wood species and wood defects are two key issues in the wood quality assessment so as to judge the physical property and commercial value of different wood products (e.g., wood veneer, lumber, or board) correctly [1]. Some visual image characteristics have been used in the wood species recognition and can be divided into two general categories: wood surface's texture analysis [2, 3] and its color analysis [4, 5]. Recently, the wood spectral reflectance characteristics are also exploited for the species classification. The more common schemes in the literature consider the vibration spectroscopy [6, 7] and the Raman spectroscopy [8]. For example, Piuri and Scotti present a scheme for the wood species classification based on the analysis of fluorescence spectra [9].

As for the wood surface's defect detection, the spectral analysis and laser scanning schemes are usually used to fulfill the qualitative detection on the wood external defects (e.g., cavity, worm tunnel, knots, or erosion) [10–12]. Researchers also have proposed some schemes for detecting wood internal defects by using X-rays, gamma rays, microwaves, and longitudinal stress waves [13–15]. For example, Wang et al.

combine the wavelet transform and neural networks to analyze and recognize different types and sizes of wood internal defects using an ultrasonic device [16].

However, simultaneous investigations on wood species and wood defect detections are scarcely performed to make an objective wood quality assessment. In fact, wood species detection and wood defect detection are usually performed with different instruments or technologies. Even if the same instrument or technology such as spectral analysis may be used for wood species and defect detection, the mutual disturbance exists so that the detection accuracy is low. For example, the wood species recognition accuracy is not good for wood veneer if there are many defects on wood surface by using spectral reflectance features. To solve this hard issue, we use a 3D laser scanning instrument to make a qualitative wood species recognition and a quantitative wood defect measurement (i.e., the precise measurement of surface area and volume of every external defect such as cavity, worm tunnel, and crack on the wood product's surface) simultaneously. Therefore, the wood detection efficacy is improved by using our scheme and it can provide a better basis for the subsequent wood quality assessment.

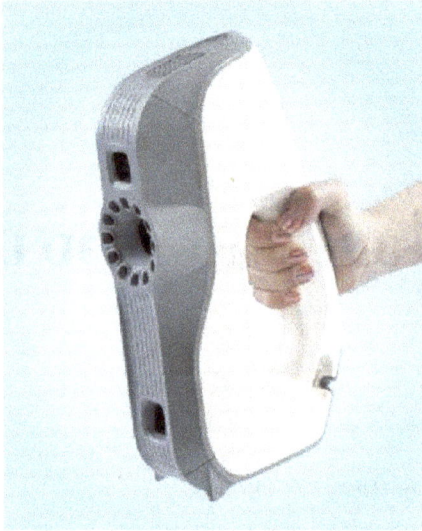

FIGURE 1: A portable Artec 3D laser scanner.

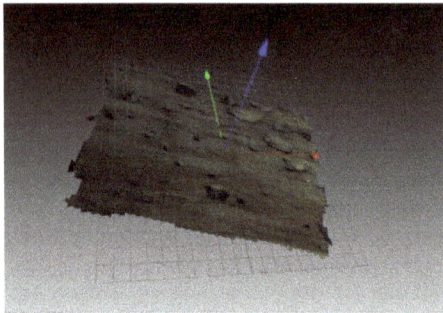

FIGURE 2: Scanned wood surface in the Artec Studio 9.

2. Materials and Methods

2.1. 3D Scan. The 3D scan is a nondestructive detection technology based on laser processing with a fast scan speed and a high scan accuracy. Therefore, it is usually used in the object's 3D reconstruction. In our scheme, a portable Artec 3D laser scanner is used to get the wood surface's 3D point cloud, as illustrated in Figure 1. This scanner is small and portable with an adjustable flash lamp and with a 3D scan resolution of 0.5 mm. It is used conveniently with a number of data storage formats. This scanner is equipped with software system Artec Studio 9 to read the scanner data and then display the scanned object's surface. The object's surface is calibrated with the instrument's *XOY* plane automatically by this software and then the object's surface is stored as a file with OBJ format. For example, Figure 2 illustrates the scanned wood surface in the Artec Studio 9. Each scanned point contains its X, Y, and Z coordinate and its RGB color information. The point's X, Y, and Z coordinates are used in the defect detection, while its RGB information is used in the species recognition.

2.2. Defect Segmentation. In this section, two steps need to be executed to segment different defects. First, every point's Z coordinate is compared with the set threshold to determine

whether this point is a depressed defect (i.e., cavity, worm tunnel, and crack) point. Second, the retained defect points are processed with a deep preferred search algorithm to segment different defects. The defect points which belong to the same defect will be marked with the same number and displayed with the same color. A detailed flow graph is illustrated in Figure 3, and different defects are displayed with different colors, as illustrated in Figure 4.

2.3. Defect Measurement. For every defect such as every cavity or worm tunnel, we propose an integration scheme to calculate its surface area and volume. To achieve this goal, we need to extend each defect point into a small surface. As illustrated in Figure 5, every three adjacent defect points form a triangle plane, while each defect point is the vertex of six triangle planes. Therefore, we should extend each defect point into a regular hexagon surface model, as illustrated in Figure 6. In Figure 6, the point O is the current processed defect point with six adjacent points $A \sim F$. We draw six midnormals on OA, OB, ..., OF, and these six midnormals intersect each other at the six vertices P, Q, R, S, N, and T. These six cross-points form a small purple regular hexagon. In fact, we finally extend each defect point into this small purple regular hexagon $PQRSNT$.

As for the surface area computation of every defect, we should firstly calculate the area of the small purple regular hexagon $PQRSNT$. Define L as the mean distance value of parallel opposing sides such as PQ and SN in $PQRSNT$, and this distance L can be approximated as the mean length value of six sides OA, OB, OC, ..., OF. After we calculate every defect point's surface area (i.e., the area of $PQRSNT$), then we can summarize all defect point's areas of one defect to get one defect's total surface area, as illustrated in

$$S = \sum_{i=1}^{n} \frac{\sqrt{3}}{2} \cdot L_i^2. \tag{1}$$

As for the volume computation of every defect, we should also similarly extend a defect point O into regular hexagon $PQRSNT$. In this way, we can summarize all defect point's volumes of one defect to get one defect's total volume, as illustrated in

$$V = \sum_{i=1}^{n} (S_i \cdot z_i) = \sum_{i=1}^{n} \left(\frac{\sqrt{3}}{2} \cdot L_i^2 \cdot z_i \right). \tag{2}$$

In (2), the corresponding volume for every defect point O is approximated as the volume of the regular hexagonal prism, as illustrated in Figure 7. The height h is approximated as z_i, which is the absolute value of Z coordinate of the ith defect point.

2.4. Species Identification. Wood species identification is performed with wood surface's color information. Each scanned point contains its X, Y, and Z coordinate and its RGB color information. The point's RGB color information is used here to calculate the color moment features for subsequent species identification. Obviously the wood surface's defect points will disturb the precise calculations of color moment features and these defect points should be deleted. Fortunately, in

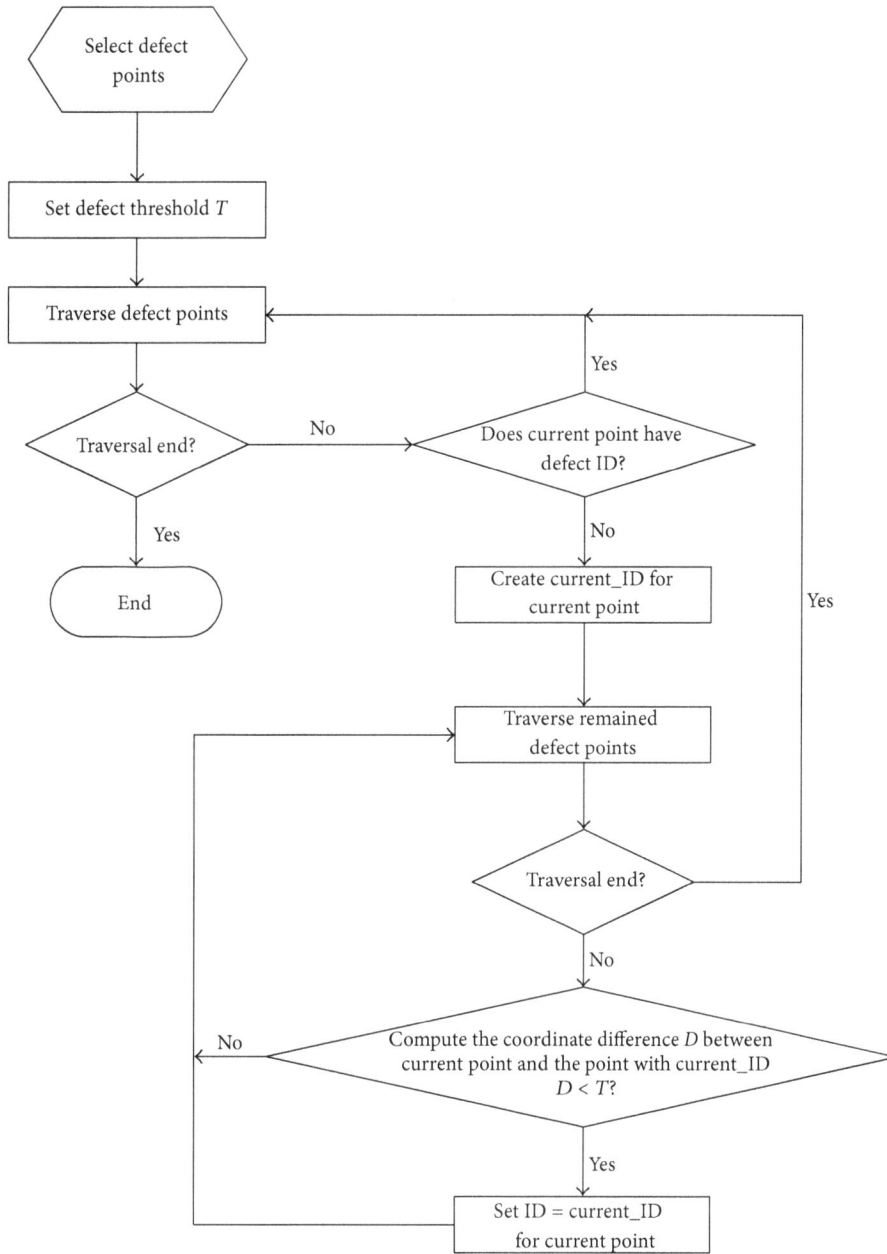

FIGURE 3: The flow graph on wood defect segmentation.

Section 2.2, we have successfully extracted these defect points and therefore the disturbance from defect points can be overcome effectively.

The following three color moments are used for RGB three channels, and we can get 9 classification features. As for the classifier design, the BP neural network is used.

$$\alpha = \frac{1}{n} \sum_{i=1}^{n} P_i,$$

$$\beta = \left(\frac{1}{n} \sum_{i=1}^{n} \left(P_i - \alpha \right)^2 \right)^{1/2},$$

$$\gamma = \left(\frac{1}{n} \sum_{i=1}^{n} \left(P_i - \alpha \right)^3 \right)^{1/3}.$$

(3)

3. Results and Discussions

3.1. Defect Detection. We select many wood products which have cavity, worm tunnel, or crack for experiments (one example is illustrated in Figure 8). The Artec 3D scanner is used to get the wood surface's point cloud data, and subsequent defect segmentation and measurement are performed by use of C language programming. One detailed defect detection result is illustrated in Figure 9.

In order to evaluate our defect detection scheme's measurement accuracy objectively, a simulation experiment is designed here. We select some standard wood boards without any defects on their smooth surfaces and then we use electric drills to drill some pores with different sizes on board surfaces. These drilled pores are the standard cylinder pores

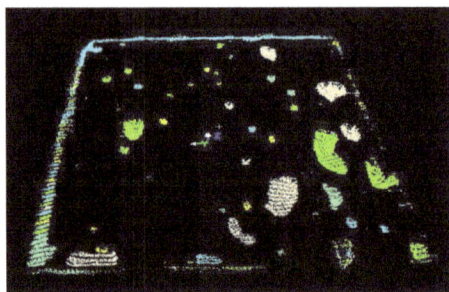

FIGURE 4: The segmented defects displayed with different colors.

FIGURE 5: Wood surface's grid graph in the Artec Studio 9 without color information.

FIGURE 6: A defect point is extended into a purple regular hexagon *PQRSNT*.

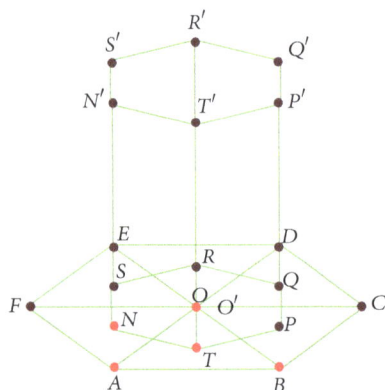

FIGURE 7: The regular hexagonal prism model for volume calculation.

FIGURE 8: The detected wood specimen.

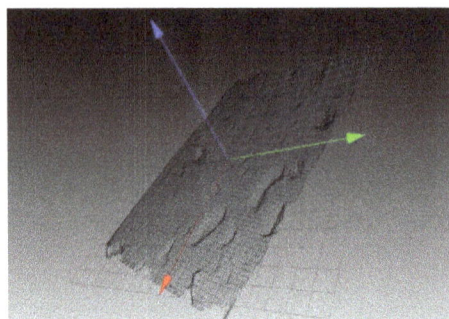

FIGURE 9: The detailed wood defect's measurement results.

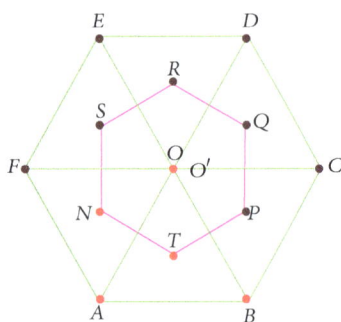

with diameter 4 mm~20 mm and height 1 mm~5 mm. We use our scheme to calculate the surface areas and volumes of these cylinder pores and the standard surface areas and volumes are calculated with cylinder's area and volume formulas. By this way, we can objectively determine our scheme's measurement errors and detailed errors are illustrated in Tables 1 and 2. By comparisons, we can see that our measurement accuracy decreases gradually when the cylinder's height increases. The main measurement error is from the scanning error of Artec scanner, which performs the surface scanning and its scanning error will increase when the depressed defect's height increases. But within the cylinder defect's height 3 mm, our scheme's measurement is good with the relative errors less than 5%.

3.2. Species Recognition. In this section, we perform the wood species identification experiments on five wood species including *Betula platyphylla, Populus davidiana, Pinus sylvestris, Picea jezoensis,* and *Larix gmelinii* (i.e., *BP, PD, PS, PJ,* and *LG*) in northeast region of China with 2000 specimens for each wood species (1000 specimens are used as training dataset and another 1000 specimens are used as test dataset). The Matlab 6.5 programming tool is used here and its neural network toolbox is used for BP network's training and identification. The training frequency is 4000 for BP network and its node number is 19 in its hidden layer. The network's training function is trainbr. For each wood species, radial section (RS) and tangential section (TS) are performed, respectively. The detailed results are illustrated

TABLE 1: Volume measurement error for cylinder cavity.

Relative error (%)	$D = 4$ mm	$D = 6$ mm	$D = 8$ mm	$D = 10$ mm	$D = 13$ mm	$D = 16$ mm	$D = 19$ mm
$H = 1.0$ mm (%)	3.0	3.0	3.0	3.1	3.0	3.0	3.0
$H = 1.5$ mm (%)	3.2	3.2	3.2	3.2	3.2	3.2	3.2
$H = 2.0$ mm (%)	3.5	3.5	3.5	3.5	3.5	3.4	3.5
$H = 2.5$ mm (%)	3.9	3.9	3.9	3.9	4.0	3.8	3.8
$H = 3.0$ mm (%)	4.8	5.0	4.8	4.8	4.8	4.9	5.0

TABLE 2: Surface area measurement error for cylinder cavity.

Relative error (%)	$D = 4$ mm	$D = 6$ mm	$D = 8$ mm	$D = 10$ mm	$D = 13$ mm	$D = 16$ mm	$D = 19$ mm
$H = 1.0$ mm (%)	2.7	2.9	2.7	2.7	2.7	2.7	2.7
$H = 1.5$ mm (%)	2.9	2.9	2.9	2.9	2.9	2.8	2.9
$H = 2.0$ mm (%)	3.3	3.5	3.3	3.3	3.6	3.3	3.3
$H = 2.5$ mm (%)	3.8	3.8	3.8	3.8	3.7	3.7	3.7
$H = 3.0$ mm (%)	4.5	4.5	4.7	4.8	4.8	4.9	5.0

TABLE 3: Pattern recognition accuracy comparisons for our scheme and ordinary color moment-based scheme (RS).

Scheme	BP	PD	PS	PJ	LG
Our scheme	95%	95%	96%	97%	96%
Ordinary scheme	87%	85%	79%	89%	89%

TABLE 4: Pattern recognition accuracy comparisons for our scheme and ordinary color moment-based scheme (TS).

Scheme	BP	PD	PS	PJ	LG
Our scheme	97%	95%	95%	94%	94%
Ordinary scheme	88%	88%	81%	87%	85%

in Tables 3 and 4, where we can see that the recognition accuracy of our scheme with defect points deleted is better than that of ordinary color moment-based scheme.

4. Conclusions

In this paper, we have proposed a simultaneous wood defect and wood species detection scheme based on 3D scanning and signal processing. It can measure the wood surface's depressed defects quantitatively in terms of area and volume. Moreover, we also find that wood defects usually disturb the wood species identification. Therefore, based on the wood defect's detection results, an improved wood species identification based on color moments is also proposed by deleting those defect points.

For example, the measurement errors for wood defect's areas and volumes can be approximately less than 5% if the defect's height is less than 3.0 mm. Moreover, the improved wood species recognition scheme can achieve a recognition accuracy of 95% approximately for 5 wood species. Therefore, our wood quality detection scheme is portable and efficient which can be applied in the practical wood processing factory. In fact, our scheme may be further used in other detection fields such as the surface detections of online industrial products.

Competing Interests

Dr. Zhao Peng declares that there is no conflict of interests regarding the publication of this paper.

Acknowledgments

This research is supported by the National Natural Science Foundation of China with Grant no. 31670717 and Heilongjiang Province Natural Science Foundation with Grant no. C2016011. It is also supported by the New Century Excellent Talents in University with Grant no. NCET-12-0809 and the Fundamental Research Funds for the Central University with Grant no. 2572014EB05-01.

References

[1] A. K. Moore and N. L. Owen, "Infrared spectroscopic studies of solid wood," *Applied Spectroscopy Reviews*, vol. 36, no. 1, pp. 65–86, 2001.

[2] M. S. Packianather and P. R. Drake, "Neural networks for classifying images of wood veneer. Part 2," *International Journal of Advanced Manufacturing Technology*, vol. 16, no. 6, pp. 424–433, 2000.

[3] M. Nakamura, M. Masuda, and K. Shinohara, "Multiresolutional image analysis of wood and other materials," *Journal of Wood Science*, vol. 45, no. 1, pp. 10–18, 1999.

[4] Q. Wei, Y. H. Chui, B. Leblon, and S. Y. Zhang, "Identification of selected internal wood characteristics in computed tomography images of black spruce: a comparison study," *Journal of Wood Science*, vol. 55, no. 3, pp. 175–180, 2009.

[5] J. E. Phelps and E. A. Mcginnes, "Growth-quality evaluation of black walnut wood, part 2-color analysis of veneer produced on different sites," *Wood and Fiber Science*, vol. 15, no. 2, pp. 177–185, 1983.

[6] S. Tsuchikawa, Y. Hirashima, Y. Sasaki, and K. Ando, "Near-infrared spectroscopic study of the physical and mechanical

properties of wood with meso- and micro-scale anatomical observation," *Applied Spectroscopy*, vol. 59, no. 1, pp. 86–93, 2005.

[7] C. R. Orton, D. Y. Parkinson, P. D. Evans, and N. L. Owen, "Fourier transform infrared studies of heterogeneity, photodegradation, and lignin/hemicellulose ratios within hardwoods and softwoods," *Applied Spectroscopy*, vol. 58, no. 11, pp. 1265–1271, 2004.

[8] B. K. Lavine, C. E. Davidson, A. J. Moores, and P. R. Griffiths, "Raman spectroscopy and genetic algorithms for the classification of wood types," *Applied Spectroscopy*, vol. 55, no. 8, pp. 960–966, 2001.

[9] V. Piuri and F. Scotti, "Design of an automatic wood types classification system by using fluorescence spectra," *IEEE Transactions on Systems, Man, and Cybernetics Part C: Applications and Reviews*, vol. 40, no. 3, pp. 358–366, 2010.

[10] P. K. Lebow, C. C. Brunner, A. G. Maristany, and D. A. Butler, "Classification of wood surface features by spectral reflectance," *Wood and Fiber Science*, vol. 28, no. 1, pp. 74–90, 1996.

[11] D. E. Kline, C. Surak, and P. A. Araman, "Automated hardwood lumber grading utilizing a multiple sensor machine vision technology," *Computers and Electronics in Agriculture*, vol. 41, no. 1-3, pp. 139–155, 2003.

[12] J. W. Funck, Y. Zhong, D. A. Butler, C. C. Brunner, and J. B. Forrer, "Image segmentation algorithms applied to wood defect detection," *Computers and Electronics in Agriculture*, vol. 41, no. 1–3, pp. 157–179, 2003.

[13] F. Longuetaud, F. Mothe, B. Kerautret et al., "Automatic knot detection and measurements from X-ray CT images of wood: a review and validation of an improved algorithm on softwood samples," *Computers and Electronics in Agriculture*, vol. 85, no. 2, pp. 77–89, 2012.

[14] S. Gao, N. Wang, L. Wang, and J. Han, "Application of an ultrasonic wave propagation field in the quantitative identification of cavity defect of log disc," *Computers and Electronics in Agriculture*, vol. 108, no. 3, pp. 123–129, 2014.

[15] V. Bucur, "Ultrasonic techniques for nondestructive testing of standing trees," *Ultrasonics*, vol. 43, no. 4, pp. 237–239, 2005.

[16] L. Wang, L. Li, W. Qi, and H. Yang, "Pattern recognition and size determination of internal wood defects based on wavelet neural networks," *Computers and Electronics in Agriculture*, vol. 69, no. 2, pp. 142–148, 2009.

10

Modulation Transfer Function of a Gaussian Beam Based on the Generalized Modified Atmospheric Spectrum

Chao Gao and Xiaofeng Li

School of Astronautics and Aeronautics, University of Electronic Science and Technology of China, 2006 Xiyuan Ave, West Hi-Tech Zone, Chengdu 611731, China

Correspondence should be addressed to Xiaofeng Li; lxf3203433@uestc.edu.cn

Academic Editor: Sulaiman Wadi Harun

This paper investigates the modulation transfer function of a Gaussian beam propagating through a horizontal path in weak-fluctuation non-Kolmogorov turbulence. Mathematical expressions are obtained based on the generalized modified atmospheric spectrum, which includes the spectral power law value of non-Kolmogorov turbulence, the finite inner and outer scales of turbulence, and other optical parameters of the Gaussian beam. The numerical results indicate that the atmospheric turbulence would produce less negative effects on the wireless optical communication system with an increase in the inner scale of turbulence. Additionally, the increased outer scale of turbulence makes a Gaussian beam influenced more seriously by the atmospheric turbulence.

1. Introduction

Optical wireless communication technology has drawn much attention for its significant technological challenges and prospective applications. It uses beams of laser propagating in the atmosphere to wirelessly transmit data at high speed. However, the atmosphere is full of numerous turbulence eddies, which has great degrading impacts on the performance of the communication system. The degrading effects of atmospheric turbulence on the communication system can be characterized statistically by the modulation transfer function (MTF) [1]. In the past few decades, various power spectrum models of refractive index have been proposed to analyze the MTF for different situations. Generally speaking, these turbulence power spectrum models can be classified into two typical categories: Kolmogorov and non-Kolmogorov models. The former have a fixed power law value of 11/3, while the latter allow the power law value to vary in the range from three to four. Most non-Kolmogorov models can be generalized from their corresponding Kolmogorov models, and thus the Kolmogorov models can be regarded as specific cases of the non-Kolmogorov models [2]. Among these models, the generalized modified atmospheric spectrum not only considers the variable spectral power law value between the ranges from 3 to 4, but also takes the finite

inner and outer scales of turbulence into account [3]. Besides, the generalized modified atmospheric spectrum features the small rise at a high wavenumber, which is clearly seen in temperature data recorded by sensors. These properties make the generalized modified atmospheric spectrum suitable and unique in the investigation of the MTF for plane and spherical waves [4].

In this study, the generalized modified atmospheric spectrum is used to investigate the MTF of a Gaussian beam in non-Kolmogorov turbulence along a horizontal path. The Gaussian beam, whose transverse electric field and intensity are normally distributed, is a typical kind of electromagnetic wave [5]. The rest of the paper is organized as follows. Section 2 introduces the generalized modified atmospheric spectrum and the MTF of a Gaussian beam. Section 3 presents a detailed expression reduction. The influences of the inner and outer scales of turbulence on the MTF of a Gaussian beam are analyzed in Section 4, followed by conclusions in Section 5.

2. Theoretical Models

2.1. Generalized Modified Atmospheric Spectrum. The generalized modified atmospheric spectrum takes the form [3]

$$\Phi_n(\kappa) = A(\alpha) C_n^2 \kappa^{-\alpha} f(\kappa, \alpha), \qquad (1)$$

where $\kappa \in [0, +\infty)$ is the angular wavenumber of the turbulence scale, $\alpha \in (3, 4)$ is the spectral power law value, C_n^2 is the generalized atmospheric structure parameter, $l_0 \geq 0$ is the inner scale of turbulence, and $L_0 \geq l_0$ is the outer scale of turbulence. $A(\alpha)$ in (1) is a function related to α:

$$A(\alpha) = \frac{\Gamma(\alpha - 1)}{4\pi^2} \sin \frac{\pi(\alpha - 3)}{2}, \qquad (2)$$

where $\Gamma(x)$ is the gamma function.

For the convenience of mathematical analysis, let

$$C_\alpha = \frac{3 - \alpha}{3} \Gamma\left(\frac{3 - \alpha}{2}\right) + a \frac{4 - \alpha}{3} \Gamma\left(\frac{4 - \alpha}{2}\right)$$
$$- b \frac{3 + \beta - \alpha}{3} \Gamma\left(\frac{3 + \beta - \alpha}{2}\right), \qquad (3)$$

where the constant coefficients a, b, and β in (3) are usually set as

$$a = 1.802,$$
$$b = 0.254, \qquad (4)$$
$$\beta = \frac{7}{6}.$$

It must be pointed out that the values of these coefficients are based on the experiments for the classic Kolmogorov turbulence but are widely used for theoretical analyses of non-Kolmogorov turbulence [1, 4]. Nevertheless, $f(\kappa, \alpha)$ in (1) takes the form

$$f(\kappa, \alpha) = \exp\left(-\frac{\kappa^2}{\kappa_l^2}\right) \times \left(1 - \exp\left(-\frac{\kappa^2}{\kappa_0^2}\right)\right)$$
$$\times \left(1 + a\left(\frac{\kappa}{\kappa_l}\right) - b\left(\frac{\kappa}{\kappa_l}\right)^\beta\right) \qquad (5)$$
$$= \sum_{i=1}^{3} \sum_{j=1}^{2} (-1)^{j-1} c_i \kappa^{p_i} \exp\left(-d_j^2 \kappa^2\right),$$

where

$$\kappa_0 = \frac{4\pi}{L_0},$$
$$\kappa_l = \frac{(\pi A(\alpha) C_\alpha)^{1/(\alpha - 5)}}{l_0}. \qquad (6)$$

And the coefficients are $c_1 = 1$, $c_2 = a/\kappa_l$, $c_3 = -b/\kappa_l^\beta$, $p_1 = 0$, $p_2 = 1$, $p_3 = \beta$, $d_1 = \sqrt{1/\kappa_l^2}$, and $d_2 = \sqrt{1/\kappa_l^2 + 1/\kappa_0^2}$.

2.2. MTF of a Gaussian Beam. The MTF is relative to the wave structure function (WSF). Based on the Rytov approximation, the WSF of Gaussian beam takes the simple form [1]

$$D(\rho) = 8Lk^2\pi^2 \int_0^1 d\xi \int_0^{+\infty} d\kappa$$

$$\times \kappa \Phi_n(\kappa) \exp\left(-\frac{L\Lambda\kappa^2\xi^2}{k}\right)$$
$$\times \left(I_0(\Lambda\rho\kappa\xi) - J_0\left(\rho\kappa\left(1 - \overline{\Theta}\xi\right)\right)\right), \qquad (7)$$

where ρ is the scalar separation between two observation points and L is the propagation optical path length. k in (7) is the angular wavenumber of Gaussian beam wave

$$k = \frac{2\pi}{\lambda}, \qquad (8)$$

where λ is the wavelength of Gaussian beam. Both Λ and $\overline{\Theta}$ in (7) are optical parameters of the Gaussian beam at the receiver

$$\Theta = \frac{\Theta_0}{\Theta_0^2 + \Lambda_0^2},$$
$$\overline{\Theta} = 1 - \Theta, \qquad (9)$$
$$\Lambda = \frac{\Lambda_0}{\Theta_0^2 + \Lambda_0^2},$$

where Θ_0 is the curvature parameter of Gaussian beam at transmitter and Λ_0 is the Fresnel ratio of Gaussian beam at transmitter

$$\Theta_0 = 1 - \frac{L}{R_0},$$
$$\Lambda_0 = \frac{2L}{kW_0}. \qquad (10)$$

In (10), R_0 is the phase front radius of Gaussian beam at transmitter, and W_0 is the radius of Gaussian beam at transmitter. $I_0(x)$ in (7) is the modified Bessel function of the first kind with zero order, and $J_0(x)$ in (7) is the Bessel function of the first kind with zero order [6]

$$I_0(x) = \sum_{n=0}^{+\infty} \frac{1}{(n!)^2} \left(\frac{x}{2}\right)^{2n},$$
$$J_0(x) = \sum_{n=0}^{+\infty} \frac{(-1)^n}{(n!)^2} \left(\frac{x}{2}\right)^{2n}. \qquad (11)$$

The atmospheric turbulence MTF takes the form [1]

$$\text{MTF}(\mu) = \exp\left(-\frac{1}{2} D(\mu d)\right), \qquad (12)$$

where μ is the normalized spatial frequency and d is the receiver aperture diameter. It is clear that the value range of the MTF is the interval from 0 to 1.

3. Expression Reduction

The calculation equation (7) will spend too much time because of its improper iterated integral. As an alternative, the closed-form expression of (7) can replace the improper iterated integral with special functions, which has corresponding packages in frequently used software. This section mainly discusses the reduction of (7).

Substituting (1) into (7), it follows that

$$D(\rho) = 8A(\alpha)C_n^2 L k^2 \pi^2 \int_0^1 d\xi \int_0^{+\infty} d\kappa$$

$$\times \kappa^{1-\alpha} f(\kappa, \alpha) \exp\left(-\frac{L\Lambda\kappa^2\xi^2}{k}\right) \tag{13}$$

$$\times \left(I_0(\Lambda\rho\kappa\xi) - J_0\left(\rho\kappa\left(1-\overline{\Theta}\xi\right)\right)\right).$$

For mathematical convenience, let

$$D_I = \int_0^1 d\xi \int_0^{+\infty} d\kappa \times \kappa^{1-\alpha} f(\kappa, \alpha) \exp\left(-\frac{L\Lambda\kappa^2\xi^2}{k}\right)$$

$$\times \left(I_0(\Lambda\rho\kappa\xi) - 1\right),$$

$$D_J = \int_0^1 d\xi \int_0^{+\infty} d\kappa \times \kappa^{1-\alpha} f(\kappa, \alpha) \exp\left(-\frac{L\Lambda\kappa^2\xi^2}{k}\right)$$

$$\times \left(J_0\left(\rho\kappa\left(1-\overline{\Theta}\xi\right)\right) - 1\right). \tag{14}$$

Thus, (13) can be presented by

$$D(\rho) = 8A(\alpha)C_n^2 L k^2 \pi^2 \times (D_I - D_J). \tag{15}$$

3.1. Reduction of D_I. Substituting (11) into (14), D_I is rewritten as

$$D_I = \int_0^1 d\xi \int_0^{+\infty} d\kappa \times \sum_{n=1}^{+\infty} \frac{1}{(n!)^2} \left(\frac{\Lambda\rho\kappa\xi}{2}\right)^{2n}$$

$$\times \kappa^{1-\alpha} f(\kappa, \alpha) \exp\left(-\frac{L\Lambda\kappa^2\xi^2}{k}\right). \tag{16}$$

In most situations, $\rho \ll 1$. This is because MTF will quickly converge to zero when ρ approaches one; that is, MTF is significantly larger than zero when ρ approaches zero. Thus, (16) could be approximated by the simpler expression

$$D_I \approx \frac{\Lambda^2\rho^2}{4} \int_0^1 \xi^2 d\xi \int_0^{+\infty} d\kappa$$

$$\times \kappa^{3-\alpha} f(\kappa, \alpha) \exp\left(-\frac{L\Lambda\kappa^2\xi^2}{k}\right). \tag{17}$$

Consider the iterated integral in (17). According to (5), there is

$$\kappa^{3-\alpha} f(\kappa, \alpha) \exp\left(-\frac{L\Lambda\kappa^2\xi^2}{k}\right)$$

$$= \sum_{i=1}^3 \sum_{j=1}^2 (-1)^{j-1} c_i \kappa^{3-\alpha+p_i} \tag{18}$$

$$\times \exp\left(-\left(d_j^2 + \frac{L\Lambda\xi^2}{k}\right)\kappa^2\right).$$

Based on the equation for $u > -1$ and $v > 0$ [7],

$$\int_0^{+\infty} x^u \exp\left(-vx^2\right) dx = \frac{1}{2} v^{-(u+1)/2} \Gamma\left(\frac{u+1}{2}\right), \tag{19}$$

we can get

$$\int_0^{+\infty} \kappa^{3-\alpha} f(\kappa, \alpha) \exp\left(-\frac{L\Lambda\kappa^2\xi^2}{k}\right) d\kappa$$

$$= \sum_{i=1}^3 \sum_{j=1}^2 (-1)^{j-1} \frac{c_i}{2} \left(d_j^2 + \frac{L\Lambda\xi^2}{k}\right)^{-(4-\alpha+p_i)/2} \tag{20}$$

$$\times \Gamma\left(\frac{4-\alpha+p_i}{2}\right).$$

Without loss of generality, the integrand in (17) takes the form

$$\xi^n \int_0^{+\infty} \kappa^p f(\kappa, \alpha) \exp\left(-\frac{L\Lambda\kappa^2\xi^2}{k}\right) d\kappa$$

$$= \sum_{i=1}^3 \sum_{j=1}^2 (-1)^{j-1} \frac{c_i}{2} \Gamma\left(\frac{1+p+p_i}{2}\right) \tag{21}$$

$$\times \left(\frac{L\Lambda}{k}\right)^{-(1+p+p_i)/2} \xi^n \left(\frac{kd_j^2}{L\Lambda} + \xi^2\right)^{-(1+p+p_i)/2},$$

where $n = 2$ and $p = 3 - \alpha$. Based on the equation for $u > 0$ and $w > 0$ [7],

$$\int_0^1 x^{u-1} \left(w^2 + x^2\right)^v dx$$

$$= \frac{1}{u} w^{2v} {}_2F_1\left(-v, \frac{u}{2}; \frac{u+2}{2}; -\frac{1}{w^2}\right), \tag{22}$$

we can get

$$\int_0^1 \xi^n d\xi \int_0^{+\infty} d\kappa \times \kappa^p f(\kappa, \alpha) \exp\left(-\frac{L\Lambda\kappa^2\xi^2}{k}\right)$$

$$= \sum_{i=1}^3 \sum_{j=1}^2 (-1)^{j-1} \frac{c_i}{2d_j^{1+p+p_i}(n+1)} \times \Gamma\left(\frac{1+p+p_i}{2}\right) \tag{23}$$

$$\times {}_2F_1\left(\frac{1+p+p_i}{2}, \frac{n+1}{2}; \frac{n+3}{2}; -\frac{L\Lambda}{kd_j^2}\right),$$

where ${}_2F_1(a, b; c; z)$ is the Gaussian hypergeometric function [6]. Thus, D_I can be computed by (17) and (23) with $n = 2$.

3.2. Reduction of D_J. Following similar procedures as presented in Section 3.1, D_J in (14) is rewritten as

$$D_J \approx -\frac{\rho^2}{4} \int_0^1 \left(1-\overline{\Theta}\xi\right)^2 d\xi \int_0^{+\infty} d\kappa$$

$$\times \kappa^{3-\alpha} f(\kappa, \alpha) \exp\left(-\frac{L\Lambda\kappa^2\xi^2}{k}\right). \tag{24}$$

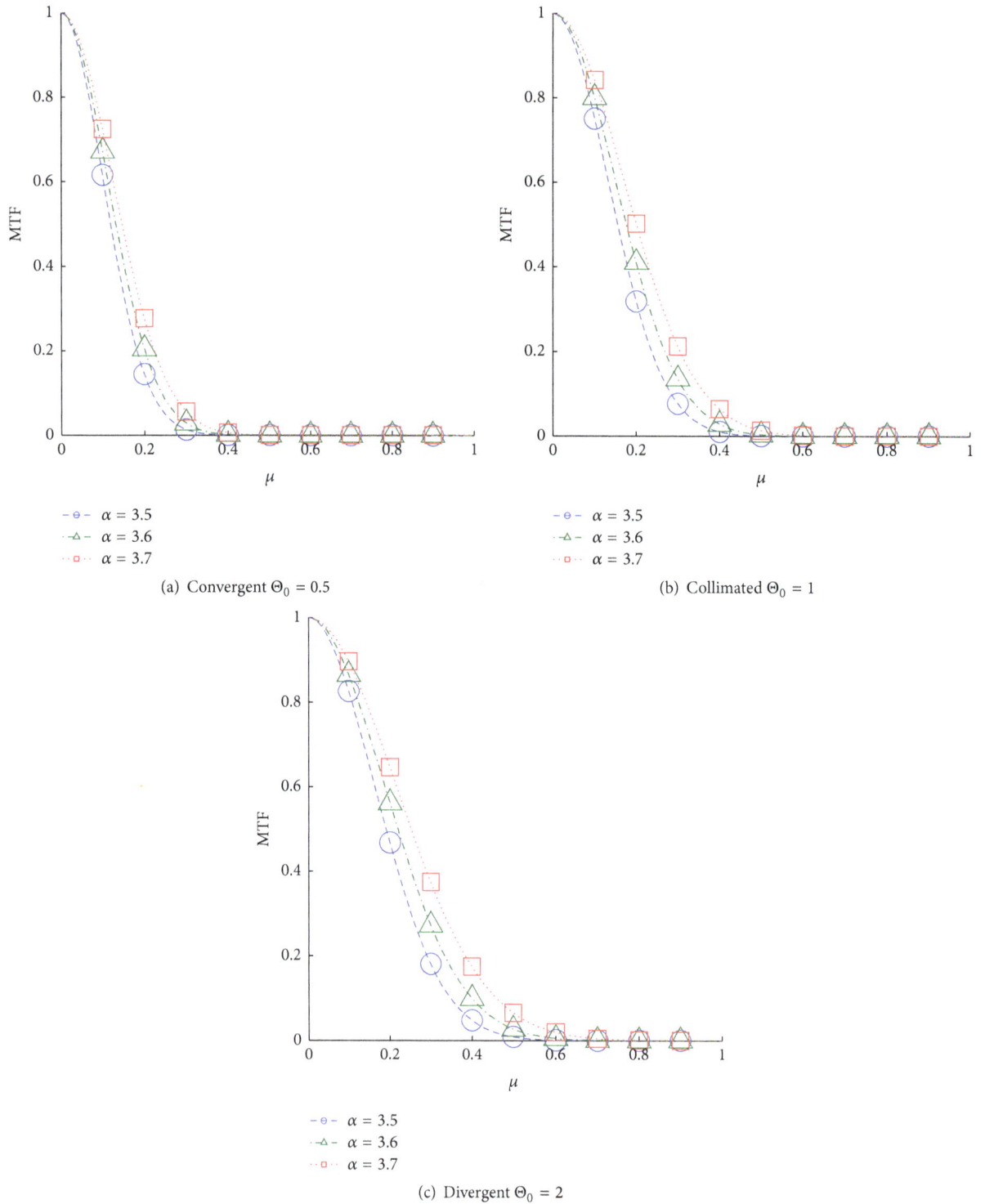

FIGURE 1: Effects of spectral power law value on MTF for different types of Gaussian beams.

Expanding (24) by the binomial theorem, it follows that

$$D_J = -\frac{\overline{\Theta}^2 \rho^2}{4} \int_0^1 \xi^2 d\xi \int_0^{+\infty} d\kappa$$

$$\times \kappa^{3-\alpha} f(\kappa, \alpha) \exp\left(-\frac{L\Lambda\kappa^2\xi^2}{k}\right)$$

$$+ \frac{\overline{\Theta}\rho^2}{2} \int_0^1 \xi \, d\xi \int_0^{+\infty} d\kappa$$

$$\times \kappa^{3-\alpha} f(\kappa, \alpha) \exp\left(-\frac{L\Lambda\kappa^2\xi^2}{k}\right)$$

$$- \frac{\rho^2}{4} \int_0^1 d\xi \int_0^{+\infty} d\kappa$$

(a) Convergent $\Theta_0 = 0.5$

(b) Collimated $\Theta_0 = 1$

(c) Divergent $\Theta_0 = 2$

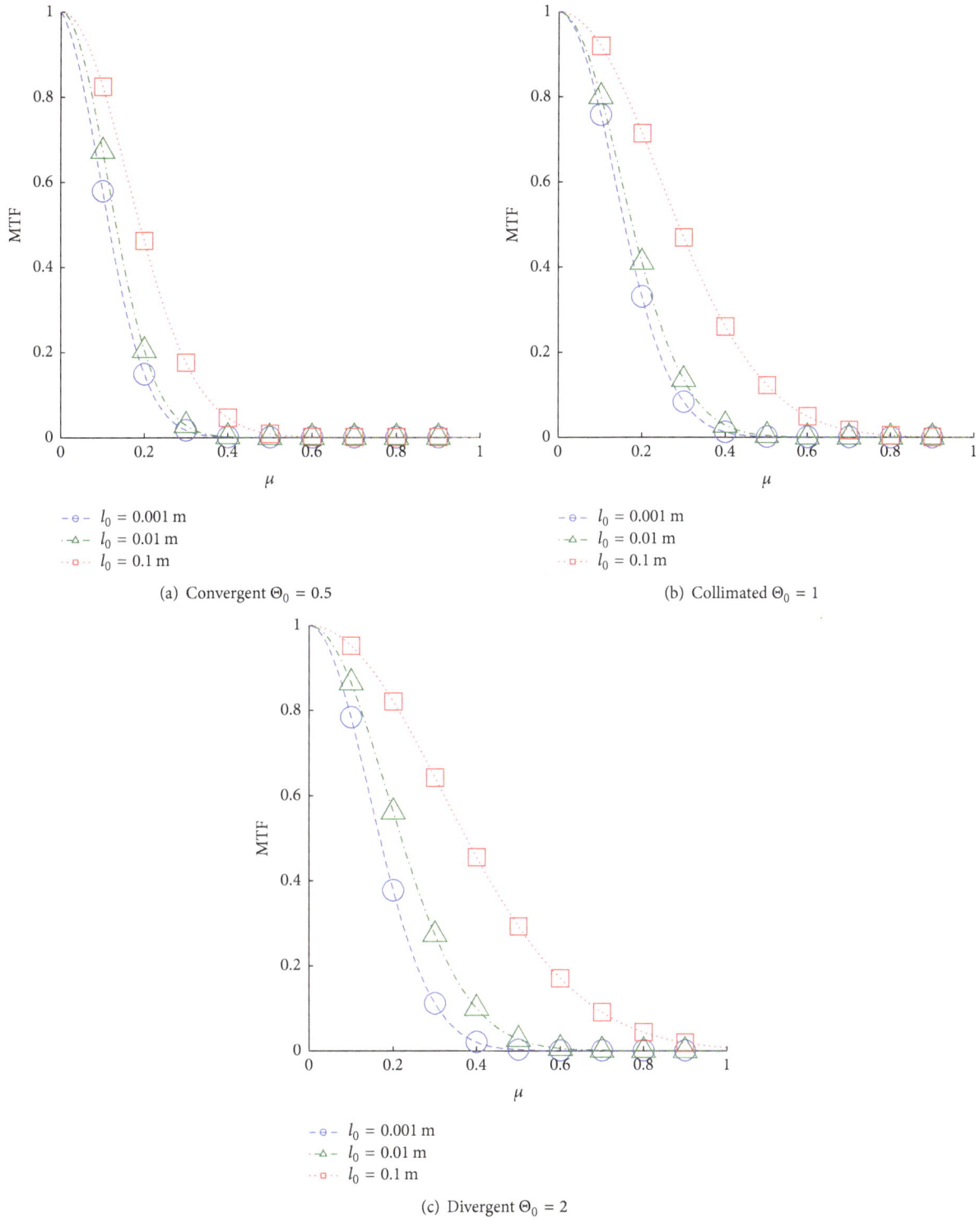

FIGURE 2: Effects of inner scale on MTF for different types of Gaussian beams.

$$\times \kappa^{3-\alpha} f(\kappa, \alpha) \exp\left(-\frac{L\Lambda\kappa^2\xi^2}{k}\right).$$

$$(25)$$

Thus, D_J could be computed by (25) and (23) with $n = 0, 1, 2$.

4. Numerical Simulations

The following simulations are conducted by the Gaussian beam with these settings: $\lambda = 1.55 \times 10^{-6}$ m, $L = 1000$ m, $k \approx 4.0537 \times 10^6$ rad/m, $C_n^2 = 1.7 \times 10^{-14}$ m$^{3-\alpha}$, $W_0 = 0.1$ m,

(a) Convergent $\Theta_0 = 0.5$

(b) Collimated $\Theta_0 = 1$

(c) Divergent $\Theta_0 = 2$

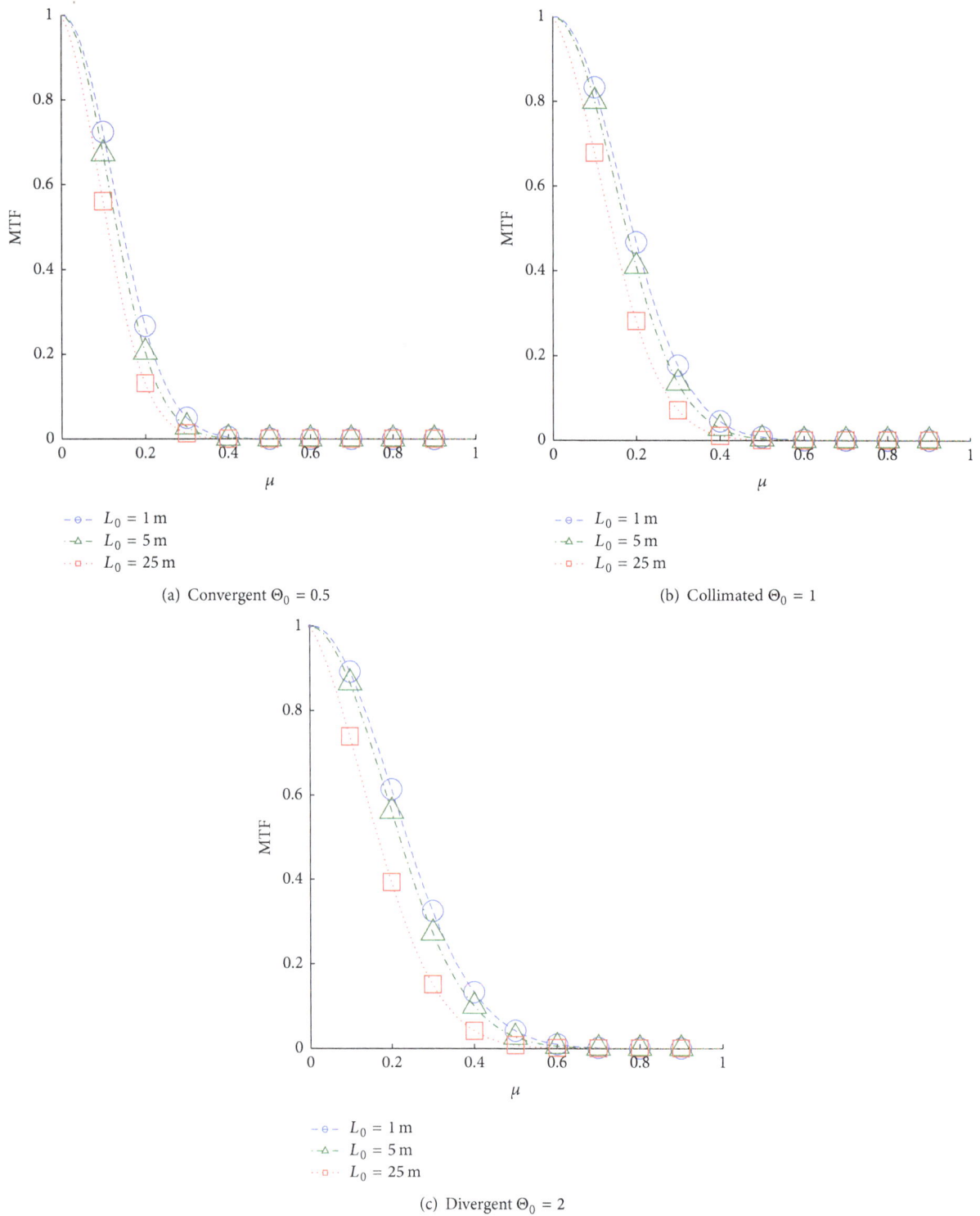

FIGURE 3: Effects of outer scale on MTF for different types of Gaussian beams.

$\Lambda_0 \approx 0.0493$, and $d = 0.1$ m. Of course, other values can also be chosen.

Figure 1 depicts the effects of spectral power law value on MTF for different types of Gaussian beams. In this calculation, the inner and outer scales of turbulence are set as $l_0 = 0.01$ m and $L_0 = 5$ m, respectively. As shown in Figure 1(a), the atmospheric turbulence apparently produces

more effects on the propagation of the convergent Gaussian beam ($\Theta_0 = 0.5$) with an increase in the normalized spatial frequency μ, which acts in accordance with common sense. Besides, from Figure 1(a), it can be found that the non-Kolmogorov atmospheric turbulence would bring more effects on the wireless optical communication system when the spectral power law value α decreases. The same trends

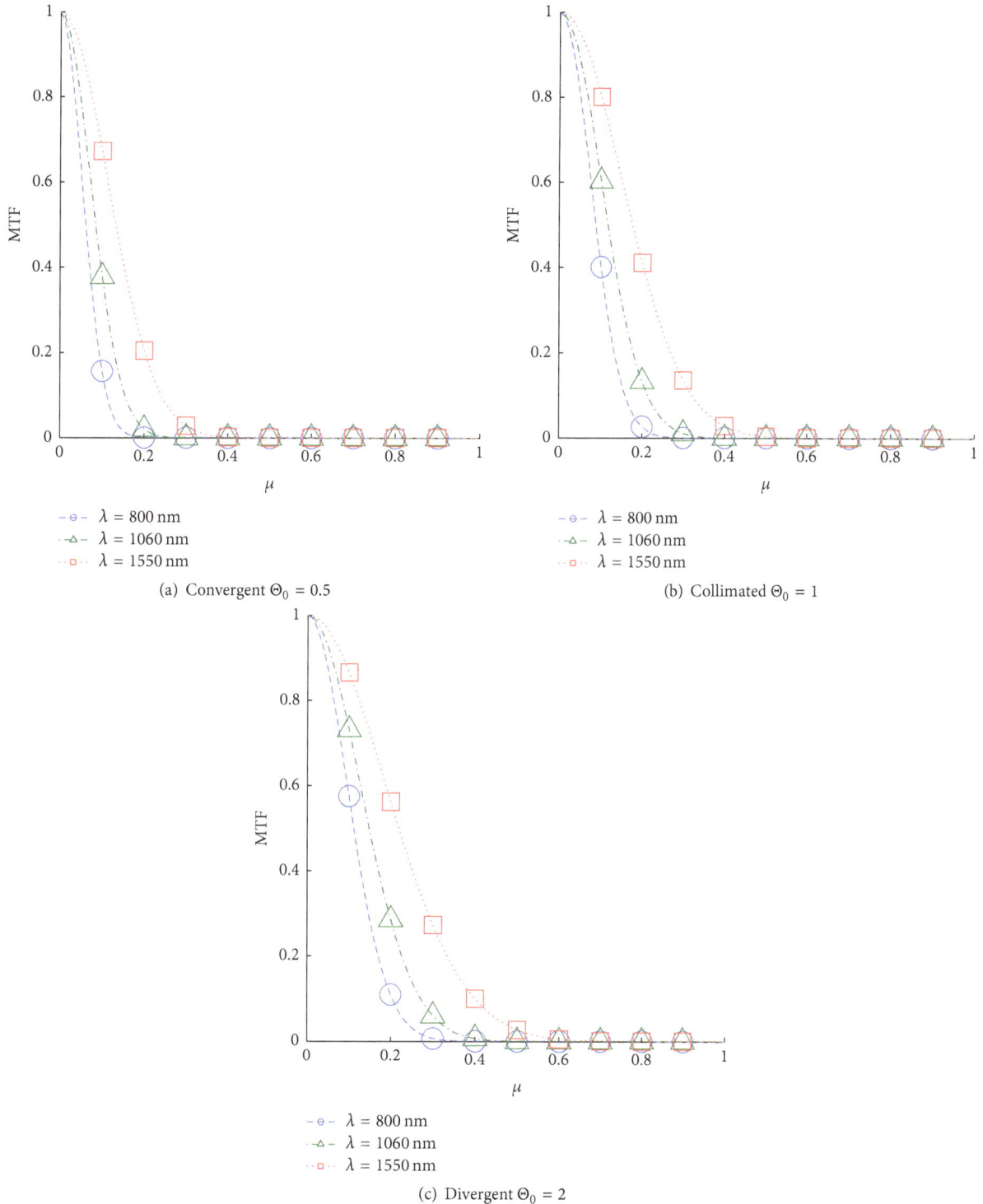

(a) Convergent $\Theta_0 = 0.5$

(b) Collimated $\Theta_0 = 1$

(c) Divergent $\Theta_0 = 2$

FIGURE 4: Effects of wavelength on MTF for different types of Gaussian beams.

are obtained for the collimated Gaussian beam ($\Theta_0 = 1$) in Figure 1(b) and the divergent Gaussian beam ($\Theta_0 = 2$) in Figure 1(c).

To analyze the effects of the turbulence inner scale on MTF, the spectral power law value and the outer scale of turbulence are fixed to constant values as $\alpha = 3.6$ and $L_0 = 5$ m. Several inner scales of turbulence are used, and calculation results are depicted in Figure 2 for different types of Gaussian beams. It can be seen that, with an increase in the inner scale of turbulence, the value of MTF also increases. This can be physically explained by the change of inertial subrange of turbulence. When the inner scale of turbulence increases, the frequency's upper bound of inertial subrange would move to a lower position, and thus the atmospheric turbulence would bring less effects on the propagation of the Gaussian beam.

The influences of outer scale of turbulence on MTF are depicted in Figure 3 for different types of Gaussian beams. For the real atmospheric turbulence, the outer scale of turbulence is usually in the order of meters. Hence, it is set to 1 m, 5 m, and 25 m, respectively. The spectral power law value and the inner scale of turbulence are set to $\alpha = 3.6$ and $l_0 = 0.01$ m as example. It can be seen that, with an increase in the outer scale of turbulence, the value of MTF decreases and thus the quality of the Gaussian beam is degraded severely by the atmospheric turbulence. This is because WSF is mostly influenced by the large-scale turbulence eddies, which are relevant to the low-frequency part of the atmospheric turbulence spectrum. A larger turbulence outer scale would lead to a larger range of inertial subrange.

For further discussions and analyses, the inner and outer scales of turbulence are assigned to constant values $l_0 = 0.01$ m and $L_0 = 5$ m, respectively. The spectral power law value still uses the default value $\alpha = 3.6$. Some typical values of wavelength in the near infrared region, $\lambda = 850$ nm, $\lambda = 1060$ nm, and $\lambda = 1550$ nm, are investigated in this simulation. Figure 4 depicts MTF for different Gaussian beams as a function of μ with different λ. It is obvious that the value of MTF increases with an increase in λ for certain type of Gaussian beam if other optical parameters are fixed. This phenomenon may be caused by the fact that the larger the beam wavelength, the more pronounced the diffraction. Thus, a laser beam with larger wavelength can be less affected by turbulence eddies.

5. Conclusions

In this paper, a theoretical expression of the MTF is derived for a Gaussian beam propagating through the non-Kolmogorov atmospheric turbulence along a horizontal path. This expression contains a variable spectral power law value, finite inner and outer scales of turbulence, and other important optical parameters of a Gaussian beam. Numerical simulations indicate that the atmospheric turbulence would produce less degrading effects on the wireless optical communication system with an increase in the spectral power law value. The decreased inner scale of turbulence makes a Gaussian beam influenced more seriously by the atmospheric turbulence. With an increase in the outer scale of turbulence, the quality of a Gaussian beam is degraded more severely by the atmospheric turbulence. A laser beam with larger wavelength can be less affected by turbulence eddies.

Competing Interests

The authors declare that there is no conflict of interests regarding the publication of this paper.

References

[1] L. C. Andrews and R. L. Phillips, *Laser-Beam Propagation through Random Media*, SPIE Optical Engineering Press, Bellingham, Wash, USA, 2nd edition, 2005.

[2] C. Gao, Y. Li, Y. Li, and X. Li, "Irradiance scintillation index for a gaussian beam based on the generalized modified atmospheric spectrum with aperture averaged," *International Journal of Optics*, vol. 2016, Article ID 8730609, 8 pages, 2016.

[3] B. Xue, L. Cui, W. Xue, X. Bai, and F. Zhou, "Generalized modified atmospheric spectral model for optical wave propagating through non-Kolmogorov turbulence," *Journal of the Optical Society of America A: Optics, Image Science, and Vision*, vol. 28, no. 5, pp. 912–916, 2011.

[4] B. Xue, L. Cao, L. Cui, X. Bai, X. Cao, and F. Zhou, "Analysis of non-Kolmogorov weak turbulence effects on infrared imaging by atmospheric turbulence MTF," *Optics Communications*, vol. 300, no. 1, pp. 114–118, 2013.

[5] C. Gao and X. Li, "An analytic expression for the beam wander of a Gaussian wave propagating through scale-dependent anisotropic turbulence," *Iranian Journal of Science and Technology, Transactions A: Science*, 2016.

[6] F. W. J. Olver, D. W. Lozier, R. F. Boisvert, and C. W. Clark, *NIST Handbook of Mathematical Functions*, Cambridge University Press, New York, NY, USA, 2010.

[7] I. S. Gradshteyn and I. M. Ryzhik, *Table of Integrals, Series, and Products*, Academic Press, Waltham, Mass, USA, 8th edition, 2014.

11

The Impact of Pixel Resolution, Integration Scale, Preprocessing, and Feature Normalization on Texture Analysis for Mass Classification in Mammograms

Mohamed Abdel-Nasser,[1] Jaime Melendez,[2] Antonio Moreno,[1] and Domenec Puig[1]

[1]Departament d'Enginyeria Informàtica i Matemàtiques, Universitat Rovira i Virgili, Avinguda Paisos Catalans 26, 43007 Tarragona, Spain
[2]Department of Radiology, Radboud University Medical Center, 6525 GA Nijmegen, Netherlands

Correspondence should be addressed to Mohamed Abdel-Nasser; egnaser@gmail.com

Academic Editor: Chenggen Quan

Texture analysis methods are widely used to characterize breast masses in mammograms. Texture gives information about the spatial arrangement of the intensities in the region of interest. This information has been used in mammogram analysis applications such as mass detection, mass classification, and breast density estimation. In this paper, we study the effect of factors such as pixel resolution, integration scale, preprocessing, and feature normalization on the performance of those texture methods for mass classification. The classification performance was assessed considering linear and nonlinear support vector machine classifiers. To find the best combination among the studied factors, we used three approaches: greedy, sequential forward selection (SFS), and exhaustive search. On the basis of our study, we conclude that the factors studied affect the performance of texture methods, so the best combination of these factors should be determined to achieve the best performance with each texture method. SFS can be an appropriate way to approach the factor combination problem because it is less computationally intensive than the other methods.

1. Introduction

Breast cancer was responsible for the largest number of cancer deaths among the EU females in 2014 [1]. Mammography is considered, in general, the most effective method for early detection of breast cancer and thus has been adopted for breast cancer screening. Computer-aided detection (CAD) systems are typically used to analyze mammograms in screening. While radiologists are generally pleased with the performance of CAD for clustered microcalcification detection, they have little confidence in CAD for mass detection. The most common complaint of radiologists is that CAD systems lead to a large number of false positives [2].

A breast cancer CAD system consists of three main stages: *segmentation* of a region of interest (ROI) from the mammogram, *feature extraction* from the ROI, and *classification*. Although mammography is a highly sensitive method for early detection of breast cancer, low specificity has been achieved in the classification of benign and malignant masses.

Texture analysis methods constitute one of the options for improving the specificity of classification algorithms applied to mammography. These methods may provide additional information in distinguishing benign and malignant masses. Although several feature extraction methods have been proposed for analyzing mammograms, improving the classification performance remains a challenging problem.

Texture analysis methods have been widely used to analyze mammographic images because they produce information about the spatial arrangement of intensities in the mammogram. Texture is one of the major mammographic characteristics for mass classification. For instance, several studies have used texture analysis methods to distinguish between normal and abnormal tissue [3–8] or to discriminate between benign and malignant masses [9–11]. Table 1 briefly summarizes some of this previous work. In addition, other studies have used texture analysis methods to estimate breast density [12] or to segment masses from mammograms [13].

TABLE 1: Summary of texture analysis methods that have been used to analyze mammograms.

Method	Extracted features	Utilized classifiers	Purpose
[6]	Local binary pattern (LBP)	Support vector machines (SVMs)	Classification of ROIs into mass/normal
[7]	Histogram of oriented gradients (HOG)	SVM	Classification of ROIs into mass/normal
[11]	Haralick's features (HAR)	k-nearest neighbour (k-NN)	Microcalcification classification
[8]	Gabor filters (GF)	Threshold-based approach	Breast cancer detection
[29]	Grey levels, texture, and features related to independent component analysis	Neural network (NN)	Classifying ROIs into normal/abnormal Classifying ROIs into benign/malignant
[30]	A set of texture features	SVM	Mass detection
[31]	Ripley's K function texture measures	SVM	Detection of breast masses
[9]	Texture features derived from concurrence matrix	NN	Microcalcification classification
[32]	A set of texture features	k-NN, SVM, random forests, logistic model trees, and Naive Bayes	Lesion classification
[10]	HAR	Bayesian classifier Fisher linear discriminant	Study the effect of pixel resolution on the performance of texture methods
[3]	LBP, robust LBP, centre symmetric LBP, fuzzy LBP, local grey level appearance, LDN, HOG, HAR, and GF	k-NN, linear SVM, nonlinear SVM random forest, and Fisher linear discriminant analysis (FLDA)	Finding the best combination among the texture methods to classify ROIs into mass/normal
[33]	Local ternary pattern and local phase quantization	SVM	Classifying tumors into benign/malignant
[34]	Novel sets of texture descriptors extracted from the cooccurrence matrix	SVM	Six medical datasets were used for validation, one of them for breast cancer
[35]	Texture analysis techniques based on the cooccurrence matrix and region-based approaches	SVM	15 datasets were used for validation, one of them for breast cancer
[36]	HOG, dense scale invariant feature transform, and local configuration pattern	SVM, k-NN, FLDA, and decision tree	Classifying ROIs into normal/abnormal Classifying ROIs into benign/malignant
[37]	Curvelet moments	k-NN	Classifying ROIs into normal/abnormal Classifying ROIs into benign/malignant

CAD systems usually focus on a ROI to study breast masses. The texture of this ROI describes the pattern of spatial variation of gray levels in a neighbourhood that is small compared to the breast area but big enough to include the masses. In other words, texture must be analyzed in a region, and the size of this region should be tuned. Thus, we should answer the question: *what is the optimal neighbourhood size (integration scale) for texture analysis*? In addition, the size of a mammogram is usually in the range of thousands of pixels. Consequently, several works have reduced the original resolution of a mammogram to reduce the computational complexity and the execution time of their algorithms [14], or to save resources (e.g., memory and storage space). However, image downsampling may also affect the performance of the texture analysis methods. Therefore, we should answer the question: *how far can we downsample the image while keeping the performance of the texture methods*?

In breast cancer CAD systems, several preprocessing operations such as image filtering or enhancement are usually applied to mammograms. Pisano et al. show that the contrast-limited adaptive histogram equalization (CLAHE) applied to a mammogram before it is displayed can make the indicative

structures of breast cancer more visible [15]. Sharpening (SH) is used to improve the detection of clustered calcifications [16]. The median filter (MF) is used to remove the noise from the mammograms [17]. *Preprocessing may affect the performance of texture analysis methods* because it effectively changes the gray levels of the images. This effect should be assessed. After extracting the texture features from a given mammogram, they are usually normalized before proceeding to the classification stage. *The utilized normalization method may also affect the final classification results.*

In this paper, we study the effect of pixel resolution, integration scale, preprocessing, and feature normalization on the performance of texture analysis methods when used to classify masses in mammograms. For that purpose, we have chosen five widely/recently used texture methods: local binary pattern (LBP), local directional number (LDN), histogram of oriented gradients (HOG), Haralick's features (HAR), and Gabor filters (GF). In order to evaluate the performance of the aforementioned methods, we extracted a set of regions of interest (ROIs) containing lesions from the mini-MIAS database [18], and we used each texture analysis method to classify the ROIs into benign or malignant. The performance of each texture method is evaluated with five pixel resolutions (200 μm, 400 μm, 600 μm, 800 μm, and 1000 μm), six integration scales (25×25, 32×32, 50×50, 64×64, 75×75, and 100×100 pixels), three preprocessing steps (CLAHE, MF, and SH), and five feature normalization methods. In addition, linear and nonlinear SVM classifiers are used.

To the best of our knowledge, only one previous study has conducted a similar evaluation. Rangayyan et al. studied the effect of pixel resolution on texture features of breast masses in mammograms [10]. However, only pixel resolution and Haralick's features were considered. In contrast, the current study takes into account a wider range of factors such as pixel resolution, integration scale, preprocessing, and feature normalization, and it considers a larger number and more powerful texture descriptors that have been successfully applied in recent relevant work. Moreover, we include linear and nonlinear SVMs; thus, both relatively simple and complex classification approaches can be assessed. Lastly, we analyze the combination of the best options for those factors using three approaches: greedy, sequential forward selection (SFS), and exhaustive search (ExS).

The rest of this paper is organized as follows. Section 2 describes the database and the methods used in this study. Section 3 shows our experimental results, which are then discussed in Section 4. Finally, Section 5 concludes our study.

2. Materials and Methods

In this study, we assess the performance of five texture analysis methods (LBP, LDN, HOG, HAR, and GF) while varying the pixel resolution, integration scale, image preprocessing algorithm, and data normalization method. To that end, we extracted a set of ROIs containing either benign or malignant masses from the mini-MIAS database. Given a certain texture analysis method, a feature vector is extracted from each ROI to be fed into a linear support vector machine (LSVM) or a nonlinear support vector machine (NLSVM). The trained models are used to determine if an unseen ROI contains a benign or a malignant mass.

2.1. Materials. The mini-MIAS database, consisting of 322 mediolateral oblique images of 161 cases, is used in our experiments. It was created from the original MIAS database by downsampling the images from 50 μm to 200 μm per pixel and clipping/padding to a fixed size of 1024×1024 pixels. A ground truth was prepared by experienced radiologists and confirmed using a biopsy procedure. The dataset is available at http://peipa.essex.ac.uk/info/mias.html. In this study 109 ROIs, 60 containing a benign mass and 49 containing a malignant mass, were used. Figure 1 shows examples of the extracted ROIs. Interested researchers can request the ROIs from the corresponding author of the paper.

The authors of the mini-MIAS database reported that they reduced the pixel resolution of the original MIAS database (digitized at 50 μm) to 200 μm by popular request. Moreover, several studies have used the pixel resolution 200 μm as a baseline resolution in their applications [14, 19]. We do the same in this work.

2.2. Texture Analysis Methods. This section explains the utilized texture analysis methods including the parameters selected for each of them.

2.2.1. Local Binary Pattern. The LBP labels the pixels of an image by comparing a 3×3-pixel neighbourhood with the value of the central pixel [20]. Pixels in this neighbourhood with a value greater than the central pixel are labelled as 1 and the rest as 0; thus, each pixel is represented by 8 bits. The size of the neighbourhood may vary on different applications (e.g., 3×3 and 5×5). A *uniform LBP* is an extension of the original LBP in which only patterns that contain at most two transitions from 0 to 1 (or vice versa) are considered. In uniform LBP mapping, there is a separate output label for each uniform pattern and all the nonuniform patterns are assigned to a single label. In this study, a 3×3 neighbourhood is used to generate the histogram of uniform LBPs for each ROI. The uniform mapping produces 59 output labels (59 dimensions) for neighbourhoods of 8 pixels. The implementation of LBP descriptor is available at http://www.cse.oulu.fi/CMV/Downloads/LBPMatlab.

2.2.2. Local Directional Number. In the LDN [21], the edge responses are computed in eight different directions by convoluting the Kirsch compass masks [22] with the ROIs. The locations of the top positive and negative edge responses are used to generate a 6-bit code for each pixel. Finally, the histogram of the LDN codes is calculated in the given ROI (64 dimensions). The implementation of LDN descriptor is available at https://gitlab.com/my-research/local-directional-number-pattern.git.

2.2.3. Histogram of Oriented Gradients. In the HOG method [23], the occurrences of edge orientations in a ROI are counted. The image is divided into blocks (small groups of cells) and then a weighted histogram is computed for each of them. The combination of the histograms of all blocks represents the final HOG descriptors. In order to get the best performance of HOG, its parameters have been empirically tuned. In this study, we used a 3×3 cell size, 8×8 cells for the block size, and a 9-bit histogram. The implementation of HOG descriptor is available at http://www.vlfeat.org/overview/hog.html.

2.2.4. Haralick's Features. The HAR features are computed from the gray level cooccurrence matrix (GLCM). In the GLCM, the distribution of cooccurring gray level values at a given offset (direction and distance) is computed [24]. A GLCM is computed from each ROI, and then 14 texture features are calculated: *angular second moment, contrast, correlation, variance, inverse difference moment, sum average, sum variance, sum entropy, entropy, difference variance, difference entropy, information measure of correlation 1, information measure of correlation 2*, and *maximal correlation coefficient* [10]. The mathematical expression of each feature can be

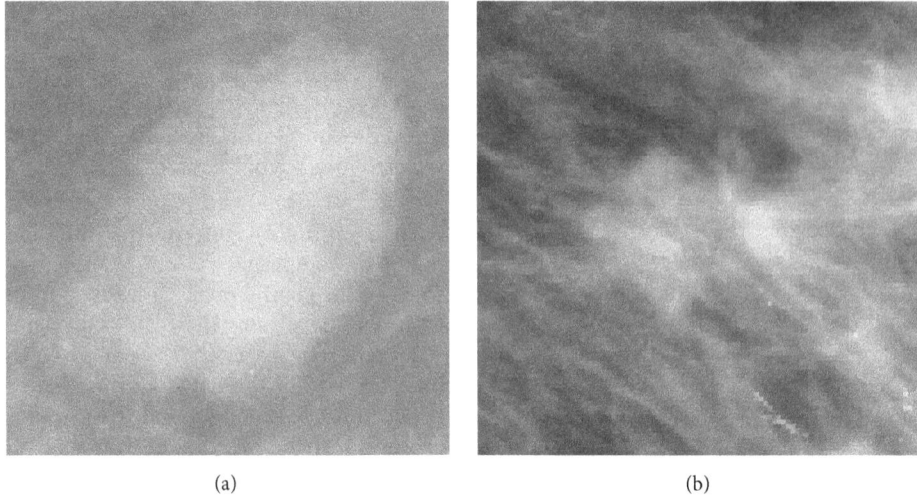

(a)   (b)

FIGURE 1: ROIs extracted from the mini-MIAS breast cancer database. A ROI containing (a) a benign mass and (b) a malignant mass.

found in the relevant previous work [25, 26]. The implementation of HAR descriptors is available at https://github.com/nutsiepully/spiff/blob/master/src/haralick.m.

2.2.5. Gabor Filters. A two-dimensional Gabor filter $g(x, y)$ can be expressed as a sinusoid with a particular frequency and orientation, modulated by a Gaussian envelope

$$g(x, y) = \exp^{(-1/2)(x^2/\sigma_x^2 + y^2/\sigma_y^2)} \exp^{-j2\pi(u_0 x + v_0 y)}, \quad (1)$$

where (u_0, v_0) is the centre of a sinusoidal function and σ_x and σ_y are the standard deviations along two orthogonal directions (which determine the width of the Gaussian envelope along the x- and y-axes in the spatial domain). Given a ROI $I(x, y)$, the filtered ROI $f(x, y)$ is the result of convoluting $I(x, y)$ and $g(x, y)$. Tuning GF to specific frequencies and directions can lead them to detect both local orientation and frequency information from an image [27]. In this study, we used 4 scales and 6 orientations to obtain these filtered ROIs. This design produces 24 responses. For each ROI, the energies of the 24 responses are calculated, and then they are aggregated in order to form the feature vector. The implementation of Gabor filters is available at https://github.com/mhaghighat/gabor.

2.3. Preprocessing. The performance of the texture analysis methods is evaluated with three preprocessing algorithms: CLAHE, median filter (MF), and sharpening (SH).

(i) *CLAHE*: it works on small regions of the input ROI (known as tiles). The contrast of each tile is enhanced; consequently the histogram of the output region approximately matches a predefined distribution [28]. In this study, the *Rayleigh* distribution is used [15].

(ii) *MF*: each pixel in the filtered ROI contains the median value of the $m \times n$ neighbourhood around the corresponding pixel in the input ROI [17]. In this study, a 3×3 neighbourhood is used.

(iii) *SH*: in order to sharpen a ROI, it is first blurred; edges are detected in the blurred ROI and added to it to produce a sharper image [16].

The preprocessing operations can be carried out using the following MATLAB functions: CLAHE (adapthisteq.m), median filter (medfilt2.m), and sharpening (imsharpen.m). Figure 2 shows examples for MF, SH, and CLAHE when they are applied to benign and malignant masses.

2.4. Feature Normalization Methods. Feature vectors are normalized in order to prevent attributes with higher numeric ranges from dominating those with lower numeric ranges. Given a feature vector $x = [x_1, x_2, x_3, \ldots, x_N]$, the normalized feature vector x_{new} is calculated using five normalization methods as follows [38, 39]:

(i) The zero mean unit variance (*zs*) method: $x_{\text{new}} = (x - \mu)/\sigma$, where μ and σ are the mean and the variance of x.

(ii) The maximum-minimum (*mn*) method: $x_{\text{new}} = (x - x_{\min})/(x_{\max} - x_{\min})$, where $x_{\max}$ and $x_{\min}$ are the maximum and minimum of x.

(iii) The ℓ^1 method scales x to unit length using the ℓ^1-norm, $x_{\text{new}} = x/\sum_{n=1}^{N} |x_n|$.

(iv) The ℓ^2 method scales x to unit length using the ℓ^2-norm, $x_{\text{new}} = x/\sqrt{\sum_{n=1}^{N} |x_n|^2}$.

(v) The *nh* method scales x to unit length as follows: $x_{\text{new}} = x/\sum_{n=1}^{N} x$.

The normalization methods can be easily implemented in MATLAB. ℓ^1- and ℓ^2-norm can be carried out using the MATLAB function *norm.m*.

2.5. Classification. Given a labelled training set of the form (x_i, y_i), $i = 1, 2, \ldots, k$, where $x_i \in \mathbb{R}^n$ are the feature values, $y_i \in \{1, -1\}$ is the class of x_i, n is the number of

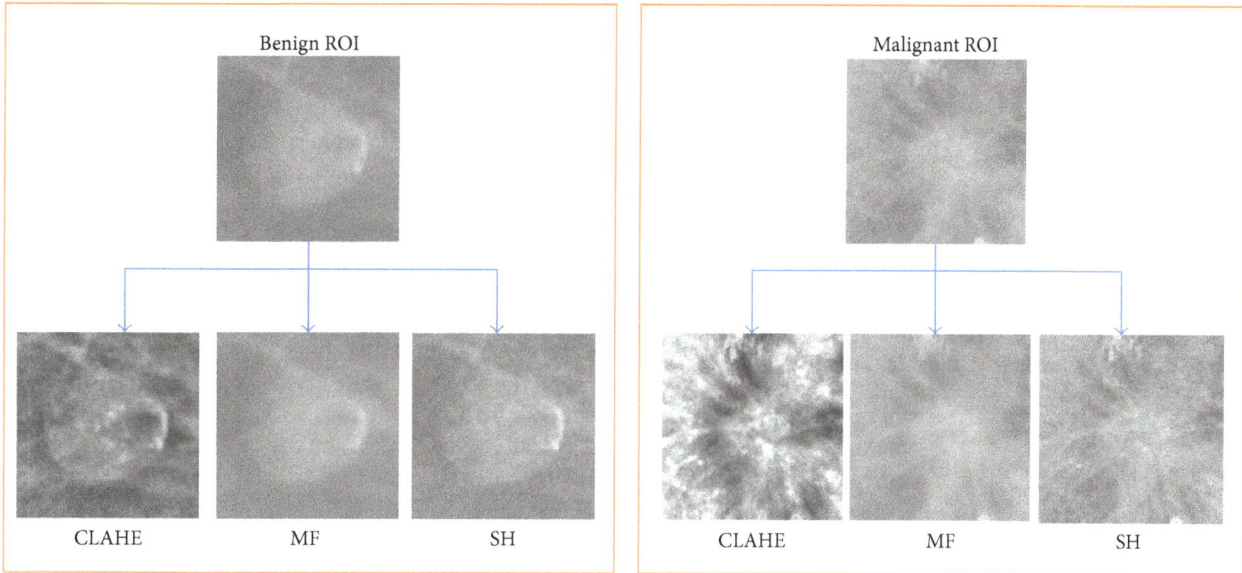

FIGURE 2: Examples of ROI preprocessing.

features, and k is the number of samples, an SVM attempts to discriminate between positive and negative classes by finding a hyperplane that separates them [40]. The SVM classifier solves the following optimization problem:

$$\|\omega\|^2_{\omega,\xi} + C\sum_{i=1}^{k}\xi_i,$$

$$\text{s.t.} \quad y_i\left(\omega^T\phi\left(x_i\right)+b\right) \geq 1-\xi_i, \quad \xi_i \geq 0, \tag{2}$$

where the soft margin parameter C controls the trade-off between the training error and the complexity of the SVM's model in order to fit the training data and to avoid overfitting. The weight vector ω is normal to the separating hyperplane. The parameter ξ is used to give a degree of flexibility for the algorithm when fitting the data and b represents the bias.

The SVM uses a kernel function to make the data linearly separable. It projects the training data x_i to a higher dimensional space as follows: $K(x_i, x_j) = (\phi^T(x_i) \cdot \phi(x_j))$. The SVM algorithm attempts to find the hyperplane with maximum margin of separation between the classes in the new higher dimensional space. In the case of a LSVM classifier, ϕ refers to a dot product. In the case of a NLSVM, the classifier function is formed by nonlinearly projecting the training data in the input space to a feature space of higher dimension by using a kernel function. In this study, we use a radial basis function (RBF) as a mapping kernel, which is defined as follows:

$$K\left(x_i, x_j\right) = \exp\left(-\gamma\left\|x_i - x_j\right\|^2_2\right), \tag{3}$$

where $\gamma = 1/2\sigma^2$, $\|x_i - x_j\|^2_2$ is the squared Euclidean distance between the two feature vectors x_i and x_j, and σ is a free parameter. In this work, we use LIBSVM [41] to implement SVM classifiers. LIBSVM is available at https://www.csie.ntu.edu.tw/~cjlin/libsvm/. A grid search algorithm is performed to find the optimal parameter of the

RBF kernel, γ, and the regularization parameter, C. For each training set, we estimated the parameters used by SVM in the classification as done in [42].

2.6. Evaluation. The performance of each texture analysis method is measured in terms of the area under the curve (AUC) of the receiver operating characteristics (ROC) curve [43]. The SVM classifier provides decision values related to the membership of each class. To generate a ROC curve, we vary a threshold over the decision values. We also use the k-fold cross validation technique to generate the training and testing data. In this procedure, the data are partitioned into k folds; thus $1/k$ of ROIs are used for testing and the rest of ROIs are used for training. In this study, $k = 10$. The mean AUC value is calculated over the cross validation process.

3. Experiments

In this section, we present the effect of pixel resolution, integration scale, preprocessing steps, and normalization methods on the performance of the texture analysis methods when they are applied to *benign/malignant* mass classification in mammograms. Moreover, we study the effect of different combinations of the aforementioned factors.

3.1. Effect of Pixel Resolution and Integration Scale. As we commented in Section 2.1, the pixel resolution 200 μm has been widely used in several studies [14, 19]. So, in this experiment we start with this pixel resolution and then the mammograms are downsampled to generate different pixel resolutions. The downsampling step includes antialiasing filtering and a bicubic interpolation. Five pixel resolutions are generated (200 μm, 400 μm, 600 μm, 800 μm, and 1000 μm), and then we use six integration scales (25 × 25, 32 × 32, 50 × 50, 64 × 64, 75 × 75, and 100 × 100 pixels) to analyze

TABLE 2: Summary of the ANOVA results of pixel resolution and integration scale with the LSVM (the value in each cell is a p value).

Method	Res	IS	Res∗IS
LBP	**0.0024**	**0.001**	0.908
LDN	0.1174	0.4035	0.8037
HOG	0.3905	0.6515	0.4636
HAR	0.7846	0.0962	0.2895
GF	0.083	0.8259	0.9864

TABLE 3: Summary of the ANOVA results of pixel resolution and integration scale with the NLSVM (the value in each cell is a p value).

Method	Res	IS	Res∗IS
LBP	0.9332	**0.0101**	**0.0095**
LDN	0.2387	0.0772	0.6451
HOG	**0.0448**	**0.0103**	0.5138
HAR	0.4253	**0.004**	0.0847
GF	0.6552	0.3109	0.2024

the texture of each ROI. In this experiment, no preprocessing is applied, and the standard zs normalization method is used to normalize the extracted feature vectors. The effect of pixel resolution and integration scale in the performance of LBP, LDN, HOG, HAR, and GF with the LSVM and the NLSVM is shown in Figure 3.

As shown in Figure 3, each texture method achieves its best AUC value at a certain pixel resolution and integration scale. Among all texture methods, LBP achieves the best AUC value (0.78) at pixel resolution 800 μm, integration scale 75 × 75.

The analysis of variance (ANOVA) test [44] has been used to examine the interaction between pixel resolutions and integration scales. The experimental design of ANOVA includes two factors: pixel resolution (Res) and integration scale (IS). Res includes five levels (200 μm, 400 μm, 600 μm, 800 μm, and 1000 μm), whereas IS includes six levels (25×25, 32 × 32, 50 × 50, 64 × 64, 75 × 75, and 100 × 100 pixels). Each combination of the levels of Res and IS produces an AUC value (response). The confidence level is set to 0.05. The results are shown in Tables 2 and 3.

As shown in Table 2, with LBP and the LSVM, the mean responses for the levels of pixel resolution are significantly different ($p = 0.0024$). Similarly, the mean responses for the levels of integration scale are significantly different. In the case of LDN, HOG, HAR, and GF, the mean responses for the levels of pixel resolution and integration scale are not significantly different. The p values indicate that the interactions between the levels of pixel resolution and integration scale (Res ∗ IS) are not significant.

As shown in Table 3, the mean responses for the levels of pixel resolution are significantly different in the case of HOG with the NLSVM. In the case of LBP, LDN, HAR, and GF, the mean responses for the levels of pixel resolution are not significantly different. The mean responses for the levels of integration scale are significantly different in the case of LBP, HOG, and HAR. With LBP and the NLSVM, the interaction

between pixel resolution and integration scale (Res ∗ IS) is significant.

3.2. Effect of Preprocessing. In this experiment, the integration scale that obtained the highest AUC value with each texture analysis method at the baseline pixel resolution of 200 μm and the standard zs normalization method are used. The effect of no preprocessing (NP), CLAHE, MF, and SH on the performance of each texture analysis method is shown in Figure 4. As can be seen, each texture method produces the highest AUC value with a certain preprocessing algorithm. In this experiment, LBP achieves the highest AUC value with SH and the NLSVM, while LDN and HAR achieve the highest AUC value with NP and the LSVM. HOG achieves the highest AUC value with CLAHE and the LSVM. In turn, GF achieves the highest AUC value with CLAHE and the NLSVM.

3.3. Effect of Feature Normalization Methods. In this experiment, we study the effect of five normalization methods (zs, mn, ℓ^1, ℓ^2, and nh) on the performance of each texture analysis method. For each texture analysis method, we use the integration scale that produces the highest AUC value at pixel resolution 200 μm. No preprocessing method is used. The effect of the normalization methods is shown in Figure 5. With the LSVM, zs normalization has led LBP and LDN to AUC values better than other normalization methods, while GF achieves its highest AUC value with ℓ^1 normalization and the NLSVM. As shown in the figure, each texture analysis method achieves its highest AUC value with a certain normalization method.

3.4. Summary of the Results. The best AUC values of each texture analysis method considering the experiments in Sections 3.1, 3.2, and 3.3 are summarized in Table 4. LBP produces the best AUC value (0.78) at pixel resolution 800 μm, integration scale 75 × 75, no preprocessing, zs normalization method, and the LSVM. In turn, HAR produces the lowest AUC value (0.61). LBP, LDN, HOG, and HAR achieve their best values with the LSVM, whereas GF achieves its best AUC value with the NLSVM.

3.5. Combining the Levels of All Factors. To find the best combination among the levels of all factors, we use three approaches: greedy, sequential forward selection (SFS), and exhaustive search (ExS). In the greedy approach, we try to combine the best options of the aforementioned factors. For each texture analysis method, we summarize the best levels of pixel resolution, integration scale, and normalization methods in Table 5.

Table 6 shows that combining the best levels of pixel resolution, integration scale, preprocessing, and feature normalization does not yield improvement on the AUC values of the texture analysis methods reported in Table 4. In fact, LBP, HOG, and GF produced substantially lower AUC values. The LSVM yields higher AUC values than the NLSVM.

Secondly, we use a SFS approach to find the best combination. It consists of two sequential steps: finding the normalization method that improves the current performance

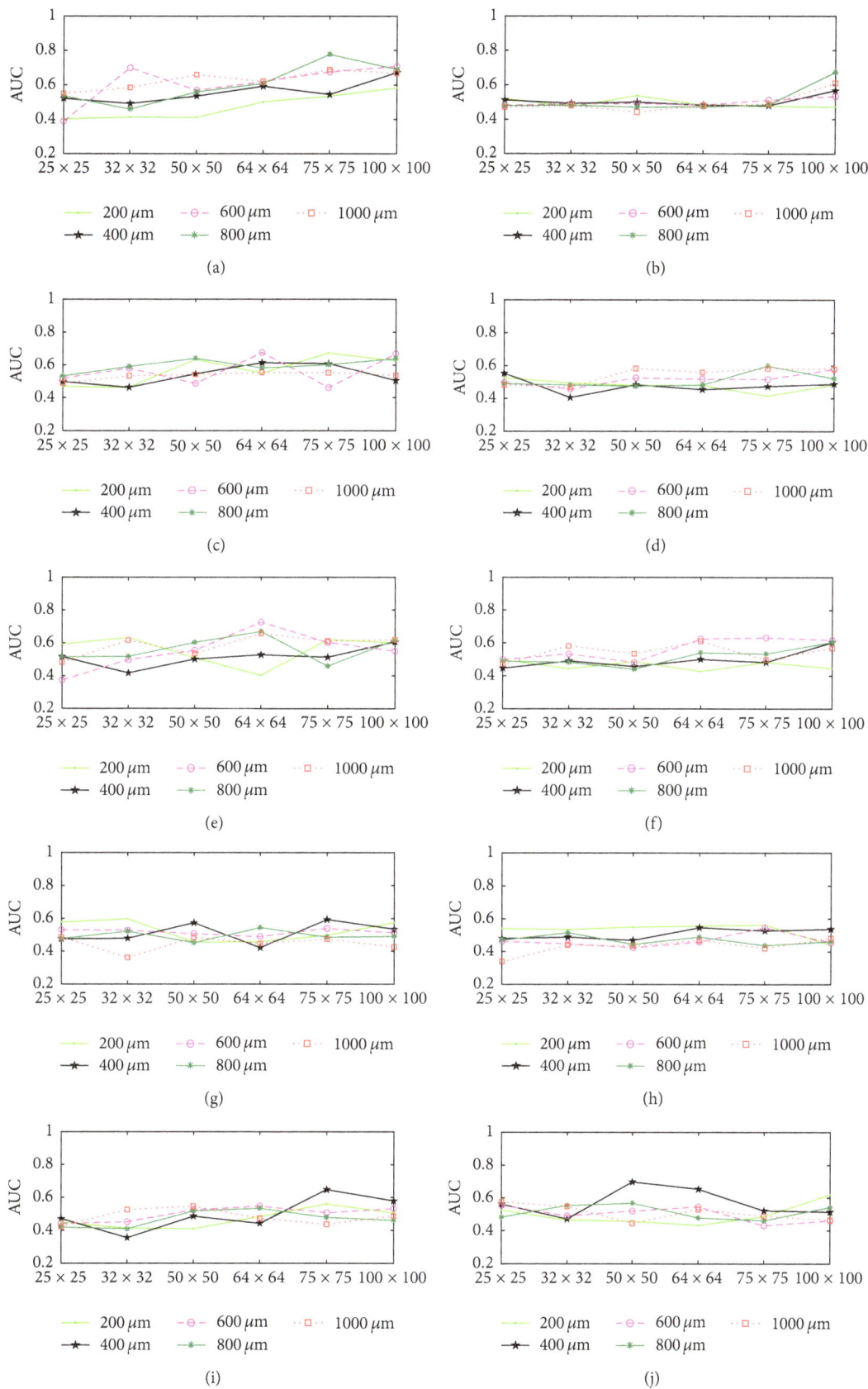

FIGURE 3: The effect of pixel resolution and integration scale on the performance of the texture methods with the LSVM (a, c, e, g, i), the NLSVM (b, d, f, h, j), (a)-(b) LBP, (c)-(d) LDN, (e)-(f) HOG, (g)-(h) HAR, and (i)-(j) GF.

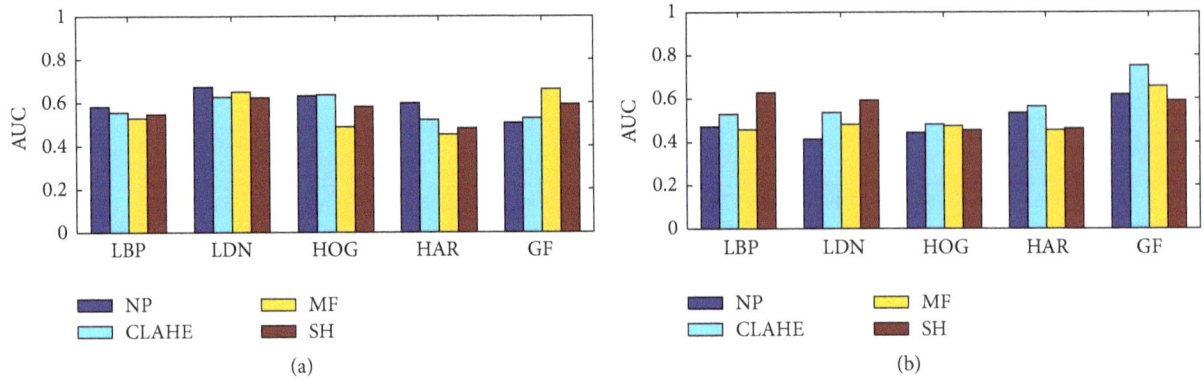

(a)

(b)

FIGURE 4: The performance of the texture analysis methods with NP, CLAHE, MF, and SH using (a) the LSVM and (b) the NLSVM.

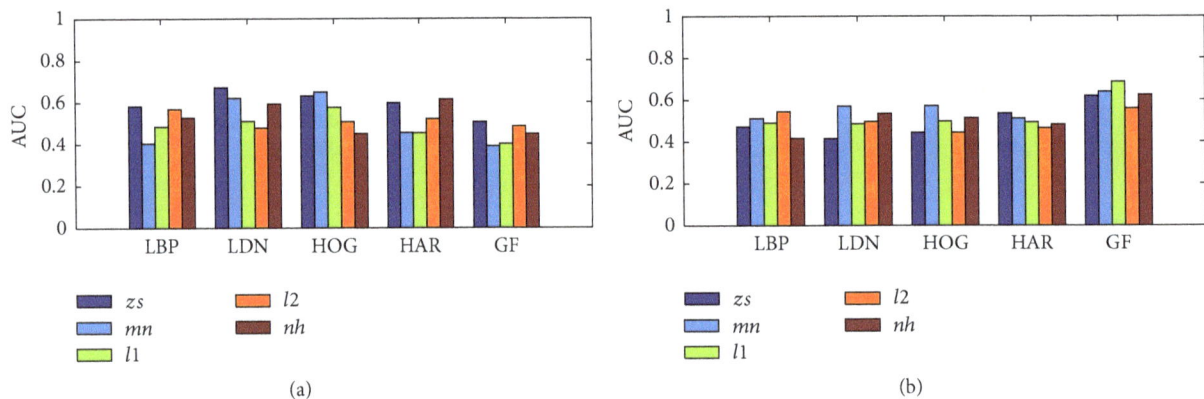

(a)

(b)

FIGURE 5: The performance of the texture analysis methods with different feature normalization methods using (a) the LSVM and (b) the NLSVM.

the most and then finding the preprocessing method that keeps improving this performance. For each texture method, in the first step, we start with the best pixel resolution and integration scale summarized in Table 5. Then, with no preprocessing, the extracted features are separately normalized by each normalization method. Then, the one that improves the performance in combination with the previous two factors is added. In the second step, we apply each preprocessing option to the ROIs (NP, CLAHE, MF, and SH). Then we extract the texture features and normalize them using the best normalization method obtained in the previous step. Both LSVM and NLSVM are used to classify the ROIs. Table 7 shows that the SFS does not improve the AUC value of GF achieved in Table 4. LBP, LDN, HOG, and HAR achieve AUC values close to the ones listed in Table 4. With all texture methods, the SFS approach achieves AUC values better than the greedy approach.

Lastly, we use an ExS algorithm, which is looking for the best combination among five pixel resolutions, six integration scales, and four preprocessing (NP, CLAHE, MF, and SH) and five data normalization methods, resulting in 600 combinations. In the previous experiments, we found that the LSVM usually achieves the best results except with GF. The NLSVM has two parameters that need to be optimized to achieve the best classification results. Adding NLSVM's

parameters optimization to the ExS substantially increases its complexity. So we decided to only use the LSVM in this final test.

As shown in Table 8, the ExS approach improves the AUC values of LDN, HOG, and HAR. The GF achieves an AUC value lower than the one listed in Table 4 because the LSVM can not perfectly separate the GF features.

4. Discussion

Many factors affect the performance of texture analysis methods when applied to benign/malignant mass classification. In this work, we study the effect of factors such as pixel resolution, integration scale, preprocessing, and feature normalization. We use the well-known mini-MIAS database in this study. We start with the original pixel resolution of the mini-MIAS database ($200\,\mu m$); then we downsample the mammograms in order to generate the pixel resolutions $400\,\mu m$, $600\,\mu m$, $800\,\mu m$, and $1000\,\mu m$. In addition, six integration scales are used (25×25, 32×32, 50×50, 64×64, 75×75, and 100×100 pixels). These integration scales cover most of the sizes of the masses in the mini-MIAS database, which range from a few pixels to tens of pixels (the mean diameter of the circle containing the masses is about 49 pixels). Several previous studies have

TABLE 4: Best AUC value for each texture analysis method and the configuration that yields it considering the experiments in Sections 3.1, 3.2, and 3.3.

Method	Best value	Res (μm)	IS	Classifier	Preprocessing	Nor.
LBP	0.78	800	75×75	LSVM	NP	*zs*
LDN	0.68	600	64×64	LSVM	NP	*zs*
HOG	0.72	600	64×64	LSVM	NP	*zs*
HAR	0.61	200	32×32	LSVM	NP	*nh*
GF	0.75	200	100×100	NLSVM	CLAHE	*zs*

TABLE 5: The best option of pixel resolution, integration scale, preprocessing, and normalization methods with each texture method.

Method	Res (μm)	IS (pixels)	Preprocessing	Normalization
LBP	800	75×75	SH	*zs*
LDN	600	64×64	NP	*zs*
HOG	600	64×64	CLAHE	*mn*
HAR	200	32×32	NP	*nh*
GF	400	50×50	CLAHE	ℓ^1

TABLE 6: Results of the greedy approach (AUC).

Method	LSVM	NLSVM
LBP	0.46	0.40
LDN	0.68	0.52
HOG	0.44	0.44
HAR	0.61	0.48
GF	0.58	0.54

TABLE 7: Results of the SFS approach.

Method	Best AUC	Best parameters
LBP	0.780	*zs*, NP, and LSVM
LDN	0.679	*zs*, NP, and LSVM
HOG	0.716	*zs*, NP, and LSVM
HAR	0.605	*nh*, NP, and LSVM
GF	0.720	*zs*, CLAHE, and NLSVM

TABLE 8: Results of the ExS approach.

Method	Best AUC	Best parameters
LBP	0.78	800, 75×75, NP, and *zs*
LDN	0.70	600, 75×75, MF, and *zs*
HOG	0.737	1000, 50×50, SH, and *mn*
HAR	0.666	800, 32×32, CLAHE, and *nh*
GF	0.691	600, 32×32, NP, and ℓ^1

used one of these integration scales to analyze the texture of mammograms [3, 6, 7]. Thus, we hypothesize that the aforementioned integration scales are able to deal with all the masses appearing in the mini-MIAS database.

The shape of breast masses is one of the powerful features that can be used to discriminate between benign and malignant masses. The boundaries of malignant masses usually have irregular shapes, while the boundaries of benign masses have regular ones. In the case of breast mass analysis, pixel resolution may be a critical factor because image downsampling may remove some fine detail from the image. However, as our results indicate, it would be possible to decrease the resolution far beyond 200 μm and obtain good classification results. A notable example is LBP, which actually achieved its best performance at 800 μm. A possible explanation is that core information such as that contained in the boundary of masses may still be preserved even after downsampling and become more useful for methods such as LBP that operate over higher order statistics of gray intensity values. Obviously, when the resolution is far too low, the classification performance degrades, as the shape of the boundaries of benign and malignant masses will be very similar. Another important factor is the integration scale, as it should be big enough to cover the masses and their boundaries and small enough to exclude other tissues. The effect of pixel resolution and integration scale on the performance of texture methods should be jointly studied.

As summarized in Table 5, each texture method achieves its highest AUC value at a certain pixel resolution and integration scale. A pixel resolution of 200 μm and an integration scale of 32×32 pixels have led HAR to its highest AUC value. In turn, a pixel resolution of 800 μm and an integration scale of 75×75 pixels have led LBP to its best AUC value. The integration scale and the pixel resolution interact with each other in a certain way. In the case of LBP, LDN, and HOG, the texture features of each method are represented in a histogram. This histogram includes the repetition of the patterns detected by each method at a certain pixel resolution and integration scale. LBP features calculated at pixel resolution 200 μm are different from those calculated at pixel resolution 400 μm. LDN and HOG also produce different patterns at different pixel resolutions. The local patterns of LBP, LDN, and HOG are usually calculated within a certain integration scale. Different integration scales will yield different histograms for the local patterns. For instance, the histograms of LBP that are calculated with the integration scales 75×75 and 100×100 are different.

ANOVA results show that the mean AUC values of the pixel resolutions are significantly different in the case of LBP with the LSVM. In addition, the mean AUC values of the integration scales are significantly different with LBP, HOG, and HAR and the NLSVM. The performance differences with respect to the pixel resolutions and the integration scales are only significantly different with the LBP and the NLSVM (p = 0.0095). These results indicate that the choice of the pixel resolution and the integration scale has a direct implication on the performance of a texture-based CAD system, because our choice substantially affects the performance of the utilized texture method.

Image preprocessing also affects the performance of the texture analysis methods. HOG and GF achieve the highest AUC values with CLAHE, while LDN and HAR perform better with NP. Indeed, CLAHE, MF, and SH change the intensities of the mammograms in different ways. As a result, each texture analysis method will produce a different AUC value with each preprocessing technique. In general, the preprocessing approach that makes the small-scale structures in the ROIs more visible would give the texture methods more discriminative power. For instance, CLAHE leads GF to its best AUC value (0.75). There is also a coherent relation between the principle of operation of some texture methods and the utilized preprocessing. For instance, the binary patterns of the LDN are calculated based on the edge responses of each pixel in the image. MF removes the outliers before calculating the edge responses. Thus, the edge responses will be properly calculated, and the discriminative power of LDN will improve.

Prior to mass classification, the calculated texture features should be normalized to prevent attributes with higher numeric ranges from dominating those with lower numeric ranges. As shown in our experiments, each texture method produces its highest AUC value with a certain normalization method. This is because each normalization method produces numerical values with different distributions. Consequently, the arrangement of the texture features in the feature space with a certain normalization method is different than with other normalization methods. Thus, the normalization technique changes the final values of the features computed by each texture method. As shown in Table 5, LBP and LDN achieve the highest AUC values with zs normalization, HOG with mn, HAR with nh, and GF with ℓ^1.

In the classification stage, we utilize two widely used classifiers in the field of mammogram analysis: the LSVM and the NLSVM. The first one tries to linearly separate the texture features in the feature space, while the second one uses a kernel function (RBF) to separate the features. As shown in Table 4, the LSVM has led LBP, LDN, HOG, and HAR to the highest AUC values. Conversely, GF achieves the best AUC value with the NLSVM, indicating that GF features are not linearly separable.

Table 4 shows a summary of the levels of pixel resolution, integration scale, preprocessing, and normalization methods that have led each texture method to its best AUC value considering the experiments in Sections 3.1, 3.2, and 3.3. HAR and GF achieve the best AUC values at pixel resolution 200 μm, while LDN and HOG give their best results at pixel resolution 600 μm. No method achieves its best AUC value with the integration scales 25 × 25 and 50 × 50 pixels.

The greedy, SFS, and ExS approaches are used to find the best combination among the levels of all factors. Although the greedy approach is the least complex approach, it yielded poor AUC values. In contrast, the ExS achieved good results, but its computational complexity is the highest. The SFS approach provides a trade-off between the accuracy and the computational complexity. It is not as complex as the ExS approach and it does not produce poor AUC values as the greedy approach. In the case of LBP, LDN, HOG, and HAR, Table 7 shows that the SFS approach produces approximately the same results as those obtained with the ExS approach. The GF achieved better AUC values with the SFS approach because it used the NLSVM, whereas using it with the ExS approach presents some additional challenges in the calculation of the optimal values of its internal parameters (γ and C).

Rangayyan et al. extracted 111 ROIs from mammograms, which were obtained from three different sources: mammographic image analysis society (MIAS), the teaching library of the Foothills Hospital in Calgary, and a screening test (the Alberta program for the early detection of breast cancer) [10]. Although using mammograms from different sources may be helpful to assess the robustness of the studied texture methods, the three mammogram sets used by Rangayyan et al. were digitized at different pixel resolutions. Thus, the characteristics of the textures extracted from the 111 ROIs may be different. This changes the characteristics of the extracted features, so the effect of pixel resolution on the performance of the texture methods may have not been properly studied. In contrast, in the current study, the ROIs were extracted from a single source (the mini-MIAS database). Rangayyan et al. extracted ROIs with different sizes (each ROI included a mass) and they did not mention the effect of the integration scale on the performance of the texture methods. Conversely, the current study has considered six integration scales. With pixel resolution 800 μm, integration scale 75 × 75, no preprocessing, zs normalization method, and the LSVM, the LBP achieves the best AUC value (0.78) compared to other texture methods, exceeding the best AUC value (0.75) achieved by Rangayyan et al. [10]. This is encouraging, so our future work will focus on improving the capabilities of an LBP-based approach by complementing it with the analysis of the fractal dimensions in multiple integration scales at different pixel resolutions.

As mentioned above, the work of [10] has some similarities to our analysis; however it obtained an AUC value less than the one of our study; in addition, the authors of [45] have studied the effect of ROI size and location on texture methods when classifying the low-risk women and the BRCA1/BRCA2 gene-mutation carriers. In turn, our study focuses on analyzing the impact of pixel resolution, integration scale, preprocessing, and feature normalization on texture methods when classifying breast tumors into benign or malignant.

In the current work we studied the impact of the abovementioned factors on the performance of texture methods, achieving the best AUC value with the LBP (0.78). However, some methods in the literature achieved better benign/malignant breast cancer classification results, such as the ones of [33–35]. For instance, the authors of [35] achieved an AUC of 0.92 because they used ROIs of different dataset (DDSM) and extracted the GLCM features from subwindows or regions (they added spatial information). We expect that the classification results of our study will be improved when utilizing the region-based approach of [35] with each texture method. One of our future research lines is to integrate the region-based approach of [35] with our analysis.

5. Conclusion

Texture analysis methods, when applied to *benign/malignant* mass classification in mammograms, are sensitive to the changes of pixel resolution, integration scale, preprocessing, and feature normalization. The best combination of the aforementioned factors should be identified to achieve the best discriminative power of each texture analysis method. We expect that the assessment performed in this study will help researchers to accomplish this task. Due to its computational cost advantage, sequential forward selection would be a suitable approach to determine a reasonable (possibly the best) factor configuration.

Competing Interests

The authors declare that they have no competing interests.

Acknowledgments

This work was partly supported by the Spanish Government through Project TIN2012-37171-C02-02.

References

[1] M. Malvezzi, P. Bertuccio, F. Levi, C. La Vecchia, and E. Negri, "European cancer mortality predictions for the year 2014," *Annals of Oncology*, vol. 25, no. 8, pp. 1650–1656, 2014.

[2] F. J. Gilbert, S. M. Astley, M. G. C. Gillan et al., "Single reading with computer-aided detection for screening mammography," *The New England Journal of Medicine*, vol. 359, no. 16, pp. 1675–1684, 2008.

[3] M. Abdel-Nasser, A. Moreno, and D. Puig, "Towards cost reduction of breast cancer diagnosis using mammography texture analysis," *Journal of Experimental & Theoretical Artificial Intelligence*, vol. 28, no. 1-2, pp. 385–402, 2016.

[4] R. Bellotti, F. De Carlo, S. Tangaro et al., "A completely automated CAD system for mass detection in a large mammographic database," *Medical Physics*, vol. 33, no. 8, pp. 3066–3075, 2006.

[5] J. Melendez, C. I. Sánchez, B. Van Ginneken, and N. Karssemeijer, "Improving mass candidate detection in mammograms via feature maxima propagation and local feature selection," *Medical Physics*, vol. 41, no. 8, Article ID 081904, 2014.

[6] A. Oliver, X. Lladó, J. Freixenet, and J. Martí, "False positive reduction in mammographic mass detection using local binary patterns," in *Medical Image Computing and Computer-Assisted Intervention—MICCAI 2007*, pp. 286–293, Springer, Berlin, Germany, 2007.

[7] V. Pomponiu, H. Hariharan, B. Zheng, and D. Gur, "Improving breast mass detection using histogram of oriented gradients," in *Medical Imaging: Computer-Aided Diagnosis*, vol. 9035 of *Proceedings of SPIE*, pp. 1–6, International Society for Optics and Photonics, San Diego, Calif, USA, March 2014.

[8] Y. Zheng, "Breast cancer detection with Gabor features from digital Mammograms," *Algorithms*, vol. 3, no. 1, pp. 44–62, 2010.

[9] H.-P. Chan, B. Sahiner, N. Patrick et al., "Computerized classification of malignant and benign microcalcifications on mammograms: texture analysis using an artificial neural network," *Physics in Medicine and Biology*, vol. 42, no. 3, pp. 549–567, 1997.

[10] R. M. Rangayyan, T. M. Nguyen, F. J. Ayres, and A. K. Nandi, "Effect of pixel resolution on texture features of breast masses in mammograms," *Journal of Digital Imaging*, vol. 23, no. 5, pp. 547–553, 2010.

[11] H. Soltanian-Zadeh, F. Rafiee-Rad, and D. S. Pourabdollah-Nejad, "Comparison of multiwavelet, wavelet, Haralick, and shape features for microcalcification classification in mammograms," *Pattern Recognition*, vol. 37, no. 10, pp. 1973–1986, 2004.

[12] A. Oliver, J. Freixenet, R. Martí et al., "A novel breast tissue density classification methodology," *IEEE Transactions on Information Technology in Biomedicine*, vol. 12, no. 1, pp. 55–65, 2008.

[13] A. Oliver, J. Freixenet, J. Martí et al., "A review of automatic mass detection and segmentation in mammographic images," *Medical Image Analysis*, vol. 14, no. 2, pp. 87–110, 2010.

[14] M. P. Sampat, A. C. Bovik, G. J. Whitman, and M. K. Markey, "A model-based framework for the detection of spiculated masses on mammography," *Medical Physics*, vol. 35, no. 5, pp. 2110–2123, 2008.

[15] E. D. Pisano, S. Zong, B. M. Hemminger et al., "Contrast limited adaptive histogram equalization image processing to improve the detection of simulated spiculations in dense mammograms," *Journal of Digital Imaging*, vol. 11, no. 4, pp. 193–200, 1998.

[16] S. Anand, R. S. S. Kumari, S. Jeeva, and T. Thivya, "Directionlet transform based sharpening and enhancement of mammographic X-ray images," *Biomedical Signal Processing and Control*, vol. 8, no. 4, pp. 391–399, 2013.

[17] T. S. Subashini, V. Ramalingam, and S. Palanivel, "Automated assessment of breast tissue density in digital mammograms," *Computer Vision and Image Understanding*, vol. 114, no. 1, pp. 33–43, 2010.

[18] J. Suckling, J. Parker, D. Dance et al., "The mammographic image analysis society digital mammogram database," in *Proceedings of the 2nd International Workshop on Digital Mammography*, pp. 375–378, York, UK, July 1994.

[19] N. Karssemeijer, "Automated classification of parenchymal patterns in mammograms," *Physics in Medicine and Biology*, vol. 43, no. 2, pp. 365–378, 1998.

[20] T. Ojala, M. Pietikäinen, and T. Mäenpää, "Multiresolution gray-scale and rotation invariant texture classification with local binary patterns," *IEEE Transactions on Pattern Analysis and Machine Intelligence*, vol. 24, no. 7, pp. 971–987, 2002.

[21] A. R. Rivera, J. R. Castillo, and O. Chae, "Local directional number pattern for face analysis: face and expression recognition," *IEEE Transactions on Image Processing*, vol. 22, no. 5, pp. 1740–1752, 2013.

[22] R. A. Kirsch, "Computer determination of the constituent structure of biological images," *Computers and Biomedical Research*, vol. 4, no. 3, pp. 315–328, 1971.

[23] N. Dalal and B. Triggs, "Histograms of oriented gradients for human detection," in *Proceedings of the IEEE Computer Society Conference on Computer Vision and Pattern Recognition (CVPR '05)*, vol. 1, pp. 886–893, IEEE, San Diego, Calif, USA, June 2005.

[24] R. M. Haralick, I. Dinstein, and K. Shanmugam, "Textural features for image classification," *IEEE Transactions on Systems, Man and Cybernetics*, vol. 3, no. 6, pp. 610–621, 1973.

[25] W. Gómez, W. C. A. Pereira, and A. F. C. Infantosi, "Analysis of co-occurrence texture statistics as a function of gray-level quantization for classifying breast ultrasound," *IEEE Transactions on Medical Imaging*, vol. 31, no. 10, pp. 1889–1899, 2012.

[26] R. P. Ramos, M. Z. do Nascimento, and D. C. Pereira, "Texture extraction: an evaluation of ridgelet, wavelet and co-occurrence based methods applied to mammograms," *Expert Systems with Applications*, vol. 39, no. 12, pp. 11036–11047, 2012.

[27] J. P. Jones and L. A. Palmer, "An evaluation of the two-dimensional Gabor filter model of simple receptive fields in cat striate cortex," *Journal of Neurophysiology*, vol. 58, no. 6, pp. 1233–1258, 1987.

[28] D. T. Puff, E. D. Pisano, K. E. Muller et al., "A method for determination of optimal image enhancement for the detection of mammographic abnormalities," *Journal of Digital Imaging*, vol. 7, no. 4, pp. 161–171, 1994.

[29] I. Christoyianni, A. Koutras, E. Dermatas, and G. Kokkinakis, "Computer aided diagnosis of breast cancer in digitized mammograms," *Computerized Medical Imaging and Graphics*, vol. 26, no. 5, pp. 309–319, 2002.

[30] P. Agrawal, M. Vatsa, and R. Singh, "Saliency based mass detection from screening mammograms," *Signal Processing*, vol. 99, pp. 29–47, 2014.

[31] L. de Oliveira Martins, A. C. Silva, A. C. De Paiva, and M. Gattass, "Detection of breast masses in mammogram images using growing neural gas algorithm and Ripley's K function," *Journal of Signal Processing Systems*, vol. 55, no. 1–3, pp. 77–90, 2009.

[32] D. C. Moura and M. A. G. López, "An evaluation of image descriptors combined with clinical data for breast cancer diagnosis," *International Journal of Computer Assisted Radiology and Surgery*, vol. 8, no. 4, pp. 561–574, 2013.

[33] L. Nanni, S. Brahnam, and A. Lumini, "A very high performing system to discriminate tissues in mammograms as benign and malignant," *Expert Systems with Applications*, vol. 39, no. 2, pp. 1968–1971, 2012.

[34] L. Nanni, S. Brahnam, S. Ghidoni, E. Menegatti, and T. Barrier, "Different approaches for extracting information from the co-occurrence matrix," *PLoS ONE*, vol. 8, no. 12, Article ID e83554, 2013.

[35] L. Nanni, S. Brahnam, S. Ghidoni, and E. Menegatti, "Region-based approaches and descriptors extracted from the co-occurrence matrix," *International Journal of Latest Research in Science and Technology*, vol. 3, pp. 192–200, 2014.

[36] S. Ergin and O. Kilic, "A new feature extraction framework based on wavelets for breast cancer diagnosis," *Computers in Biology and Medicine*, vol. 51, pp. 171–182, 2014.

[37] S. Dhahbi, W. Barhoumi, and E. Zagrouba, "Breast cancer diagnosis in digitized mammograms using curvelet moments," *Computers in Biology and Medicine*, vol. 64, pp. 79–90, 2015.

[38] S. Aksoy and R. M. Haralick, "Feature normalization and likelihood-based similarity measures for image retrieval," *Pattern Recognition Letters*, vol. 22, no. 5, pp. 563–582, 2001.

[39] P. Juszczak, D. Tax, and R. Duin, "Feature scaling in support vector data description," in *Proceedings of the 8th Annual Conference of the Advanced School for Computing and Imaging (ASCI '02)*, pp. 95–102, 2002.

[40] C. Cortes and V. Vapnik, "Support-vector networks," *Machine Learning*, vol. 20, no. 3, pp. 273–297, 1995.

[41] C.-C. Chang and C.-J. Lin, "LIBSVM: a library for support vector machines," *ACM Transactions on Intelligent Systems and Technology*, vol. 2, article 27, 2011.

[42] G. B. Junior, A. C. Cardoso de Paiva, A. Corrêa Silva, and A. C. M. de Oliveira, "Classification of breast tissues using Moran's index and Geary's coefficient as texture signatures and SVM," *Computers in Biology and Medicine*, vol. 39, no. 12, pp. 1063–1072, 2009.

[43] T. Fawcett, "An introduction to ROC analysis," *Pattern Recognition Letters*, vol. 27, no. 8, pp. 861–874, 2006.

[44] P. Armitage, G. Berry, and J. N. Matthews, *Statistical Methods in Medical Research*, John Wiley & Sons, New York, NY, USA, 2002.

[45] H. Li, M. L. Giger, Z. Huo et al., "Computerized analysis of mammographic parenchymal patterns for assessing breast cancer risk: effect of ROI size and location," *Medical Physics*, vol. 31, no. 3, pp. 549–555, 2004.

12

Theoretical and Experimental Demonstration on Grating Lobes of Liquid Crystal Optical Phased Array

Xiangru Wang,[1,2] Liang Wu,[3] Man Li,[2] Shuanghong Wu,[1] Jiyang Shang,[4] and Qi Qiu[1]

[1]*School of Optoelectronic Information, University of Electronic Science and Technology of China, Chengdu 610054, China*
[2]*Science and Technology on Electro-Optical Information Security Control Laboratory, Sanhe 065201, China*
[3]*School of Physical Electronics, University of Electronic Science and Technology of China, Chengdu 610054, China*
[4]*Shanghai Aerospace Electronic Technology Institute, Shanghai 201109, China*

Correspondence should be addressed to Xiangru Wang; xiangruwang@uestc.edu.cn

Academic Editor: Venkata S. R. Jampani

High deflection efficiency is one of the urgent requirements for practical liquid crystal optical phased array (LC-OPA). In this paper, we demonstrate that high order grating lobes induced from fringe effect are the most important issue to reduce occupation of main lobe. A novel theoretical model is developed to analyze the feature of grating lobes when the device of LC-OPA is working on the scheme of variable period grating (VPG) or variable blazing grating (VBG). Subsequently, our experiments present the relevant results showing a good agreement with the theoretical analysis.

1. Introduction

Phase controlled steering effects are most often produced by using an array of unit antennas, the amplitude and initial phase of which are individually controllable. Thereby, the technique of phase array (PA) has been suggested to be a wave director from mechanical wave to electromagnetic wave, from microwave to optical domain [1–3]. Meanwhile, phased array has two categories: active PA and passive PA [2]. Active PA has been more widely deployed on the microwave domain because of its easy control and advanced machining accuracy. However, on the optical domain, the wavelength is almost close to 1 micron. To achieve the feature size of submicron, nanoscale processing has already dramatically developed on the optical domain applications, such as adaptive reshaping [4], optical tweezers [5], and optical Yagi-Uda antenna [6]. Even more, an encouraging letter published in Nature in 2013 reported that the first large scale optical phased array (OPA) made by CMOS technology has 64×64 units [7].

Comparing with active OPA, passive OPA does not have any other heat deposition, such as quantum loss, nonradiation jump, and Joule heat. Since the promotion on the concept of OPA, there have been at least three methods to realize it

[3, 8, 9]: LiNbO3, PLZT, and liquid crystal (LC). Therein, the OPA using nematic LC usually called liquid crystal optically phased array (LC-OPA) has a great potential to achieve practical OPA system for steering optical laser beam inertia less, nonmechanical, and low SWaP (Size Weight and Power consumption).

LC-OPA has already demonstrated its properties of light weight and high precision on steering laser beam to generate an optical space-time division multiple access network [10]. After the survival experiment in the equivalent space environment, modest effects were observed, but none were deemed significant enough to impact the performance of the device for beam steering applications on space communication [11]. After the invention of LC-OPA by Dr. McManamon et al. in 1993 [12], he published an all-thing-considered review paper to summarize its nearly 30 years of development [13] including liquid crystal polarization gratings to achieve 99.5% diffraction efficiency [14], simulation on high efficiency improvement [15], volume holographic to amplify steering angle [16], and the scattering free polymer network liquid crystal invented by Dr. Sun et al. to overcome transparency reduction on the submillisecond liquid crystal devices [17, 18]. Meanwhile, the concept of LC-OPA has been

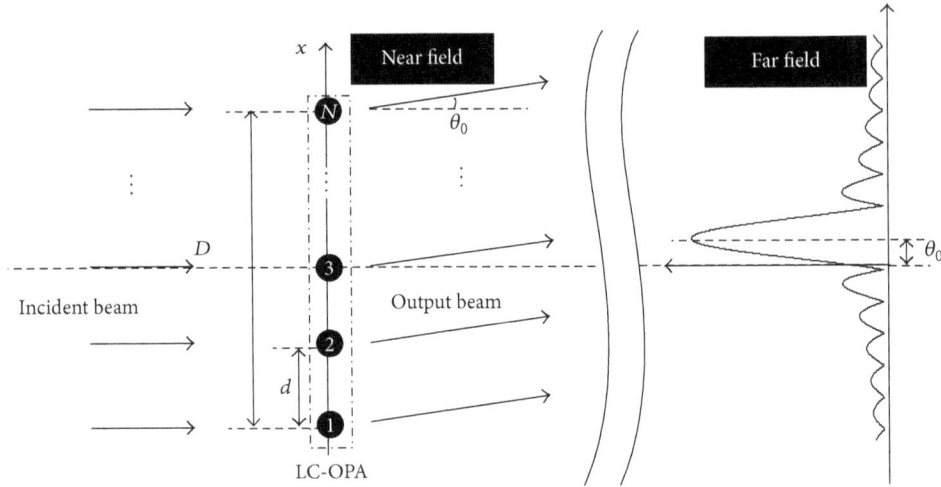

FIGURE 1: The sketch of liquid crystal optical phased array.

suggested into the midinfrared domain [19] and even some alternative methods for steering laser beam such as lenslet array [20] and MEMS-based [21].

Although the performance of LC-OPA has been improved on a variety of features, to date, there has not been a complete analysis on the high order grating lobes. In this paper, it derives a theoretical model on the main lobe and grating lobes and being verified by experimental result.

2. Theory

Because of the passive feature of LC-OPA, it has a uniform radiative laser source propagating through the LC film shown in Figure 1. Thereby, the wave front of the polarization dependent near field will be modulated by the liquid crystal film with a gradient refractive index distribution. For the device of LC-OPA, modulation on the amplitude of near field could be neglected because the absorption of working medium is very tiny. After long distance propagation, far field could be generated according to the principle of diffraction so as to a steering peak on the given angular position if a suitable near field phase modulation is given.

Steering can be accomplished by a physical prism. If the refractive index of a prism is changeable by loading different voltages, light could be steered as well at different angles. Meanwhile, we can take advantage of the fact that there is no difference for light waves if they have 0, 2π, 4π, or $2k\pi$ phase shift. The phase can be made as a 2π subtraction when it exceeds full round 2π in the case of large aperture. According to this property of diffraction, it generates the first scheme, shown in Figure 2(a), to form a deflective beam. and the scheme has already been widely used in the current microwave phased array called variable period grating (VPG), where the steering angle θ_s is governed by $\sin\theta_s = \Delta\phi_s/k_0 d$, where $\Delta\phi_s$ is the phase step between two adjacent electrodes, k_0 is vacuum wavenumber, and d is width of electrode and gap. Because of the proportional relationship between $\Delta\phi_s$ and θ_s, it is easy to realize a continuous and full coverage scan on a given domain.

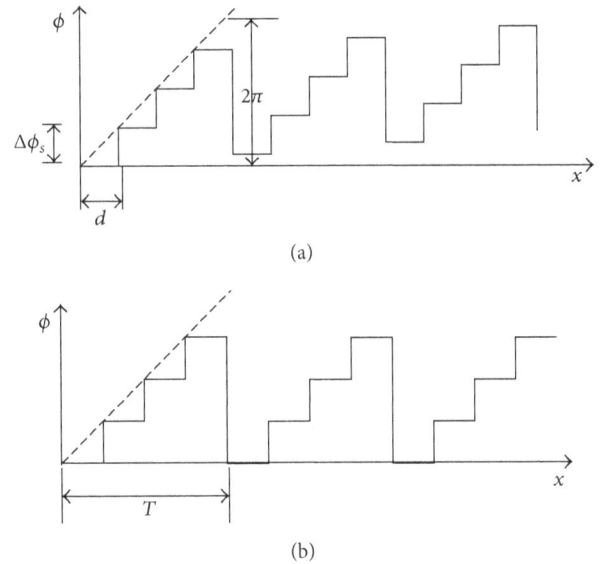

(a)

(b)

FIGURE 2: Ideal phase modulation of LC-OPA on the scheme of VPG (a) and VBG (b).

Meanwhile, another scheme to steer wave is blazing grating, shown in Figure 2(b). When an in-plane wave is modulated by periodic medium, tens of periodic spots are generated on the far field that is called grating lobes. Therein, the profile of those grating lobes is determined by the modulation function in one period, and the angular position of the mth order grating lobe θ_m is determined by the grating equation $\sin\theta_m = m\lambda_0/T$, where T is still the modulation period. When the device is a discrete one, the first order lobe we desire is determined by $\sin\theta_s = \lambda_0/Nd$, where λ_0 is vacuum wavelength and N is the number of electrodes in one period T. Comparing with the scheme of VPG, VBG has a constant period T. It also always starts at the same initial phase for each period. However, because of the inverse proportional relationship between θ_s and adjustable integer

N, it is impossible to realize a continuous and full coverage scan on one device.

After long distance propagation through the free space, electric field component of the modulated beam is governed by free space Helmholtz equation. If the distance L meets Fraunhofer's approximation condition, that is, $4L\lambda_0 \gg D^2$, where D is the diameter of transmitting aperture, E field in the far field E_{far} is completely determined by the near field E_{near} according to the Fraunhofer's equation

$$E_{far}(\theta_x) = A \int_{-\infty}^{+\infty} E_{near}(x) \exp\left(-jk_0 \sin\theta_x x\right) dx, \quad (1)$$

where the constant A is determined by the principle of energy conservation and θ_x is the angular spectrum position. Thereby, the far field distribution can be obtained from a given complicated near field by the numerical method of FFT (Fast Fourier Transformation).

E field on the near field is the transmitted beam after the liquid crystal film,

$$E_{near} = E_{in} \cdot T_a \cdot T_p, \quad (2)$$

where the incident beam E_{in} is usually configured as a standard TEM00 in-plane wave or Gaussian mode with a half beam waist of ω, $E_{in} = \exp(-x^2/\omega^2)$. The transmission function includes two parts: amplitude factor T_a and phase factor T_p. The amplitude factor T_a is only a gate function owing to the limited transmitting aperture with a full width of D_t, $T_a = \text{rect}(x/D_t)$. The phase factor T_p is determined by the electric controlled liquid crystal film, $T_p = \exp[j\phi(x)]$, where $\phi(x)$ is additional phase retardation. According to the characteristic of Fourier transformation, far field distribution E_{far} can be rewritten as the convolution of two parts that result from gate limited Gaussian beam and phase modulated transmission function, respectively,

$$E_{far} \sim \text{FFT}\{E_{in} \cdot T_a\} \otimes \text{FFT}\{T_p\}. \quad (3)$$

The shape of main lobe and side lobe is determined by $\text{FFT}\{E_{in} \cdot T_a\}$, having the same result as previous theories. That is, the full width of main lobe is $\Delta\theta_{main} = 2\lambda_0/D_t$ for an in-plane wave through an aperture with a width of D_t. If the beam waist ω of incident Gaussian beam is much smaller than the aperture, the full width of main lobe is $\Delta\theta_{main} = 4\lambda_0/\pi\omega$.

Because of the property of transmission function T_p, a pure phase modulation, if it is a periodic function in the ideal model of LC-OPA, its Fourier transformation must be a group of delta functions, the center position of which is determined by $\text{FFT}\{T_p\}$.

The periodic transmission factor T_p can be expanded on the Fourier series

$$T_p = \sum_{n=-\infty}^{+\infty} A_m \exp\left(j \cdot m \cdot \frac{2\pi}{T} \cdot x\right), \quad (4)$$

where the integer m is the order number and A_m is determined by the equation

$$A_m = \frac{1}{T} \int_0^T T_p(x) \exp\left(j \cdot m \cdot \frac{2\pi}{T} \cdot x\right) dx. \quad (5)$$

Because T_p is a phase modulation function on the form of $T_p = \exp[j\phi(x)]$, the coefficient A_m is rewritten that

$$A_m = \frac{1}{T} \int_0^T T_p(x) \exp\left[-j\left(m \cdot \frac{2\pi}{T} \cdot x - \phi(x)\right)\right] dx. \quad (6)$$

To evaluate the property clearly, a group of parameters are assumed for simulation: electrode width $a = 4\,\mu m$, electrode gap $b = 1\,\mu m$, electrode length $l = 10\,mm$, and electrode number $M = 2000$, that is, electrode period $d = a+b = 5\,\mu m$, and transmitting aperture width $D_t = Md = 10\,mm$.

In the case of VPG, phase modulation ϕ is linear with position x that can be written $\phi = kx$; then, after a distance of $2\pi/k$, the phase increase 2π, thereby, $\exp[j \cdot k(x+2\pi/k)] = \exp(j \cdot kx)$, so that the factor of phase modulation T_p is periodic and its period T is $2\pi/k$, where k is variable. Substituting $\phi(x)$ into (6), the amplitude of each order is coefficient A_m and it is obtained that $A_1 = 1$ and $A_m = 0$ for other orders. So the peak angular position θ_s on the far field can be obtained by $\sin\theta_s = k/k_0$.

In the case of VBG, a periodic transmission function is that the phase shift is periodic; that is, $\phi(x+T) = \phi(x)$, where T is variable. Owing to the periodic property on phase ϕ with a period of T_p it is not because of wave property of incident beam, the amplitude of each order is coefficient A_m obtained from (6), and the peak angular position θ_s on the far field can be obtained by $\sin\theta_s = \lambda_0/T$.

Therefore, no matter VPG or VBG, these two schemes can be unified because they have the same reason to steer the incident beam to the given angle. The phase factor of their transmission function has a common property that it is periodic with a period of $T = \lambda_0/\sin\theta_s$. The difference is that on the scheme of VBG, the period T is a user defined variable. On the scheme of VPG, we configure the slope of phase modulation k, and the corresponding period $T = 2\pi/k$.

In the practical LC-OPA, fringe effect has already been evaluated in many papers. Owing to the no source boundary condition on the gap between electrodes, there is not only a flyback area for each 2π reset, but also phase dropdown on each electrode gap. Therein, the width of flyback is L_b. Thereby, the phase shift can be written on the form of $\phi(x) = \phi_{ideal}(x) + \phi_{fringe}(x)$, where $\phi_{ideal}(x)$ is ideal phase modulation, the function of which is linear with position like a saw tooth so as to a periodic transmission function T_p

$$\phi'_{ideal}(x)$$

$$= \begin{cases} \phi_{init} + k_0 \sin\theta_s(x - x_1) & x \in D_0 \quad (7) \\ \phi'_{init} + \left(k_0 \sin\theta_s - \frac{2\pi}{T}\right)(x - x'_1) & x \in D_f, \end{cases}$$

where the subdomain D_0 indicates the ordinary domain and D_f indicates the flyback domain. The width of flyback L_b is determined by the thickness of liquid crystal device. x_1 and x'_1 are the starting position and ending position of this period, respectively. ϕ_{init} and ϕ'_{init} are their phase at x_1 and x'_1, respectively. On the scheme of VBG, the distance $|x_1 - x'_1| = T$, $\phi_{init} = \phi'_{init} = 0$. However, on the scheme of VPG, the phase on the finite width electrode is loaded step by step, phase shift

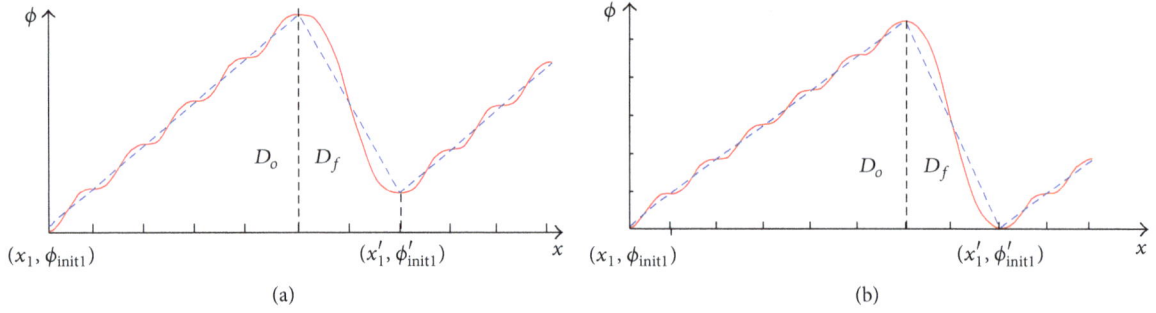

FIGURE 3: Fringe effect of LC-OPA on the scheme of (a) VPG and (b) VBG.

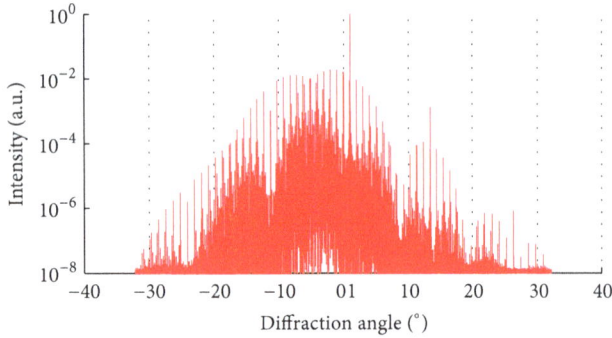

FIGURE 4: Far field of phase modulated by fringe effected LC-OPA.

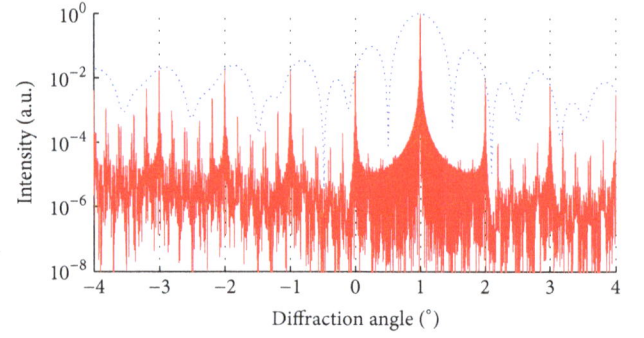

— Far field

···· Profile

FIGURE 5: Zoom-in figure of normalized intensity of far field of VPG scheme.

on the ith electrode ϕ_i is determined by $\phi_i = k(i-1)d$, when it exceeds 2π, and it would make a reset into $\phi_{\text{init}} = \phi_i - 2\pi$ to start another new 2π round where the initial phase is ϕ_{init}, as shown in Figure 2(a). During the range of one quasi period with N electrodes, the fringe effect induced item $\phi_{\text{fringe}}(x)$ is assumed to be

$$\phi_{\text{fringe}}(x) = \begin{cases} -p \cdot \sin\left[\dfrac{2\pi}{d} \cdot (x - x_1)\right] & x \in D_0 \\ q \cdot \sin\left[\dfrac{2\pi}{d} \cdot (x - x_1)\right] & x \in D_f, \end{cases} \quad (8)$$

and because of continuous and desirable feature on phase function, the coefficients p, q should be on the condition of $p : (1 - q) = d : L_b$. Then, if we assume $p = \Delta\phi_s/2\pi$, so $q = 1 - L_b/d \cdot \Delta\phi_s/2\pi$. In the case of VPG, the phase modulation is shown in Figure 3(a); in the case of VBG, it is shown in Figure 3(b). Meanwhile, the values of coefficient and flyback range L_b are fully governed by the group of EM equations and liquid crystal molecular director equation.

After the numerical FFT operation on the near field E_{near}, if we suppose steering angle θ_s is 1 degree, the far field is as shown in Figure 4. Comparing with the ideal model with only one lobe on the target angle, there are more grating lobes owing to the additional fringe item. These high order grating lobes are on the position of $\theta = n\theta_s$, when n is order number of arbitrary integer.

Therein, the zoom-in figure of Figure 4 is shown in Figure 5; the profile in blue dot line is the FFT of E_{near} in the range of $[0, T]$, where T is calculated by $2\pi/k_0 \sin\theta_s$. The intensity of each grating lobe and main lobe at the position

of $\theta = n\theta_s$ is governed by the blue dot line. Besides the defined higher order grating lobes, there are also some other neglectable peaks like noise. The intensity of most of them is below 30 dB; those are generated by the overlap of multiside lobes.

This property would give us two important issues: first, the phase distribution on the near field can be inversely derived by detecting the intensity and position of each grating lobe; second, by optimizing the phase delay of small account of electrodes in the range of $[0, T]$ can optimize the deflection efficiency. In other words, the phase distribution during one period $[0, T]$ is the major factor to influence the grating distribution or the deflection efficiency.

In the case of VPG, owing to the fringe effect, the transmission factor $T_p(x)$ is not perfectly periodic. However, for the most of given value T_{p0}, T_p has a property of repetition on x meeting the equation $\phi(x) = \phi(x + X_{qp})$. Meanwhile, the repetition distance X_{qp} is $2\pi/k_0 \sin\theta_s$; that is, the transmission factor $T_p(x)$ is a quasi-periodic function of position, where the period $X_{qp} = 2\pi/k_0 \sin\theta_s$. For the most given value of modulated phase, they would generate a group of grating lobes on the diffraction of $\theta = n\theta_s$. And for the phase value which does not have periodic overlap, it would generate the noise lobes. This can be verified in the case of VBG. In contrast, the transmission factor $T_p(x)$ is perfectly periodic because of the periodic phase modulation. Then, the far field it generates has a very perfect grating lobes shape

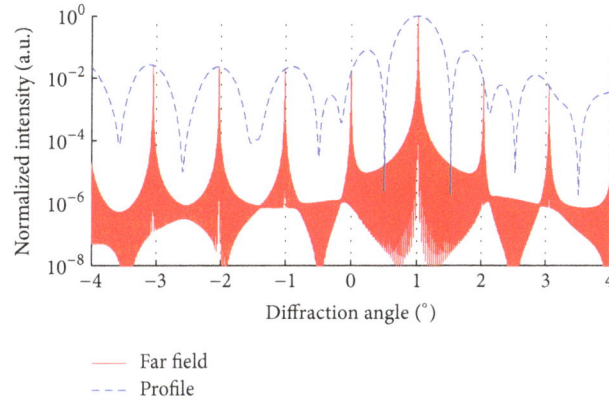

FIGURE 6: Zoom-in figure of normalized intensity of far field of VBG scheme.

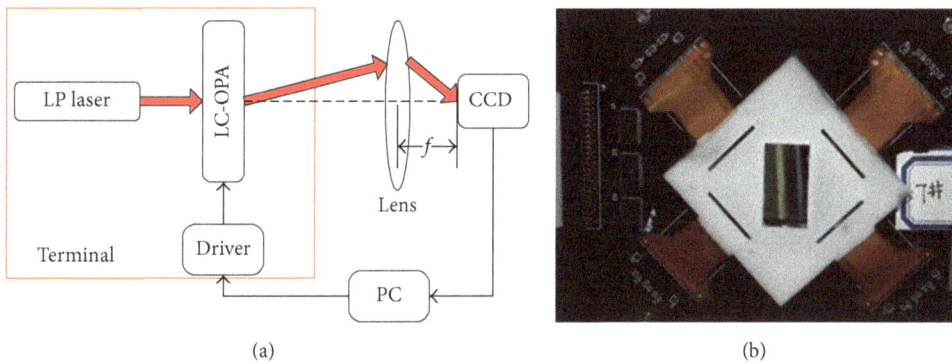

FIGURE 7: (a) Experimental setup; (b) LC-OPA phase shifter.

without noise as the VPGs as shown in Figure 6 when the steering angle is 1.02 degrees that is the most close to 1 degree by the grating function $\theta_s = \sin^{-1}(\lambda_0/Nd)$.

3. Experiment

Figure 7(a) depicts the experimental setup for studying the properties of far field. The incident beam is linear polarized Nd:YAG laser. After phase modulation by the LC-OPA phase shifter device, the output laser beam has a coherent far field observed by a high resolution CCD on the focal plane of Fourier lens. Meanwhile, the data loaded on LC-OPA is generated by PC computer and translated by driver module.

Meanwhile, the device of LC-OPA we used is developed from grating electrodes fabrication to circuit design and liquid crystal filling. The widths of electrode and gap are $4\,\mu m$ and $1\,\mu m$, respectively. The thickness of LC cell is maintained by spacers with a diameter of $10\,\mu m$. The effective optical aperture is $10\,mm \times 15\,mm$ as shown in Figure 7(b). Meanwhile, the LC-OPA module is driven by four parallel ordinary liquid crystal display drivers IC controlled by FPGA.

For a given group of steering angle as examples from $-3°$ to $3°$ with an increasing step of $0.5°$, the intensity distributions of far field at those steering angles are captured by CCD and combined together, as shown in Figure 8 from right to left. In order to display the low intensity grating lobe, we adjust the exposure time to the maximum value to detect the grating

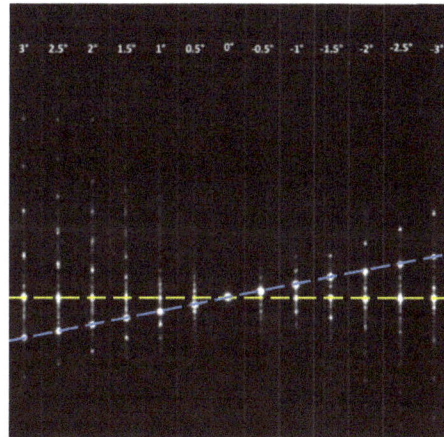

FIGURE 8: Experimental result on the far fields of different steering angles.

lobes and side lobes. The main lobes are lined up by a blue dashed line. They all have good accuracy to form good order on one line. And, the zero-order positions are lined up by the yellow dashed line.

Meanwhile, in Figure 8, grating lobes of higher order are shown with the same distance between each of them. Its distance is the same as the distance between main lobe and original position. It has good agreement with the theoretical analysis that $\theta_n = n \cdot \theta_s$. Besides the grating lobes, Figure 8 also illustrates few relatively dark spots between grating lobes. Although they look very clear in this snapped picture, they do not occupy too much energy when we measured using power meter, because of overexposure.

4. Conclusion

Theoretical and experimental evidence is presented showing that grating lobes and side lobes are generated by fringe effect in the far field. On both working schemes of variable period grating (VPG) and variable blazing grating (VBG), the angular positions of grating lobes are determined by $\theta_n = n \cdot \theta_s$; they have the same physical reason to steer incident beam that their phase factor of transmission function is periodic or quasi-periodic. The normalized intensity of grating lobes is determined by phase modulation in one period or quasi-period. This theoretically and experimentally verified conclusion has been obtained to explain the principle reason on grating lobes.

The principle reason of grating lobes would give us another important issue: optimization on deflection efficiency can be accomplished by optimizing phase delays of small account of electrodes in the range of $[0, T]$, not only VBG scheme but also VPG scheme; higher order and zero order grating lobes can be suppressed as much as possible. Meanwhile, the method of optimization and experimental demonstrations on the steering efficiency improvement is going to be presented in detail in the coming work.

Competing Interests

The authors declare that they have no competing interests.

Acknowledgments

This work is sponsored by NSFC Contracts 61405029 and 91438108 and funded by Open Foundation of National Defense Key Laboratory and SAST 2015087.

References

[1] R. Underbrink J, "Aeroacoustic phased array testing in low speed wind tunnels," in *Aeroacoustic Measurements*, Experimental Fluid Mechanics, pp. 98–217, Springer, Berlin, Germany, 2002.

[2] R. J. Mailloux, *Phased Array Antenna Handbook*, Artech House, Boston, Mass, USA, 2005.

[3] P. F. Mcmanamon, T. A. Dorschner, D. L. Corkum et al., "Optical phased array technology," *Proceedings of the IEEE*, vol. 84, no. 2, pp. 268–298, 1996.

[4] M. Aeschlimann, M. Bauer, D. Bayer et al., "Adaptive subwavelength control of nano-optical fields," *Nature*, vol. 446, no. 7133, pp. 301–304, 2007.

[5] M. L. Juan, M. Righini, and R. Quidant, "Plasmon nano-optical tweezers," *Nature Photonics*, vol. 5, no. 6, pp. 349–356, 2011.

[6] T. Kosako, Y. Kadoya, and H. F. Hofmann, "Directional control of light by a nano-optical Yagi-Uda antenna," *Nature Photonics*, vol. 4, no. 5, pp. 312–315, 2010.

[7] J. Sun, E. Timurdogan, A. Yaacobi, E. S. Hosseini, and M. R. Watts, "Large-scale nanophotonic phased array," *Nature*, vol. 493, no. 7431, pp. 195–199, 2013.

[8] R. A. Meyer, "Optical beam steering using a multichannel lithium tantalate crystal," *Applied Optics*, vol. 11, no. 3, pp. 613–616, 1972.

[9] J. A. Thomas and Y. Fainman, "Programmable diffractive optical element using a multichannel lanthanum-modified lead zirconate titanate phase modulator," *Optics Letters*, vol. 20, no. 13, pp. 1510–1512, 1995.

[10] W. J. Miniscalco and S. A. Lane, "Optical space-time division multiple access," *Journal of Lightwave Technology*, vol. 30, no. 11, pp. 1771–1785, 2012.

[11] S. A. Lane, J. A. Brown, M. E. Tremer et al., "Radiation testing of liquid crystal optical devices for space laser communication," *Optical Engineering*, vol. 48, no. 11, Article ID 114002, 11 pages, 2009.

[12] P. F. McManamon, E. A. Watson, T. A. Dorschner, and L. J. Barnes, "Nonmechanical beam steering for active and passive sensors," in *Infrared Imaging Systems: Design, Analysis, Modeling, and Testing IV, 2*, vol. 1969 of *Proceedings of SPIE*, August 1993.

[13] P. F. McManamon, P. J. Bos, M. J. Escuti et al., "A review of phased array steering for narrow-band electrooptical systems," *Proceedings of the IEEE*, vol. 97, no. 6, pp. 1078–1096, 2009.

[14] J. Kim, C. Oh, M. J. Escuti, L. Hosting, and S. Serati, "Wide-angle nonmechanical beam steering using thin liquid crystal polarization gratings," in *Proceedings of the Advanced Wavefront Control: Methods, Devices, and Applications VI*, vol. 7093, San Diego, Calif, USA, August 2008.

[15] X. Wang, B. Wang, P. J. Bos, J. E. Anderson, J. J. Pouch, and F. A. Miranda, "Finite-difference time-domain simulation of a liquid-crystal optical phased array," *Journal of the Optical Society of America A: Optics and Image Science, and Vision*, vol. 22, no. 2, pp. 346–354, 2005.

[16] "Raytheon steered agile beams," STAB, Final Rep. AFRL-SN-WP-TR-2004-1078, 2005.

[17] J. Sun, S. Xu, H. Ren, and S.-T. Wu, "Reconfigurable fabrication of scattering-free polymer network liquid crystal prism/grating/lens," *Applied Physics Letters*, vol. 102, no. 16, Article ID 161106, 2013.

[18] J. Sun, Y. Chen, and S.-T. Wu, "Submillisecond-response and scattering-free infrared liquid crystal phase modulators," *Optics Express*, vol. 20, no. 18, pp. 20124–20129, 2012.

[19] Y. Chen, H. Xianyu, J. Sun et al., "Low absorption liquid crystals for mid-wave infrared applications," *Optics Express*, vol. 19, no. 11, pp. 10843–10848, 2011.

[20] E. A. Watson, W. E. Whitaker, C. D. Brewer, and S. R. Harris, "Implementing optical phased array beam steering with cascaded microlens arrays," in *Proceedings of the IEEE Aerospace Conference*, vol. 3, pp. 1429–1436, Big Sky, Mont, USA, March 2002.

[21] P. J. Gilgunn and G. K. Fedder, "Flip-chip integrated SOI-CMOS-MEMS fabrication technology," in *Proceedings of the Technical Digest of the Solid-State Sensor, Actuator and Microsystems Workshop*, pp. 10–13, Hilton Head Island, SC, USA, June 2008.

13

Modified Three-Dimensional Multicarrier Optical Prime Codes

Rajesh Yadav and Gurjit Kaur

Department of ECE, School of ICT, Gautam Buddha University, Greater Noida, India

Correspondence should be addressed to Rajesh Yadav; raj_opyadav@yahoo.co.in

Academic Editor: Gang-Ding Peng

We propose a mathematical model for novel three-dimensional multicarrier optical codes in terms of wavelength/time/space based on the prime sequence algorithm. The proposed model has been extensively simulated on MATLAB for prime numbers (P) to analyze the performance of code in terms of autocorrelation and cross-correlation. The simulated outcome resembles the mathematical model and gives better results over other methods available in the literature as far as autocorrelation and cross-correlation are concerned. The proposed 3D optical codes are more efficient in terms of cardinality, improved security, and providing quality of services.

1. Introduction

Optical Code Division Multiple Access (OCDMA) network has a great potential to cater the needs of the fast access communication system. OCDMA is a suitable multiple access technique for local area network where traffic is bursty in nature [1]. Optical CDMA is an efficient multiplexing scheme, as it does not require synchronization between the transmitter and receiver and there is the efficient utilization of available bandwidth [2]. The performance of optical CDMA networks depends on the optimal selection of the optical code and coding configuration. It is desired to have code set that can correctly decode the desired user signal in the presence of other users [3]. The motivation for designing more dimensional code is to have increased code set and also have significant code weight to support quality of services for the given code length. The length of a code plays an important role in system performance and system complexity [4], although it is possible to improve the correlation property by using long code words. However, the code length cannot be increased much because of the power and processing time restrictions for encoder and decoder [5]. The weight of code represents the signal power. A code sequence with larger code weight is less sensitive to interference than that with a lower code weight. In this paper, novel modified three-dimensional (3D) multicarrier prime codes are proposed having sufficient code weight along with significant code length for supporting QoS in optical CDMA communication systems. The construction of proposed three-dimensional codes is conceptualized on the basis of prime sequence algorithm.

The rest of paper is organized as follows: Section 2 discusses the OCDMA coding theory that gives an insight of the optical codes for optical CDMA communication system. A concise introduction to optical prime codes for OCDMA systems is also discussed. Section 3 presents the mathematical modeling of proposed wavelength/time/space modified 3D multicarrier optical prime codes. Section 4 discusses the result of the proposed 3D optical prime codes and its performance estimation in terms of the autocorrelation and cross-correlation function. In Section 5 the current findings along with the future directions are concluded. The paper ends with the references studied and cited in the paper.

2. Optical Codes

A lot of researchers have studied optical codes like one-dimensional (1D), two-dimensional (2D), and now three-dimensional codes to provide efficient codes. Prucnal has proposed one-dimensional optical code based on the prime number P over a Galois field GF(P). The codes have an autocorrelation value of $P - 1$ and cross-correlation value of two [6]. The construction of prime codes is based on the congruence codes. Later on, one-dimensional prime code

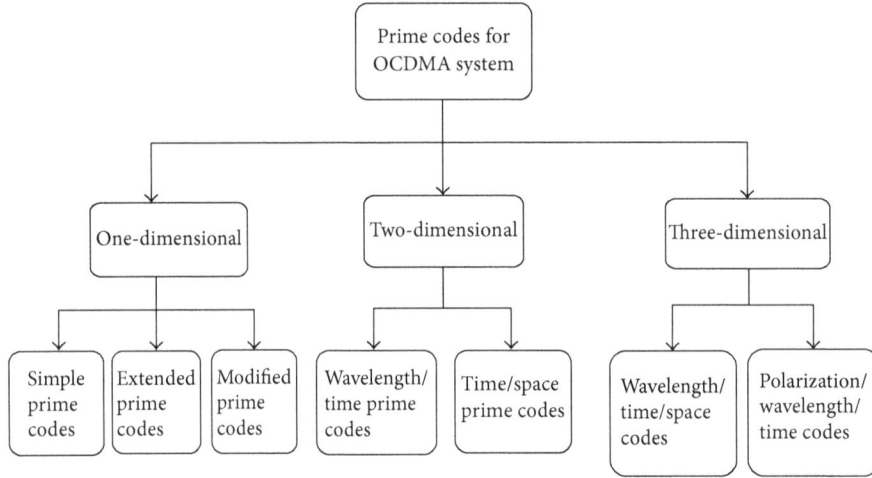

FIGURE 1: Prime code word family hierarchy [11].

with improved correlation property is proposed by many researchers [7–10]. In the one-dimensional code to increase the code set size, the code length is increased significantly to keep the correlation property satisfactory. To overcome this limitation two-dimensional codes are developed with better code set size and good correlation properties. Mendez et al. describe the 2D codes constructed from 1D code based on Golomb ruler to increase the number of code set sizes [7]. Yu and Park have proposed an algorithm to construct 2D code based on 1D prime sequence [8]. Yim et al. proposed a design for wavelength/time 2D code for O-CDMA and carried out the performance analysis of the 2D codes [9]. The performance analysis of the two-dimensional optical codes for bit error rate and a number of users supported carried out by Shivaleela and Kaur has revealed that there is a significant improvement in optical CDMA system using 2D codes over 1D code [10, 11].

The successful performance of two-dimensional optical codes over the one-dimensional codes has led the researcher to construct the three-dimensional codes with better code set size and improved system performance. McGeehan presented a time-wavelength-polarization 3D optical code for increasing the number of users in OCDMA LAN's application. This coding technique has limitation due to the polarization mode dispersion and complex polarization control at all the stages in the network [12, 13]. Kumar et al. present multiple pulse per plane codes using a row-wise orthogonal pairs (RWOP) algorithm for wavelength and spatial channel allocation [14]. The design offers the very low probability of error due to multiple access interference at lower cardinality. The RWOP algorithm based 3D code has a limitation imposed by multipath interference. The hierarchy of prime sequence codes is represented in Figure 1 [11]. Here we proposed wavelength/time/space spread 3D optical codes based on prime sequence algorithm derived from multicarrier prime codes for OCDMA [14, 15]. The comparative analysis of 1D, 2D, and 3D optical codes based on prime sequence is carried out by Yadav and Kaur [16]. The generation algorithm of 3D codes is discussed in detail and the orthogonality of the 3D

optical codes is demonstrated mathematically and simulated on the MATLAB.

3. Generation of 3D Optical Code

In this research work, we have presented new modified three-dimensional multicarrier optical code for OCDMA network based on prime sequence algorithm. The 3D codes family can be represented as $(W * T * S, w, \lambda_a, \lambda_c)$ where W, T, and S denote the number of wavelengths, time, and spatial channels domain used, respectively. "w" signifies the weight of the code. λ_a and λ_c are the autocorrelation and cross-correlation of the optical code, respectively. Figure 2 illustrates the algorithm of modified three-dimensional multicarrier prime code constructions.

Three-dimensional optical code construction starts with the prime number P over the Galois field (GF). A prime sequence is constructed for a prime number P as [6, 17]

$$Z_r = \left(z_{r,0}, z_{r,1}, \ldots, z_{r,s}, \ldots, z_{r,(P-1)} \right),$$
$$r = 0, 1, \ldots, P - 1, \tag{1}$$

where the element in this prime sequence is given by

$$Z_{r,s} = r \cdot s \pmod{P}, \tag{2}$$

where $r = (0, 1, \ldots, P - 1)$ and $s = (0, 1, \ldots, P - 1)$ are the elements over the Galois field GF(P). For prime number $P = 3$, the corresponding code set matrix $Z_{r,s}$ is shown by (3), where r and s are the elements over the Galois field GF(3). The following 3×3 matrix will be constructed:

$$Z_{r,s} = \begin{bmatrix} Z_{0,0} & Z_{0,1} & Z_{0,2} \\ Z_{1,0} & Z_{1,1} & Z_{1,2} \\ Z_{2,0} & Z_{2,1} & Z_{2,2} \end{bmatrix}. \tag{3}$$

This matrix represents a code set for prime number $P = 3$. In this 3D prime code set, every code word is represented by a matrix as given by

$$Z_{r,s} = \begin{bmatrix} C_{r,s,0,0} & C_{r,s,0,1} & \cdots & C_{r,s,0,P^2-1} \\ & \vdots & \ddots & \vdots \\ C_{r,s,P-1,0} & C_{r,s,P-1,1} & \cdots & C_{r,s,P-1,P^2-1} \end{bmatrix}. \quad (4)$$

This matrix represents a code word whose element can be derived as $C_{m,k} = C_{r,s,m,0} \cdots C_{r,s,m,k} \cdots C_{r,s,m,P^2-1}$, where $r = (0, 1, \ldots, P - 1)$, $s = (0, 1, \ldots, P - 1)$, $m = (0, 1, \ldots, P - 1)$, and $k = (0, 1, \ldots, P^2 - 1)$. With the prime sequence matrix $Z_{r,s}$ the value of $C_{m,k}$ will be mapped as

$$C_{m,k} = \begin{cases} W_{r \cdot n \oplus t \cdot s \cdot m, r \cdot m \oplus s \cdot n} & \text{if } k = ((r \cdot n \oplus t \cdot s \cdot m)\, P + n), \text{ for } m, n = \{0, 1, \ldots, P - 1\} \\ 0 & \text{elsewere.} \end{cases} \quad (5)$$

This mapping will define the value of matrix elements or position of wavelength in the code word matrix of size 3×9 for $P = 3$. A matrix is constructed by setting each element according to the rule $C_{m,k} = W_{r \cdot n \oplus t \cdot s \cdot m, r \cdot m \oplus s \cdot n}$, where $n = (0, 1, \ldots, P - 1)$, $m = (0, 1, \ldots, P - 1)$, $t = \{0, 1, \ldots, P - 1\}/[B]$, and $\oplus$ denotes a modulo addition. $[B]$ is a set of numbers that contains every integer designed with $1/s^2 \pmod{P}$ for $s = \{1, 2, \ldots, P - 1\}/2$. The genuine value of t can be calculated using this formula and that can be used to obtain $Z_{r,s}$ for various prime numbers. Each element $W_{r \cdot n \oplus t \cdot s \cdot m, r \cdot m \oplus s \cdot n}$ represents a single transmitting wavelength, among P^2 different wavelengths in the mth space slot and nth time period position in $Z_{r,s}$. A binary 3D multicarrier prime code can be derived as having the exact position of wavelength in the space slot and time period.

In $Z_{r,s}$ code word matrix, each row will have one set of disjoint wavelength and in total, there are P distinct wavelengths per row. The space group is represented by $m = (0, 1, \ldots, P - 1)$. In this way, we have $P^2 - 1$ prime code words of length P^2 and weight P^2 and have P^2 different wavelengths where $m \leq P$. These P^2 distinct wavelengths are divided into "m" space groups with P different wavelengths per space/channel. Every time slot can have a minimum of one wavelength. The optical code words can then be transmitted over "m" different channels having P different wavelengths each distributed over P^2 time slots thus acting as a 3D code. In this way, the position of P different wavelengths can be calculated in every m spatial channel according to the formula given in (5). The cardinality of 3D optical prime code is given by $|C| = P^2 - 1$; therefore for $P = 3$ it will be 8 and $Z_{0,0}$ for $(a, b) = (0, 0)$, respectively, will not exist.

For prime number $P = 3$ the valid value of t we have is 2. In the codeword $Z_{0,1}$, $r = 0$ and $s = 1$ are in GF(3). The elements of the code word matrix are calculated according to the rule $C_{m,k} = W_{r \cdot n \oplus t \cdot s \cdot m, r \cdot m \oplus s \cdot n}$ as shown in

$$Z_{0,1} = \begin{bmatrix} W_{0,0} W_{0,1} W_{0,2} & 000 & 000 \\ 000 & 000 & W_{2,0} W_{2,1} W_{2,2} \\ 000 & W_{1,0} W_{1,1} W_{1,2} & 000 \end{bmatrix}. \quad (6)$$

The corresponding code matrix is of size 3×9 and weight 9. Each "W" in the code word matrix represents the position of "1" in the code word that will be transmitted using an optical pulse of certain wavelength over the space channel in the corresponding time slot. It can be seen that each code word matrix of a three-dimensional code is divided into P rows in the space domain having length P^2 for each row in the time domain; the P^2 column has been divided into P groups each having P subcolumns.

The code word has P^2 wavelengths distributed over the P^2 time slot with each column having only one wavelength and each space channel will have P wavelengths. The P^2 wavelengths are in the shape of 3 groups and 3 wavelengths per group. The length of the 3D optical code word is 9 for $P = 3$. Figure 3 shows the visualization of a wavelength/time/space spread 3D code word $Z_{0,1}$ for prime sequence $(P = 3)$ and $t = 2$. Similarly, for the code word $Z_{2,2}$ the elements are calculated according to the rule $C_{m,k} = W_{r \cdot m \oplus t \cdot s \cdot n, r \cdot n \oplus s \cdot m}$. The corresponding code matrices $Z_{r,s}$ for $r = 2$ and $s = 2$ are of size 3×9 and weight 9 as shown in Figure 4. Equation (7) shows the resulting code word matrix as follows:

$$Z_{2,2} = \begin{bmatrix} W_{0,0} 00 & 00 W_{1,1} & 0 W_{2,2} 0 \\ 0 W_{0,1} 0 & W_{1,2} 00 & 00 W_{2,0} \\ 00 W_{0,2} & 0 W_{1,0} 0 & W_{2,1} 00 \end{bmatrix}. \quad (7)$$

In OCDMA system using modified 3D multicarrier prime code, each subscriber is allocated a code matrix as its code word. To transmit information bit "1" subscriber sends out a sequence of the optical signal in the space, time, and wavelength domain as per the intended receiver code word matrix. At the receiving end, the optical signal in every spatial channel is segregated and routed to correlate the every wavelength separately in the time domain. After successfully correlating the received optical signals are combined together, thus resulting in a high peak optical signal for the proposed receiver. This is considered as autocorrelation function, whereas for all others, it gives a series of low peak optical signal. This is considered as cross-correlation function. The received peak optical pulse is further passed

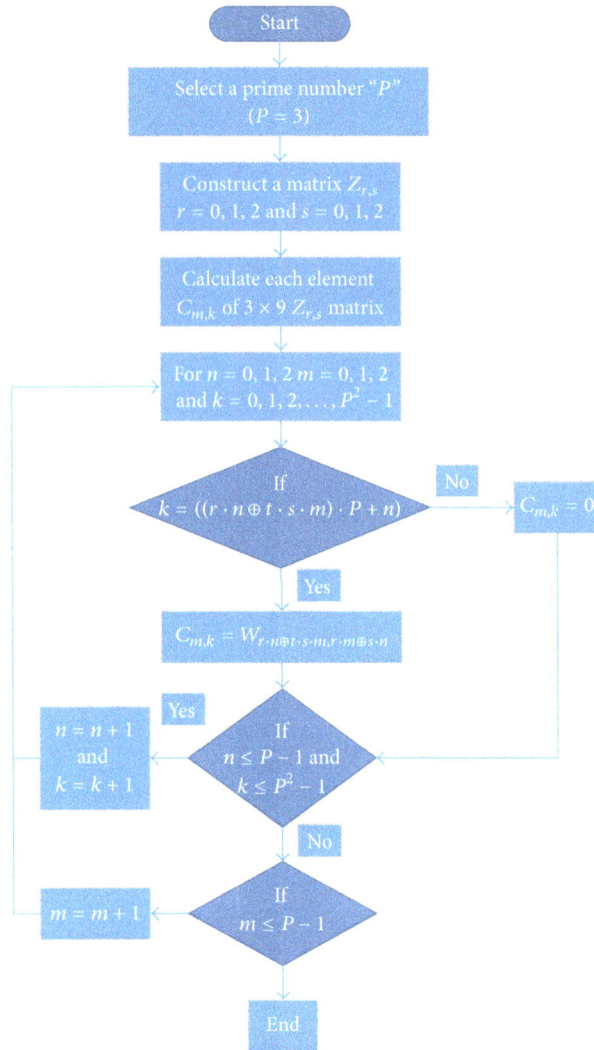

FIGURE 2: Flowchart of three-dimensional code construction algorithm.

Space	K	0	1	2	3	4	5	6	7	8
$m = 2$		W_{00}	W_{01}	W_{02}	0	0	0	0	0	0
$m = 1$		0	0	0	0	0	0	W_{20}	W_{21}	W_{22}
$m = 0$		0	0	0	W_{10}	W_{11}	W_{12}	0	0	0

$n=0$ $n=1$ $n=2$ $n=0$ $n=1$ $n=2$ $n=0$ $n=1$ $n=2$

Time $\longrightarrow$

FIGURE 3: Visualization of 3D optical prime code word $Z_{0,1}$ over GF($P = 3$) and $t = 2$.

to photodetector to detect the bit "1." For transmitting information bit "0" nothing is sent.

4. Result and Discussion

From optical CDMA network point of view, there is a need to design more secure codes with sufficient code weight, optimal code length, and higher code set size. The proposed code is compared with the 1D and 2D optical codes based on prime sequence algorithm for analysis as presented in Figure 5. It

Space	K	0	1	2	3	4	5	6	7	8
$m = 2$		W_{00}	0	0	0	0	W_{11}	0	W_{22}	0
$m = 1$		0	W_{01}	0	W_{12}	0	0	0	0	W_{20}
$m = 0$		0	0	W_{02}	0	W_{10}	0	W_{21}	0	0

$n=0$ $n=1$ $n=2$ $n=0$ $n=1$ $n=2$ $n=0$ $n=1$ $n=2$

Time $\longrightarrow$

FIGURE 4: Visualization of 3D optical prime code word $Z_{2,2}$ over GF($P = 3$) and $t = 2$.

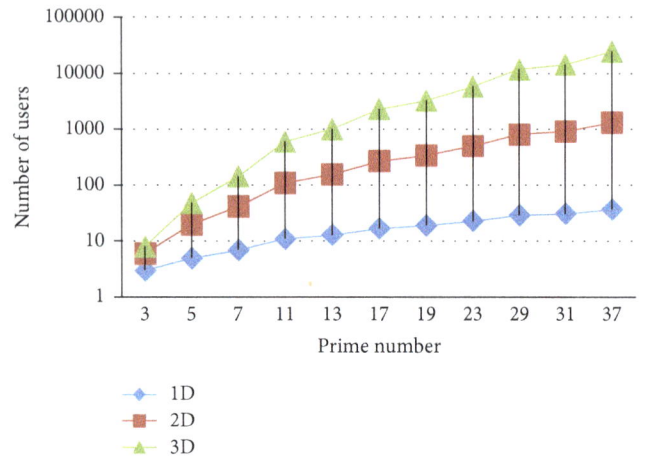

FIGURE 5: Number of users versus prime number for different optical prime codes.

can be observed that the proposed 3D codes are better in terms of a number of available codes (Cardinality) whereas they have comparable code length for a given prime number. However, the proposed 3D codes have higher code weight than 1D and 2D codes.

In optical CDMA, the interference among the users sharing the common fiber channel known as the multiple access interference is usually the dominant source of bit errors [6]. The performance of optical codes in OCDMA networks is mainly measured in terms of the autocorrelation and cross-correlation function values. The cross-correlation value depicts the interference between users. In contrast, the autocorrelation value depicts signal power and differentiates the users in the presence of other users. The intelligent design of the code word sequence is important to reduce the contribution of MAI to the total received signal. The address code words must satisfy two conditions; that is, all address code words should be distinguishable from shifted versions of it and all address code words should be distinguished from a possibly shifted version of every other code word in a code set [16].

The autocorrelation and cross-correlation value of the developed 3D optical code are evaluated on MATLAB tool. The wavelength/time/space spread 3D optical code is represented by $(W * T * S, w, \lambda_a, \lambda_c)$. "$W$" signifies the wavelengths, "T" denotes the time slots, and "S" specifies the space channels. "w" signifies the weight of the code. λ_a and λ_c are the autocorrelation and cross-correlation of the code,

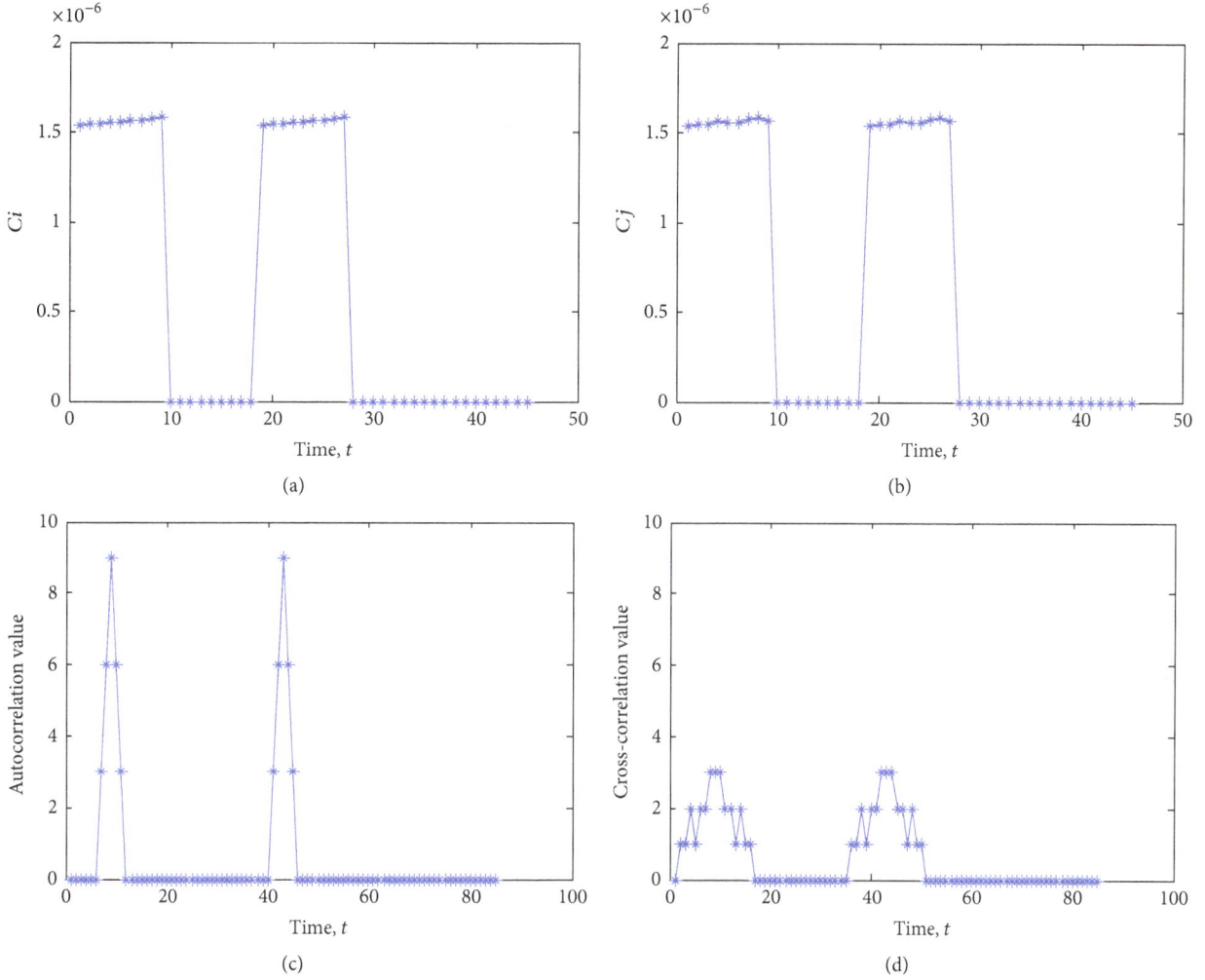

FIGURE 6: (a) Code word $Z_{0,1}$ of the code set of 3D optical code over GF(3). (b) Code word $Z_{2,2}$ of the code set of 3D optical code over GF(3). (c) Autocorrelation function of optical code $Z_{0,1}$ for the data bit stream 10100. (d) Cross-correlation function of optical codes $Z_{0,1}$ and $Z_{2,2}$, for the data bit stream 10100.

respectively. In (8a) and (8b) the correlation property for a pair of prime code words C_k and B_k with discrete data stream format are calculated.

$$\lambda_a = \sum_{k=0}^{P-1}\sum_{m=0}^{P-1}\sum_{n=0}^{P-1}C_{k,m,n}C_{k\cdot m\cdot n\oplus\tau} = P^2 \quad \text{for } 0 \le \tau \ge P^2, \quad (8a)$$

$$\lambda_c = \sum_{k=0}^{P-1}\sum_{m=0}^{P-1}\sum_{n=0}^{P-1}C_{k,m,n}B_{k\cdot m\cdot n\oplus\tau} \le 3 \quad \text{for } 0 \le \tau \ge P^2, \quad (8b)$$

where $C_{k,m,n}$ and $B_{k,m,n} \in (0,1)$ are an element of matrix Z, P is the prime number over Galois field, and $\oplus$ denotes the modulo addition.

It can be observed from the above correlation function that the peak autocorrelation is P^2 which occurs when $C_{k,m,n} = C_{k,m,n}$. Also, the cross-correlation value is "3" occurring at all synchronized time T when $C_{k,m,n} \ne B_{k,m,n}$. Figure 6 shows the two code words $Z_{0,1}$ and $Z_{2,2}$ of the code set of wavelength/time/space spread modified 3D multicarrier codes over GF(3) and of the length 9 and weight

9. It also illustrates that the autocorrelation property of 3D code sequence $Z_{0,1}$ having maximum peak value equals 9 at all synchronized times when the sequence follows the data stream 10100. Similarly, the cross-correlation property of 3D code sequence $Z_{0,1}$ with $Z_{2,2}$ is equal to 3 at all synchronized time T for the same data stream 10100. Similarly, Figure 7 represents the code words $Z_{2,0}$ and $Z_{1,1}$ and their autocorrelation and cross-correlation property, respectively, for the input data stream 10100.

Yen and Chen carried out the study of three-dimension Optical Code Division Multiple Access for optical fiber sensor networks. For efficient communication using OCDMA system, it is preferred to maximize the autocorrelation peak and minimize the cross-correlation function for sorting the correct signal and interference [18]. J. Singh and M. L. Singh present a new family of 3D wavelength/time/space codes named Golomb ruler-with-zero-insertions balanced codes for differential detection (GRZI-BCDD) using the unique interpulse distance property of Golomb rulers. The 3D code based on GRZI-BCDD requires fiber ribbons that

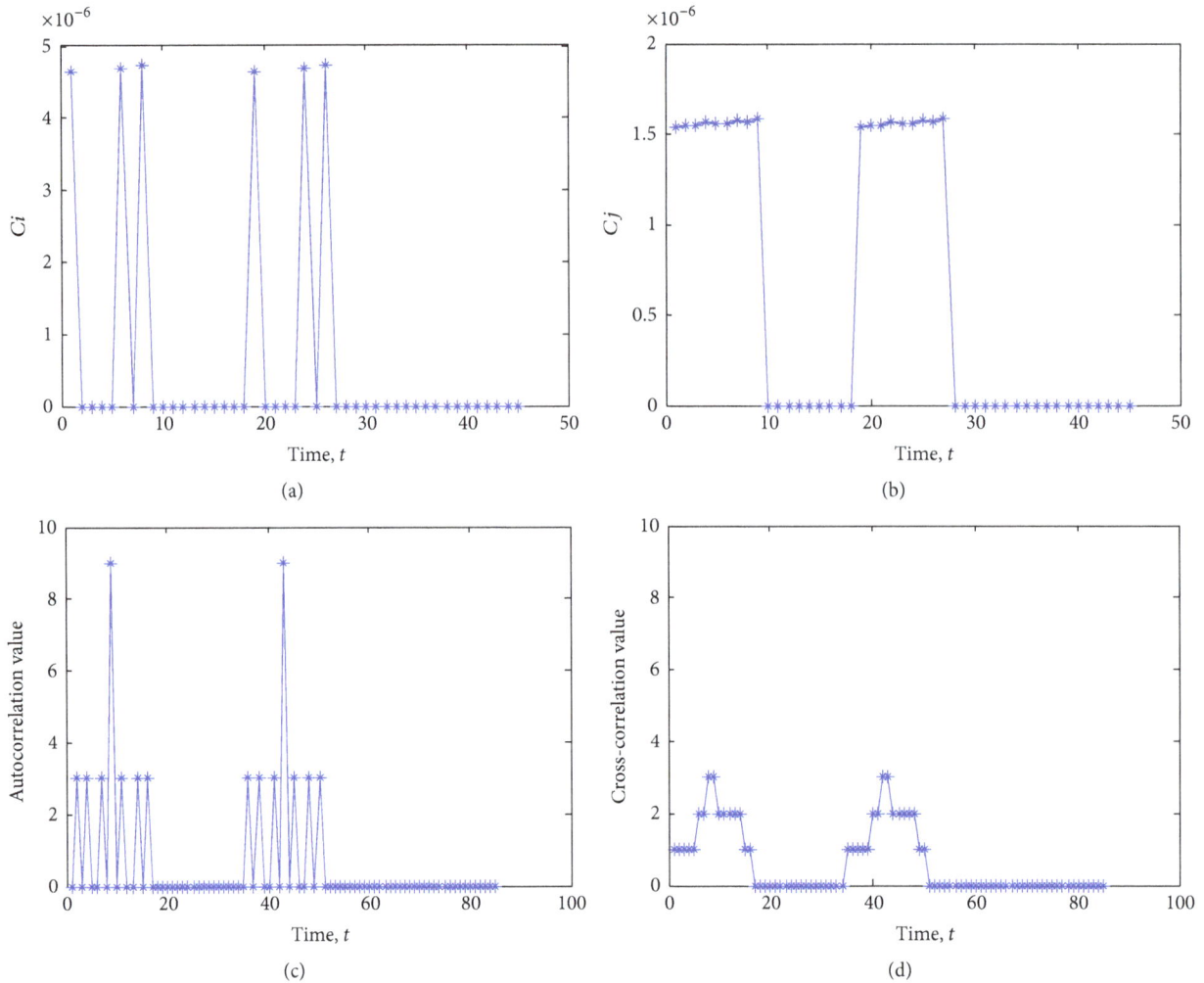

FIGURE 7: (a) Code word $Z_{2,0}$ of the code set of 3D optical code over GF(3). (b) Code word $Z_{1,1}$ of the code set of 3D optical code over GF(3). (c) Autocorrelation function of optical code $Z_{2,0}$ for the data bit stream 10100. (d) Cross-correlation function of optical codes $Z_{2,0}$ and $Z_{1,1}$, for the data bit stream 10100.

increase the system complexity [19]. To improve the ability of suppressing the multiple access interference, the 3D code set must have a sufficient code length of each code word. The proposed code incorporates the third dimension to improve the code cardinality, security, and coding flexibility that will help in providing multimedia services and quality of services. The limiting factor is its implementation because the complexity of system hardware is increased as the dimension increases. However do not require fiber ribbons for practical implementation.

5. Conclusion

This work presents the proposed wavelength/time/space spread modified 3D multicarrier codes using the prime sequence algorithm. The code discussed in the text has been designed for the prime number ($P = 3$) having code length 9; however, it can be increased for the assumed prime number. Thus, by introducing more dimensions, this coding approach not only overcomes the problem of a sufficient number of users for a given prime number but also makes the proposed

code more secure and support quality of services. However, a marginal complexity is increased from an implementation point of view due to additional dimension as compared to 1D and 2D codes. In addition, the generation algorithm of code is simple and has optimum orthogonal properties. Therefore, it reduces the cost involved. The simulation results on MATLAB depict that these 3D optical codes have maximum autocorrelation peak ($\lambda_a = P^2$) with cross-correlation value ($\lambda_c = 3$).

Competing Interests

The authors declare that they have no competing interests.

References

[1] A. Stok and E. H. Sargent, "The role of optical CDMA in access networks," *IEEE Communications Magazine*, vol. 40, no. 9, pp. 83–87, 2002.

[2] J. A. Salehi, "Emerging optical CDMA techniques and applications," *International Journal of Optics and Photonics*, vol. 1, no. 1, pp. 15–32, 2007.

[3] G.-C. Yang and W. C. Kwong, "Performance analysis of optical CDMA with prime codes," *Electronics Letters*, vol. 31, no. 7, pp. 569–570, 1995.

[4] R. Yadav and G. Kaur, "Optical CDMA: technique, parameters and applications," in *Proceedings of the 4th International Conference on Emerging Trends in Engineering & Technology*, pp. 90–98, Kurukshetra, India, October 2013.

[5] S. Park, B. K. Kim, and B. W. Kim, "An OCDMA scheme to reduce multiple access interference and enhance performance for optical subscriber access networks," *ETRI Journal*, vol. 26, no. 1, pp. 13–20, 2004.

[6] P. R. Prucnal, M. A. Santoro, and T. R. Fan, "Spread spectrum fiber-optic local area network using optical processing," *Journal of Lightwave Technology*, vol. 4, no. 5, pp. 547–554, 1986.

[7] A. J. Mendez, R. M. Gagliardi, V. J. Hernandez, C. V. Bennett, and W. J. Lennon, "Design and performance analysis of wavelength/time (W/T) matrix codes for optical CDMA," *Journal of Lightwave Technology*, vol. 21, no. 11, pp. 2524–2533, 2003.

[8] K. Yu and N. Park, "Design of new family of two-dimensional wavelength-time spreading codes for optical code division multiple access networks," *Electronics Letters*, vol. 35, no. 10, pp. 830–831, 1999.

[9] R. M. H. Yim, L. R. Chen, and J. Bajcsy, "Design and performance of 2-D codes for wavelength-time optical CDMA," *IEEE Photonics Technology Letters*, vol. 14, no. 5, pp. 714–716, 2002.

[10] E. S. Shivaleela and T. Srinivas, "Construction of wavelength/time codes for fiber-optic CDMA networks," *IEEE Journal on Selected Topics in Quantum Electronics*, vol. 13, no. 5, pp. 1370–1377, 2007.

[11] R. Yadav and G. Kaur, "Optical CDMA codes: a review," in *Proceedings of the 7th International Conference Advanced Computing and Communication Technologies*, pp. 263–269, Panipat, India, 2013.

[12] J. E. McGeehan, S. M. R. Motaghian Nezam, P. Saghari et al., "3D time-wavelength-polarization OCDMA coding for increasing the number of users in OCDMA LANs," in *Proceedings of the Optical Fiber Communication Conference (OFC '04)*, pp. 444–446, February 2004.

[13] S. Amiralizadeh and K. Mehrany, "Analytical study of multiple access interference and beat noise in polarization-wavelength-time optical CDMA systems," in *Proceedings of the 18th International Conference on Telecommunications (ICT '11)*, pp. 206–210, IEEE, Ayia Napa, Cyprus, May 2011.

[14] M. Ravi Kumar, S. S. Pathak, and N. B. Chakrabarti, "Design and performance analysis of code families for multi-dimensional optical CDMA," *IET Communications*, vol. 3, no. 8, pp. 1311–1320, 2009.

[15] C.-F. Hong and G.-C. Yang, "Multicarrier FH codes for multicarrier FH-CDMA wireless systems," *IEEE Transactions on Communications*, vol. 48, no. 10, pp. 1626–1630, 2000.

[16] R. Yadav and G. Kaur, "Design and performance analysis of 1D, 2D and 3D prime sequence code family for optical CDMA network," *Journal of Optics*, 2016.

[17] G. C. Yang and W. C. Kwong, "Prime codes with applications to CDMA optical and wireless networks," in *Prime Codes with Applications to CDMA Optical and Wireless Networks*, pp. 43–108, Artech House, Norwood, Mass, USA, 2002.

[18] C.-T. Yen and C.-M. Chen, "A study of three-dimensional optical code-division multiple-access for optical fiber sensor networks," *Computers and Electrical Engineering*, vol. 49, pp. 136–145, 2016.

[19] J. Singh and M. L. Singh, "Design of 3-D wavelength/time/space codes for asynchronous fiber-optic CDMA systems," *IEEE Photonics Technology Letters*, vol. 22, no. 3, pp. 131–133, 2010.

14

Development of Combinatorial Pulsed Laser Deposition for Expedited Device Optimization in CdTe/CdS Thin-Film Solar Cells

Ali Kadhim,[1,2] Paul Harrison,[1] Jake Meeth,[1,3] Alaa Al-Mebir,[1,2] Guanggen Zeng,[1,4] and Judy Wu[1]

[1] Department of Physics and Astronomy, University of Kansas, Lawrence, KS 66046, USA
[2] Departments of Physics, College of Science, University of Thi-Qar, Nasiriya, Thi-Qar, Iraq
[3] Electrical Engineering Division, Department of Engineering, University of Cambridge, Cambridge CB3 OFA, UK
[4] College of Materials Science and Engineering, Sichuan University, Chengdu 610064, China

Correspondence should be addressed to Ali Kadhim; phy_ali82@yahoo.com and Judy Wu; jwu@ku.edu

Academic Editor: Martin Kröger

A combinatorial pulsed laser deposition system was developed by integrating a computer controlled scanning sample stage in order to rapidly screen processing conditions relevant to CdTe/CdS thin-film solar cells. Using this system, the thickness of the CdTe absorber layer is varied across a single sample from 1.5 μm to 0.75 μm. The effects of thickness on CdTe grain morphology, crystal orientation, and cell efficiency were investigated with respect to different postprocessing conditions. It is shown that the thinner CdTe layer of 0.75 μm obtained the best power conversion efficiency up to 5.3%. The results of this work shows the importance that CdTe grain size/morphology relative to CdTe thickness has on device performance and quantitatively exhibits what those values should be to obtain efficient thin-film CdTe/CdS solar cells fabricated with pulsed laser deposition. Further development of this combinatorial approach could enable high-throughput exploration and optimization of CdTe/CdS solar cells.

1. Introduction

CdTe solar cells have shown great promise in competing with Si solar cells, which currently dominate the photovoltaic (PV) market. CdTe solar cells have a higher theoretical limiting efficiency than Si solar cells due to CdTe's nearly optimal band gap for our Sun and high absorption coefficient [1–4]. Recent improvements seen in CdTe solar cells make it reasonable for CdTe to take over a significant portion of the PV market [5]. However, the champion CdTe cells for power conversion efficiency have all been thick film (~5–8 μm) devices. If these kinds of cells are manufactured on a large scale the cost will eventually increase significantly considering the limited amount of Te available. As such, recent research has been focused on thin-film CdTe solar cells (~1 μm), which in addition to protecting Te reserves would reduce the overall material cost of device fabrication. Altering the structural parameters of these thin-film cells, in addition

to the compositional and postprocessing parameters, can easily lengthen and complicate the optimization process. Therefore, these conditions are typically optimized with respect to a specific thickness which conventionally calls for many separate samples to be fabricated.

Combinatorial processing and characterization are the method of producing a sample with varied material properties across a single sample [6]. This effectively allows a continuum of device performances to be measured as a function of the varied property with a single sample. Combinatorial pulsed laser deposition (cPLD) has been used before to vary chemical compositions across C-MOS transistors, for example, [6]. PLD is a relatively new fabrication technique applied for CdTe solar cells. While its application to solar cells has proven effective in past research, a systematic study of the effect of the PLD processing conditions is lacking [7, 8]. PLD is advantageous for thin-film depositions due to its highly controllable deposition rate and also its many easily

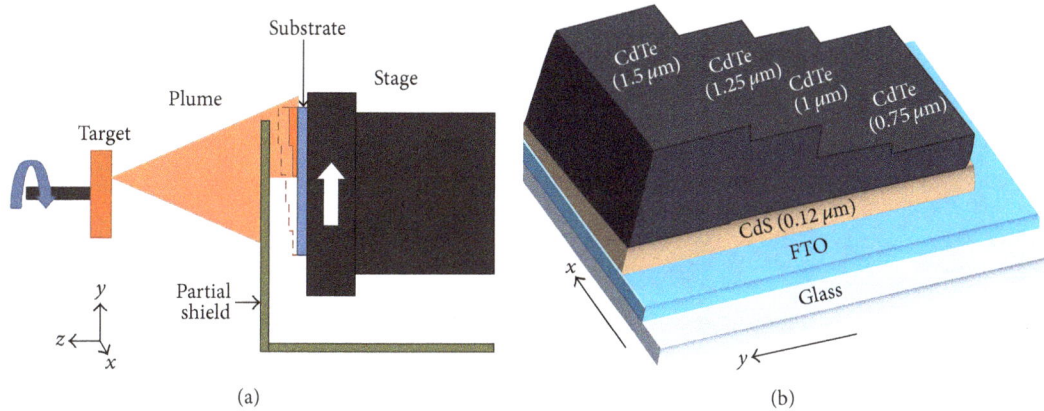

FIGURE 1: A depiction of the combinatorial PLD process which allows different thicknesses to be deposited on a single substrate (a) and a schematic drawing of the samples made with varying CdTe thickness (b).

adjustable ablation parameters including laser repetition rate, pulse length, energy density, target-substrate distance, and chamber atmosphere and temperature [9]. In combinatorial PLD fabrication of CdS/CdTe thin-film solar cells, additional advantages are provided in generating different device structures for expedited optimization of the device performance.

In thin-film CdS/CdTe solar cells, the CdTe grain size and microstructure relative to its thickness are extremely important to device performance. In order to investigate the effects of CdTe thickness on CdTe microstructure and cell performance and establish a method of probing for device optimization, a programmable scanning sample stage was implemented into the PLD system for combinatorial fabrication of CdTe. In particular, this work analyzed the properties of cPLD made CdTe and studied cell performance when varying the CdTe thickness (1.5, 1.25, 1, and 0.75 μm) on the same sample. A reference sample (denoted in this work by sample A) was fabricated without the typical CdCl$_2$ annealing treatment and tested to extract the effect of the annealing on the microstructure and crystallinity of CdTe and the resulting CdS/CdTe cell performance. It is demonstrated that the PLD conditions used in this work result in the highest efficiency being obtained by the thinnest CdTe layer of 0.75 μm with an overall maximum efficiency of 5.3%.

2. Experimental Method

The PLD system consists of a 248 nm KrF excimer laser from Lambda Physik with a 20 ns laser pulse duration. The absorber and window layer targets are mounted inside a vacuum chamber on rotating stands that allow for both targets to be moved into the path of the laser and rotation about the axis of the targets facilitates uniform ablation of the target surface [7, 8, 10–12]. The CdS and CdTe targets were obtained from ACI Alloys and had a 99.99% purity. The substrate stage was mounted into the chamber across from the targets with a distance of 5.5 cm separating the two. The scanning stage has two axes of movement and it can move in any motion desired that is perpendicular to the laser plume axis. Two external stepper motors drive the stage under computer control. The computer control is achieved

by a custom computer program which makes it possible for the stage to undergo very complicated motions. Located in the path of the ablated laser plume, in between the target and substrate, is a partial shield that covers half of the stage when the stage is in its neutral position. This allows for precise control over which area of the substrate is deposited on, granting a combinatorial way to control in situ thickness variations on a single substrate. As the substrate was moved in the y-direction to make the different CdTe thicknesses it was also constantly scanning in the x-direction (in and out of the page in Figure 1(a)) to achieve uniform thickness across the entire substrate. All vacuum and stage equipment is custom and home-built. Thicknesses of 1.5, 1.25, 1.0, and 0.75 μm were chosen to be deposited on top of the 120 nm thick CdS window layer. A depiction of the cPLD setup and the solar cell design is shown in Figure 1.

TEC 15 (from Pilkington North America) soda lime glass, which has a fluorine-doped tin oxide layer with a sheet resistance of 15 $\Omega/\square$, was used as the conductive substrate and serves as the front contact to the devices. Before any deposition the substrates were thoroughly cleaned by first a boiling in deionized (DI) water followed by sonication in DI water, acetone, and IPA for 5 minutes each. The CdS window layer was heated to 200°C with the sample stage heater and deposited at this temperature in 1.5 mTorr of Argon gas flow with a laser repetition rate of 10 Hz, spot size of 8.7 mm^2, and pulse energy of 150 mJ with a 0.3 OD filter reducing the energy to 70 mJ. The CdTe absorber layer parameters are identical except the laser spot size is decreased to 7.6 mm^2 and the thickness is varied in 250 nm steps from 750 nm to 1500 nm using the partial shield and moving stage. After ablation is complete, the sample was annealed in the vacuum chamber at 400°C for 10 min in 20 Torr Ar and cooled naturally overnight to room temperature.

The CdCl$_2$ anneal was carried out by placing the samples on top of a piece of glass that had been coated with CdCl$_2$ by dropping a supersaturated methanol/CdCl$_2$ solution onto it and letting it dry in air. The sample was kept ~3 mm from the CdCl$_2$ coated glass during the annealing process which took place in a tube furnace with 100 sccm Ar and 25 sccm O$_2$ flow at 360°C for varying times. Four different cells were

Layer 1 (1.5 μm)   Layer 2 (1.25 μm)   Layer 3 (1 μm)   Layer 4 (0.75 μm)

(a) First row: sample A (no CdCl$_2$ treatment)

Layer 1 (1.5 μm)   Layer 2 (1.25 μm)   Layer 3 (1 μm)   Layer 4 (0.75 μm)

(b) Second row: sample B3 (15 min CdCl$_2$ treatment)

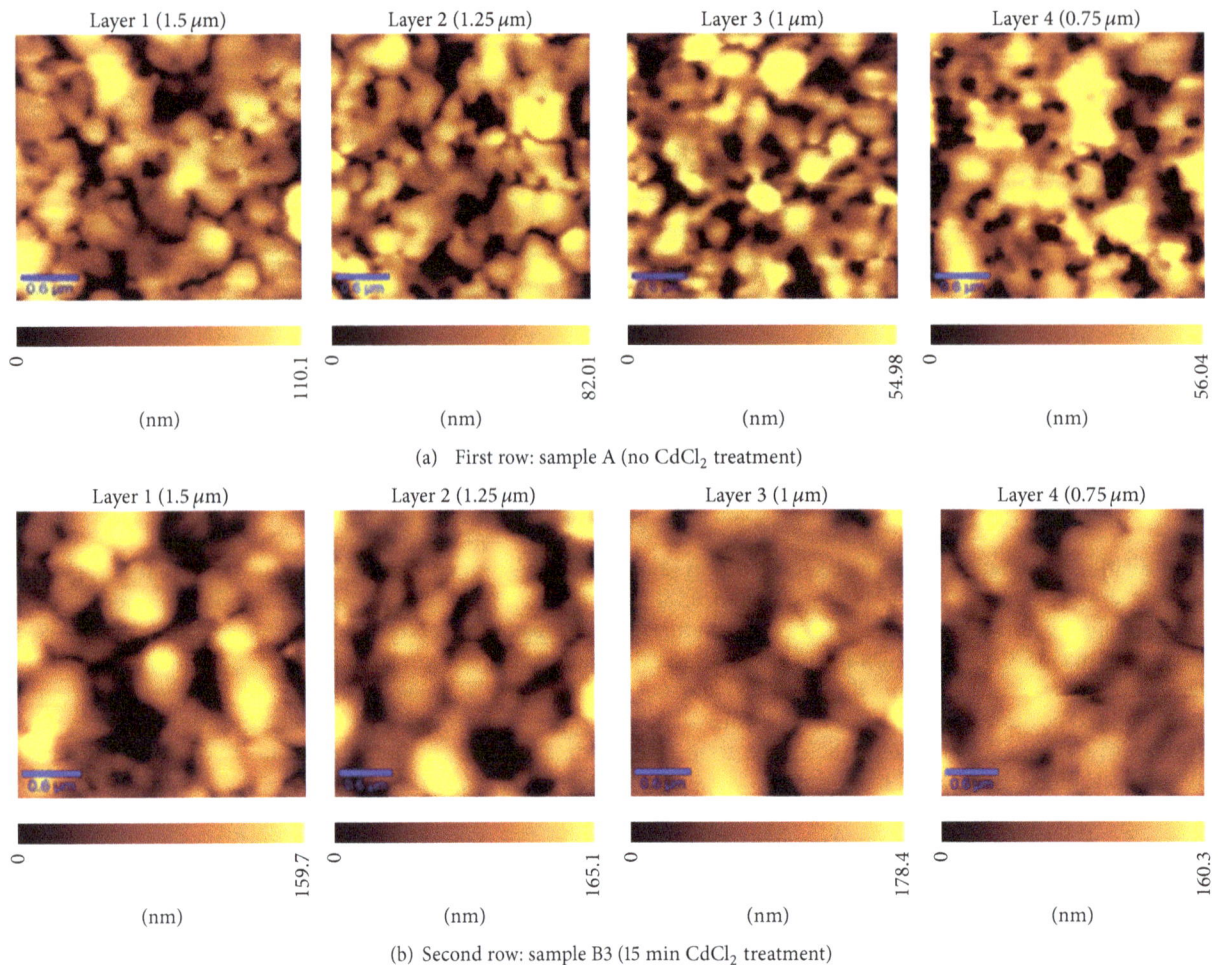

FIGURE 2: First row of AFM images is corresponding to sample A (No CdCl$_2$ annealing). The second row is corresponding to the sample B3 (with 15 min CdCl$_2$ annealing). The scale bar for all images is 0.6 μm.

made each with a varying CdCl$_2$ annealing time of 10, 12, 15, and 17 minutes. These samples are denoted by samples B1, B2, B3, and B4, respectively. All samples were submerged for four seconds in a Bromine etchant produced by mixing 0.2 mL Bromine with 40 mL methanol. The samples were then immediately rinsed with methanol, acetone, and IPA. The etching process is used to remove contaminants from the surface as well as make a Te rich surface for better ohmic contact. A Cu doped HgTe/graphite paste (0.017 g Cu, 4 g HgTe, and 10 g graphite paste) was then made to be applied to the etched CdTe layer as the back contact. Small contacts with average areas of roughly 1.25 mm^2 are then applied across the entire sample. Four contacts were placed on each layer for device and uniformity testing. After application, the sample is again baked in the tube furnace with 100 sccm Ar flow at 280°C for 30 minutes. Finally, Ag electrodes were carefully applied to the back contacts with a toothpick and baked in air at 150°C for 1 hour.

J-V characterizations are carried out using a CHI660D electrochemical workstation and a Newport 50–500 W Xenon lamp solar simulator at 1.5 AM (100 mW/cm^2). The CHI 660D electrochemical workstation and a Newport monochromator were used for external quantum efficiency

(EQE) measurements. Atomic force microscopy (AFM) and Raman spectroscopy (488 nm excitation wavelength) were performed using a WiTec Alpha 300 confocal MicroRaman system to obtain surface roughness and phase orientation for the different growth and annealing conditions.

3. Results and Discussion

Atomic force microscopy (AFM) was applied to characterize the CdTe surface morphology. The results are compared in Figure 2 for the CdTe variable thickness sample (sample B3) and the reference sample (sample A), which was fabricated in the same cPLD process but did not experience the CdCl$_2$ anneal treatment. Calculations from the AFM analysis software indicate that the average roughness of the CdTe in sample A is in the range of 13–16 nm while, by visual inspection, the CdTe grain size is roughly 140–160 nm. On the other hand, in sample B3, the average roughness for all layers was approximately 30 nm and the grain size is much larger. There is a correlation between the grain size and the thickness of CdTe. This is most easily seen in sample B3 where it is obvious that the grains get bigger as the thickness gets smaller. The thinnest layer of CdTe at 0.75 μm has grains that vary in

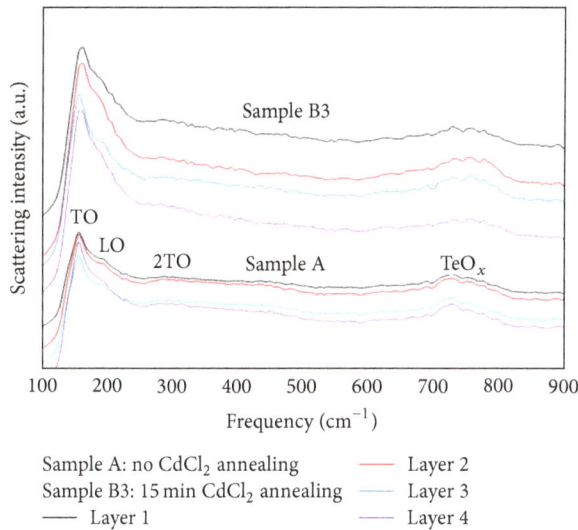

Sample A: no CdCl$_2$ annealing   —— Layer 2
Sample B3: 15 min CdCl$_2$ annealing   —— Layer 3
—— Layer 1   —— Layer 4

FIGURE 3: The Raman spectra of samples A and B3 before and after the CdCl$_2$ treatment.

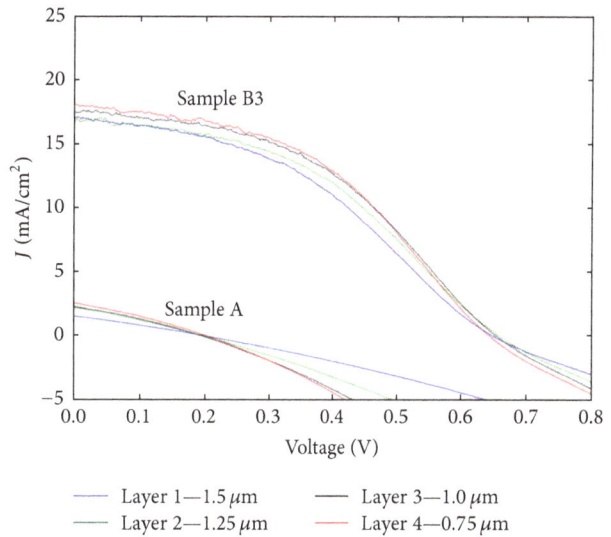

—— Layer 1—1.5 μm   —— Layer 3—1.0 μm
—— Layer 2—1.25 μm   —— Layer 4—0.75 μm

FIGURE 4: J-V curves for the best performing cells from the two samples made without (A) and with (B3) CdCl$_2$ annealing.

size from 300 to 700 nm while the thickest layer at 1.5 μm has grains in the range of 150–550 nm. The average grain size for the thinnest layer is approximately 65% of the total thickness. For the thicker layer the average grain size is closer to 20% of the total thickness. This difference seen in grain size per thickness is most likely because the recrystallization is more easily achieved in the thinner layers as there is less material.

Raman spectroscopy was performed in order to examine the structural properties of the different CdTe layers and the data can be seen in Figure 3. The data for each measurement has been shifted on the y-axis (arbitrary units) for better visibility. The transverse optical phonon mode (TO) for CdTe is known to be located at a Raman shift of 141 cm^{-1}, which can clearly be seen in all of the plots. The peak at the 169 cm^{-1} shift is assigned to the longitudinal optical phonon (LO) of CdTe [13]. The peaks at 292 cm^{-1} and 750 cm^{-1} are attributed to the 2TO mode of CdTe and tellurium oxide (TeO$_x$), respectively [14, 15]. The TeO$_x$ signature found in both samples is possibly attributed to oxygen residues which might exist during the fabrication of the samples. The thickness of CdTe appears to have no role on crystal orientation as the Raman spectra between layers of the same sample are nearly identical. The selection rules for CdTe illustrate that the TO and LO modes of CdTe are allowed from (110) and (100), respectively. Also, both modes can be allowed from (111) [16–18]. The LO mode is not highly pronounced in the layers of sample B3 suggesting that the polycrystalline structure is predominantly in the (110) crystal orientation.

Figure 4 shows the J-V curves that obtained the highest performance for the two samples made with and without CdCl$_2$ treatment. The JV data from sample B3 shows that the best performance came from the thinnest CdTe layer of 0.75 μm. This layer achieved a maximum efficiency of 5.3% with a J_{SC}, V_{OC}, and FF of 17.6 mA/cm^2, 664 mV, and 46%, respectively. Obviously, with CdTe having an absorption coefficient approaching 10^5 cm^{-1} in the visible spectrum, this layer will absorb the fewest photons due to its thickness

in accordance with the Beer-Lambert Law. However, the shortest travel distance for charges to be collected at the electrodes may outweigh the loss in photon absorption as compared to its thicker counterparts. The data from all layers can be seen in Tables 1(a) and 1(b), which contain the average values for the four cells made on each layer. All of the layers in sample A have extremely low performance as expected from its poor crystallinity and unoptimized microstructure. In addition, the extremely small V_{OC} is attributed to the weak electric field established by the p-n junction, due to the lack of doping in CdTe which is achieved during the CdCl$_2$ annealing process and quite possibly pinhole formation. The CdTe for this sample is possibly nanocrystalline, which causes huge amounts of recombination due to the small grain sizes [19, 20]. This would affect not only the fill factor, but the J_{SC} as well. The order of magnitude increase in the J_{SC} of sample B3 is due to the increased number of majority carriers, improvements in the grain connectivity, and increases in the grain size, which becomes large enough to eliminate a significant portion of the grain boundaries. However, the rollover seen in sample B3 indicates that there is a Schottky barrier present which occurs due to the mismatch in work function between the CdTe and back contact [21]. Studies are already underway to resolve this issue.

For comparison five total samples were fabricated using cPLD by which the variable thickness of CdTe has been obtained 1.5, 1.25, 1.0, and 0.75 μm in correspondence to layers 1, 2, 3, and 4, respectively, on the same substrate. Sample A regards those not treated with ex situ CdCl$_2$ annealing after the cPLD deposition, while samples B were exposed to the vapors of CdCl$_2$ at 360°C, and they were fabricated with different durations of CdCl$_2$ annealing 10, 12, 15, and 17 minutes corresponding to samples B1, B2, B3, and B4, respectively, as mentioned previously. Table 2 and Figure 5 show the highest efficiencies of the five samples and indicate that the best performance has been obtained at 15 minutes CdCl$_2$ treatment for all thicknesses. Therefore, comparisons

TABLE 1: (a) Thicknesses and electrical properties of sample A without CdCl$_2$ annealing. (b) Thicknesses and electrical properties of sample B3 with 15 min. CdCl$_2$ annealing.

(a)

Sample A	Layer 1	Layer 2	Layer 3	Layer 4
Thickness of CdS (μm)	0.12	0.12	0.12	0.12
Thickness of CdTe (μm)	1.5	1.25	1	0.75
V_{OC} (mV)	214	198	203	197
J_{SC} (mA/cm^2)	1.7	2.5	2.6	3.0
FF (%)	27	28	29	29
Efficiencies (%)	0.10	0.14	0.15	0.18

(b)

Sample B3	Layer 1	Layer 2	Layer 3	Layer 4
Thickness of CdS (μm)	0.12	0.12	0.12	0.12
Thickness of CdTe (μm)	1.5	1.25	1	0.75
V_{OC} (mV)	638	639	645	651
J_{SC} (mA/cm^2)	16.5	16.7	17.0	18.0
FF (%)	36	39	42	45
Efficiencies (%)	4.0	4.2	4.6	5.1

TABLE 2: Efficiencies for five samples processed at different CdCl$_2$ annealing times.

Layer thickness	No CdCl$_2$	10 min.	12 min.	15 min.	17 min.
Layer 1 (1.5 μm)	0.08	1.73	2.39	4.46	2.90
Layer 2 (1.25 μm)	0.12	1.95	4.35	4.79	3.51
Layer 3 (1.0 μm)	0.13	2.30	4.54	5.09	3.99
Layer 4 (0.75 μm)	0.15	2.95	5.16	5.34	4.08

of physical properties are made between samples A and B to elucidate the effect of the annealing on the solar cells with variable thicknesses of CdTe.

Figure 6 shows the key parameters of the solar cells as a function of thickness. The V_{OC} is essentially unchanged regardless of the CdTe thickness indicating that the same or comparable charge doping level, and thus the built-in bias voltage, was achieved in all layers. The J_{SC} values are also nearly identical between the layers of different thickness regardless of whether they were processed with or without the CdCl$_2$ annealing. However, at least a sixfold increase in the J_{SC} value was observed in all layers in samples with CdCl$_2$ annealing with respect to their counterparts without such annealing. The biggest factor contributing to differences in performance between layers was the FF. There is a linear decrease in FF observed with increasing CdTe thickness. The higher FF values seen at smaller CdTe thicknesses indicate the benefit of less recombination at smaller CdTe thickness as expected.

The efficiency of the solar cell can be further characterized by measuring the external quantum efficiency EQE, which represents the ratio between the numbers of electrons collected by the solar cell to the number of photons incidents on the solar cell [22]. Figure 7 shows the extreme difference in the EQE between samples A and B3 and allows us to gain

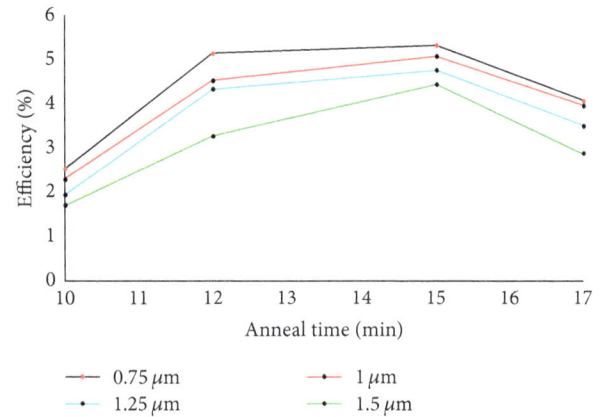

FIGURE 5: Plot of maximum solar cell efficiency versus CdCl$_2$ anneal time, which were all obtained at a CdTe thickness of 0.75 μm.

more insight into the behavior between the different CdTe thicknesses. The two thickest CdTe layers in sample B3 have nearly identical EQE curves implying that the recombination relative to CdTe thickness is the same for these two layers. When reducing the CdTe thickness from 1.25 μm to 1 μm an improvement in EQE was observed. An even further increase in EQE was observed when reducing the thickness by another 250 nm to 0.75 μm. The entire increase in collection efficiency from layer 3 to layer 4 happens above the CdS band edge. This observation combined with the comparable grain size analysis for these two layers clearly indicates that the thinnest layer is the better performer because of less recombination in the CdTe due to the decreased number of grain boundaries encountered by electrons/holes in this cell.

4. Conclusion

CdTe/CdS thin-film solar cells with variable CdTe thickness in the range of 0.75 μm to 1.5 μm on the same sample were fabricated using a cPLD system. This combinatorial approach of device fabrication has allowed for an expedited optimization of the CdTe microstructure, crystallinity, and CdS/CdTe heterojunction in CdCl$_2$ annealing. The grain size of the CdTe was found dependent on its thickness and an average grain size of ~450 nm in the cPLD CdTe yields the best power conversion efficiency (5.3%) in the solar cells of the thinnest CdTe layer of 0.75 μm, most probably due to the benefit of reduced charge recombination outweighing the reduced optical absorption. The cPLD method provides an efficient approach for exploration of device structures and can be used to further optimize these devices by changing a variety of PLD and CdTe/CdS solar cell parameters on the same device as well as allowing for quick exploration of devices with more complex cell structures.

Conflict of Interests

The authors declare that there is no conflict of interests regarding the publication of this paper.

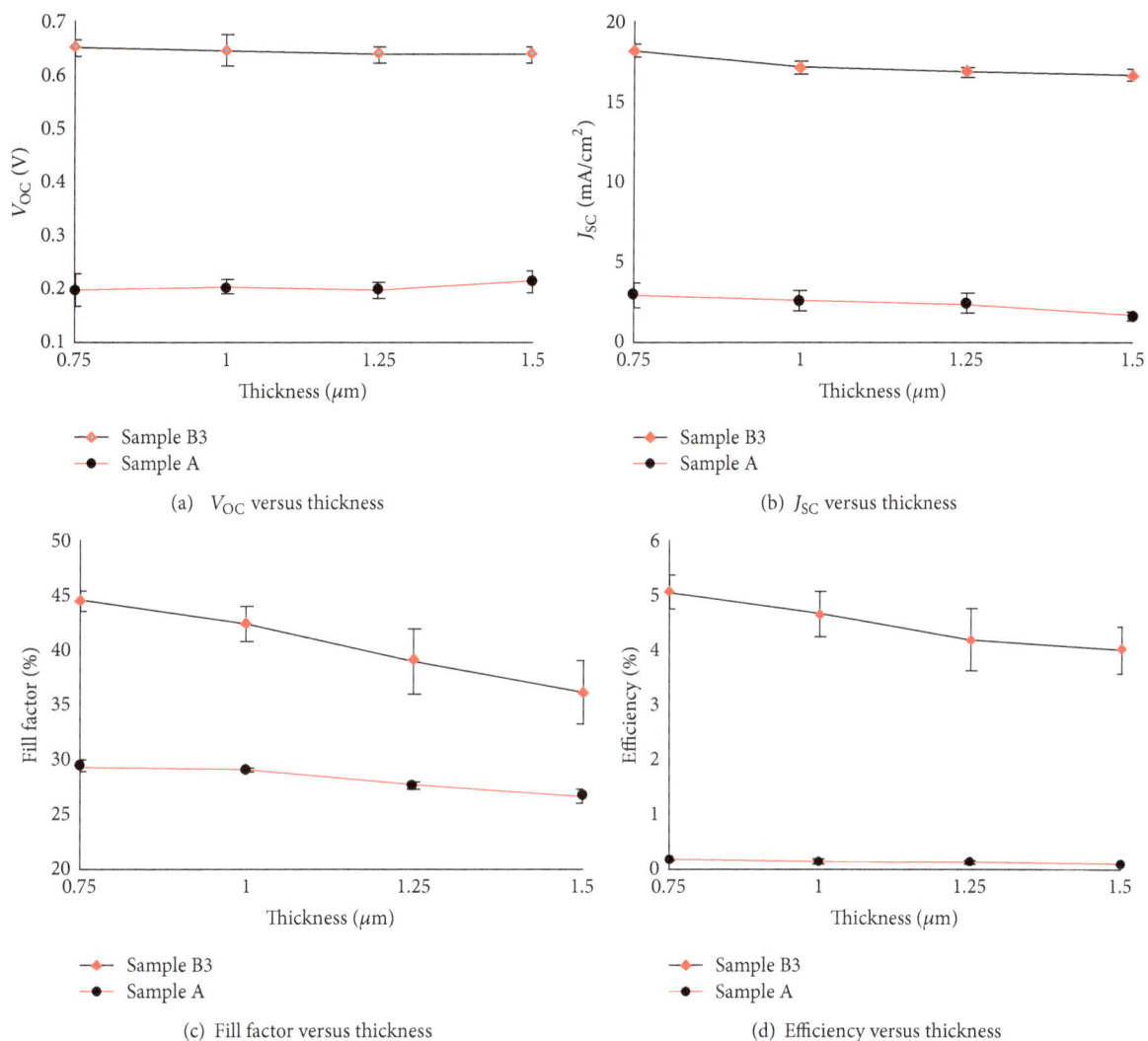

(a) V_{OC} versus thickness

(b) J_{SC} versus thickness

(c) Fill factor versus thickness

(d) Efficiency versus thickness

FIGURE 6: The values of open circuit voltage (a), short circuit current (b), fill factor (c), and efficiency (d) as a function of CdTe thickness for sample A made without any $CdCl_2$ annealing and sample B3 that underwent a 15 min $CdCl_2$ annealing.

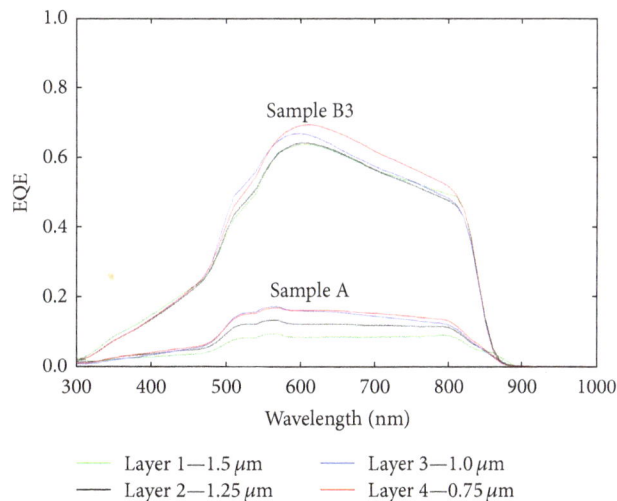

FIGURE 7: The quantum efficiencies for the best performing cells from the CdS/CdTe solar cells made without (sample A) and with (sample B3) $CdCl_2$ annealing at 360°C for 15 min.

Acknowledgments

The authors acknowledge support in part by NASA Contract no. NNX13AD42A, ARO Contract no. W911NF-16-1-0029, and NSF Contracts nos. NSF-DMR-1105986, NSF-DMR-1337737, and NSF-DMR-1508494. Paul Harrison would like to acknowledge the support from the Redeker Scholarship made possible by Maynard and Carol Redeker.

References

[1] M. Gloeckler, I. Sankin, and Z. Zhao, "CdTe solar cells at the threshold to 20% efficiency," *IEEE Journal of Photovoltaics*, vol. 3, no. 4, pp. 1389–1393, 2013.

[2] J. L. Cruz-Campa and D. Zubia, "CdTe thin film growth model under CSS conditions," *Solar Energy Materials and Solar Cells*, vol. 93, no. 1, pp. 15–18, 2009.

[3] B. McCandless and J. Sites, "Cadmium telluride solar cells," in *Handbook of Photovoltaic Science and Engineering*, chapter 14, John Wiley & Sons, Hoboken, NJ, USA, 2003.

[4] X. Wu, "High-efficiency polycrystalline CdTe thin-film solar cells," *Solar Energy*, vol. 77, no. 6, pp. 803–814, 2004.

[5] M. A. Green, K. Emery, and Y. Hishikawa, "Solar cell efficiency tables (version 44)," *Progress in Photovoltaics*, vol. 22, no. 7, pp. 701–710, 2014.

[6] V. Craciun, D. Craciun, I. N. Mihailescu et al., "Combinatorial pulsed laser deposition of thin films," in *High-Power Laser Ablation VII*, vol. 7005 of *Proceedings of SPIE*, May 2008.

[7] J. Meeth, P. Harrison, J. Liu et al., "Pulsed laser deposition of thin film CdTe/CdS solar cells with CdS/ZnS superlattice windows," in *Proceedings of the 39th IEEE Photovoltaic Specialists Conference (PVSC '13)*, pp. 1996–1999, Tampa, Fla, USA, June 2013.

[8] P. Harrison, J. Meeth, J. Liu, R. Lu, L. Feng, and J. Wu, "The effects of fabrication pressure on the physical properties of CdS/CdTe thin film solar cells made via pulsed laser deposition," in *Proceedings of the IEEE 39th Photovoltaic Specialists Conference (PVSC '13)*, pp. 2590–2593, Tampa, Fla, USA, June 2013.

[9] H.-U. Krebs, M. Weisheit, J. Faupel et al., "Pulsed laser deposition (PLD)—a versatile thin film technique," in *Advances in Solid State Physics*, B. Kramer, Ed., vol. 43 of *Advances in Solid State Physics*, pp. 505–518, Springer, Berlin, Germany, 2003.

[10] B. Li, J. Liu, G. Xu, R. Lu, L. Feng, and J. Wu, "Development of pulsed laser deposition for CdS/CdTe thin film solar cells," *Applied Physics Letters*, vol. 101, no. 15, Article ID 153903, 2012.

[11] F. d. Moure-Flores, J. G. Quiñones-Galván, A. Guillén-Cervantes et al., "Physical properties of CdTe:Cu films grown at low temperature by pulsed laser deposition," *Journal of Applied Physics*, vol. 112, Article ID 113110, 2012.

[12] P. Hu, B. Li, L. Feng et al., "Effects of the substrate temperature on the properties of CdTe thin films deposited by pulsed laser deposition," *Surface and Coatings Technology*, vol. 213, pp. 84–89, 2012.

[13] M. Kimata, T. Suzuki, K. Shimomura, and M. Yano, "Interdiffusion of In, Te at the interface of molecular beam epitaxial grown CdTe/InSb heterostructures," *Journal of Crystal Growth*, vol. 146, no. 1–4, pp. 433–438, 1995.

[14] N. Dewan, V. Gupta, K. Sreenivas, and R. S. Katiyar, "Growth of amorphous TeO$_x$ ($2 \leq x \leq 3$) thin film by radio frequency sputtering," *Journal of Applied Physics*, vol. 101, no. 8, Article ID 084910, 2007.

[15] D. Wang, Z. Hou, and Z. Bai, "Study of interdiffusion reaction at the CdS/CdTe interface," *Journal of Materials Research*, vol. 26, no. 5, pp. 697–705, 2011.

[16] P. M. Amirtharaj and F. H. Pollak, "Raman scattering study of the properties and removal of excess Te on CdTe surfaces," *Applied Physics Letters*, vol. 45, no. 7, pp. 789–791, 1984.

[17] S. S. Islam, S. Rath, K. P. Jain, S. C. Abbi, C. Julien, and M. Balkanski, "Forbidden one-LO-phonon resonant Raman scattering and multiphonon scattering in pure CdTe crystals," *Physical Review B*, vol. 46, no. 8, pp. 4982–4985, 1992.

[18] J. Menéndez, A. Pinczuk, J. P. Valladares, R. D. Feldman, and R. F. Austin, "Resonance Raman scattering in CdTe-ZnTe superlattices," *Applied Physics Letters*, vol. 50, article 1101, 1987.

[19] L. A. Kosyachenko, E. V. Grushko, and V. V. Motushchuk, "Recombination losses in thin-film CdS/CdTe photovoltaic devices," *Solar Energy Materials and Solar Cells*, vol. 90, no. 15, pp. 2201–2212, 2006.

[20] B. Maniscalco, A. Abbas, J. W. Bowers et al., "The activation of thin film CdTe solar cells using alternative chlorine containing compounds," *Thin Solid Films*, vol. 582, pp. 115–119, 2015.

[21] Z. Fang, X. C. Wang, H. C. Wu, and C. Z. Zhao, "Achievements and challenges of CdS/CdTe solar cells," *International Journal of Photoenergy*, vol. 2011, Article ID 297350, 8 pages, 2011.

[22] J. Nelson, *The Physics of Solar Cells*, Imperial College Press, 2003.

15

Synthesis, Optical Characterization, and Thermal Decomposition of Complexes Based on Biuret Ligand

Mei-Ling Wang, Guo-Qing Zhong, and Ling Chen

School of Material Science and Engineering, Southwest University of Science and Technology, Mianyang 621010, China

Correspondence should be addressed to Guo-Qing Zhong; zgq316@163.com

Academic Editor: Marek Samoc

Four complexes were synthesized in methanol solution using nickel acetate or nickel chloride, manganese acetate, manganese chloride, and biuret as raw materials. The complexes were characterized by elemental analyses, UV, FTIR, Raman spectra, X-ray powder diffraction, and thermogravimetric analysis. The compositions of the complexes were $[Ni(bi)_2(H_2O)_2](Ac)_2 \cdot H_2O$ (**1**), $[Ni(bi)_2Cl_2]$ (**2**), $[Mn(bi)_2(Ac)_2] \cdot 1.5H_2O$ (**3**), and $[Mn(bi)_2Cl_2]$ (**4**) (bi = $NH_2CONHCONH_2$), respectively. In the complexes, every metal ion was coordinated by oxygen atoms or chlorine ions and even both. The nickel and manganese ions were all hexacoordinated. The thermal decomposition processes of the complexes under air included the loss of water molecule, the pyrolysis of ligands, and the decomposition of inorganic salts, and the final residues were nickel oxide and manganese oxide, respectively.

1. Introduction

Biuret contains two acylamino groups and one imino group, the structure of which determines its value on the synthesis of some complexes as a neutral ligand. In medicine, biuret can be used as pharmaceutical intermediates to preparation hypnotics, sedatives, and some special drugs which have the functions of diuresis and lowering the blood pressure. In chemical industry, biuret plays an important role in the produce of the flame retardants of papers, the fiber bleaching agent, the paint of textiles, the foaming agent of foamed plastics, and the additive agent of the paint, adhesives, resins, plastics, dyes, and lubricating oils, and so on. In agriculture, biuret can be utilized as long-effective fertilizers rich in nitrogen. In animal husbandry, biuret is an excellent nonprotein nitrogen feed additive, and it has a better palatability and higher-usage comparing with urea which is usually used as feed additive. Several teams had synthesized and characterized the complexes of rare earth metals [1], actinide metals [2, 3], and alkaline earth metals [4] based on biuret ligand. However, the complexes of transition metals with biuret ligand have been rarely reported [5, 6], particularly in the comparison between different metal ions on the synthesis of the biuret complexes. With the rapid development of animal industry in China, the prospects of biuret complexes which are used as feed additives for ruminants are considerable. The complexes of trace elements with biuret which are added in the feed of ruminants can play a dual role in supplementing both trace elements and nonprotein nitrogen, which can promote the growth of animals and improve the economic efficiency. Here we report the synthesis of four biuret complexes, study their optical properties, and characterize them by elemental analyses, UV, FTIR, Raman spectra, X-ray powder diffraction, and thermogravimetric analysis.

2. Experimental

2.1. Materials and Physical Measurements. All chemicals purchased were of analytical reagent grade and used without further purification. Nickel acetate, nickel chloride, manganese acetate, manganese chloride, and biuret were purchased from Sinopharm Chemical Reagent Co. Ltd. of Shanghai.

Elemental analyses for C, H, N, and O in the complexes were measured on a Vario EL CUBE elemental analyzer, and the content of nickel and manganese was determined by EDTA complexometric titration with murexide and chrome black T as indicators, respectively. UV spectra were performed on a UV-3150 spectrophotometer. IR spectra were

TABLE 1: Elemental analysis results of the complexes (calculated values are in brackets).

Complex	Formula	M_r	$w(M)/\%$	$w(C)/\%$	$w(H)/\%$	$w(O)/\%$	$w(N)/\%$
1	$NiC_8H_{22}O_{11}N_6$	436.97	13.29 (13.43)	22.03 (21.97)	5.17 (5.09)	40.36 (40.28)	19.15 (19.23)
2	$NiC_4H_{10}O_4N_6Cl_2$	335.75	17.62 (17.48)	14.48 (14.30)	2.96 (3.01)	18.99 (19.06)	24.92 (25.04)
3	$MnC_8H_{19}O_{9.5}N_6$	406.19	10.79 (13.53)	23.19 (23.63)	3.69 (4.72)	43.34 (37.42)	28.79 (20.70)
4	$MnC_4H_{10}O_4N_6Cl_2$	332.00	16.42 (16.55)	14.39 (14.46)	2.98 (3.04)	19.35 (19.28)	24.41 (25.32)

obtained with KBr pellets on a Nicolet 5700 FT-IR spectrophotometer in the range of 4000–400 cm^{-1}. Raman spectra were recorded on an InVia Laser Raman spectrometer. The powder X-ray diffraction data were collected on a D/max-II X-ray diffractometer with Cu $K_{\alpha 1}$ radiation, the voltage of 35 kV, the current of 60 mA, and the scanning speed of 8° min^{-1}, in the diffraction angle range of 10–80°. The thermogravimetric analysis data were obtained using a SDT Q600 thermogravimetry analyzer in the air atmosphere in the temperature range of 25–800°C with a heating rate of 10°C min^{-1}.

2.2. Synthesis of [Ni(bi)$_2$(H$_2$O)$_2$](Ac)$_2$·H$_2$O (1). Ni(Ac)$_2$·4H$_2$O (2.49 g, 10 mmol) and biuret (2.06 g, 20 mmol) were weighed and dissolved in 80 mL methanol, and the solution was green. The mixed solution was stirred on a magnetic stirrer for about 6 h under reflux reaction. After the solution cooling, the resultant was separated from the reaction mixture by filtration and washed by some methanol and dried in the phosphorus pentoxide desiccator for 1 week. The product was green powder (3.59 g) and the yield was about 82.2%.

2.3. Synthesis of [Ni(bi)$_2$Cl$_2$] (2). Complex 2 was synthesized by the same procedure as that for the synthesis of complex 1 except for using NiCl$_2$·6H$_2$O (2.38 g, 10 mmol) instead of Ni(Ac)$_2$·4H$_2$O as the start reactant. The product was green powder (2.18 g) and the yield was about 64.9%.

2.4. Synthesis of [Mn(bi)$_2$(Ac)$_2$]·1.5H$_2$O (3). The synthesis of complex 3 was similar to that of 1 except that Mn(Ac)$_2$·4H$_2$O (2.45 g, 10 mmol) was used to replace Ni(Ac)$_2$·4H$_2$O. There was a difference that the reaction mixture was cooling in the refrigerator. The solubility of complex 3 at room temperature was larger than that of complexes 1 and 2, and the yield was higher at low temperature. The product was pale pink powder (2.37 g) and the yield was about 58.3%.

2.5. Synthesis of [Mn(bi)$_2$Cl$_2$] (4). The synthetic method of 4 was the same as 3 other than that Mn(Ac)$_2$·4H$_2$O was replaced by MnCl$_2$·4H$_2$O (1.98 g, 10 mmol). The product was pale pink powder (2.00 g) and the yield was about 60.2%.

3. Result and Discussion

3.1. Composition and Property. The results of elemental analyses for the complexes are shown in Table 1. The experimental results coincide with the theoretical calculation, and the composition of the complexes is [Ni(bi)$_2$(H$_2$O)$_2$](Ac)$_2$·H$_2$O, [Ni(bi)$_2$Cl$_2$], [Mn(bi)$_2$(Ac)$_2$]·1.5H$_2$O, and [Mn(bi)$_2$Cl$_2$] (bi

= NH$_2$CONHCONH$_2$), respectively. In order to make sure whether chlorine atoms were coordinated or ionic, a qualitative test was conducted; namely, a few drops of AgNO$_3$ solution were added into the aqueous solution containing complexes 2 and 4; there was no precipitation formation. This indicates that the chlorine atoms are coordinated to the metal ions rather than ionic. The solid complexes are stable in the air, easily dissolved in water, and not easy to absorb moisture. Every Ni(II) ion in complex 1 is coordinated by six oxygen atoms from two biuret molecules and two coordinated water molecules, while Ni(II) and Mn(II) ions in complexes 2 and 4 are coordinated with two chloride ions and four oxygen atoms from two biuret molecules. In contrast, the Mn(II) ion in complex 3 is coordinated by six oxygen atoms from two biuret molecules and two acetate anions. Several complexes in which the acetate anions are coordinated had been reported [7–9]. In the four complexes, six coordinated atoms form an octahedral geometry, and four oxygen atoms presenting quadrilateral from two biuret are in the same plane with the metal ions. The other two coordinated atoms are on both sides of the plane. The metal ions are in the centre of the octahedron. And the octahedron is symmetric; thus, the structures of the complexes are stable.

3.2. UV Spectroscopy Analysis. The UV spectra of the complexes and biuret are shown in Figure 1. It is not difficult to find that there is an absorption peak around 200 nm for biuret as well as the four complexes we have synthesized, which shows a close spectral similarity between the four complexes and biuret. In other words, the UV absorption of the complexes derived from biuret. The very strong absorption at short wavelengths is attributed to π-π^* transitions, originating from the carbonyl groups of biuret ligand. The locations of the UV absorption peaks for these complexes are found to be a bit different, indicating that some changes in the π-electron system of biuret have taken place. The outer electronic structure of Ni^{2+} is 3d^8, and there is no empty 3d orbit. Compared with Ni^{2+}, the outer electronic structure of Mn^{2+} is 3d^5, and there are two empty 3d orbits. As a result, the electrons of Cl$^-$ in 3p orbits can be filled into the empty 3d orbits of Mn^{2+}. Therefore, the UV spectrum of complex 4 differs from those of other studied complexes and the free biuret ligand. In conclusion, the oxygen atoms of the carbonyl groups in biuret molecules are coordinated to the metal ions.

3.3. IR Spectroscopy Analysis. The IR spectra of the four complexes and biuret are shown in Figures 2–4, and the main infrared spectral data of biuret and its complexes are listed in Table 2. In the spectra of the four complexes, two

TABLE 2: Infrared spectra of the ligand and the complexes (cm^{-1}).

bi	[Ni(bi)$_2$(H$_2$O)$_2$](Ac)$_2$·H$_2$O	[Ni(bi)$_2$Cl$_2$]	[Mn(bi)$_2$(Ac)$_2$]·1.5H$_2$O	[Mn(bi)$_2$Cl$_2$]	Vibration type
3415	3433, 3379	3383	3406	3406	$\nu_{as}(NH_2) + \nu(H_2O)/\nu_{as}(NH_2)$
3254	3196, 3115	3268, 3198	3216	3264, 3196	$\nu_s(NH_2)$
1719	1686	1693	1693	1685	$\nu(C=O)$
1585	1530	1582	1551	1571	$\delta(NH_2)$
1499, 1423	1445, 1411	1490	1512	1485	$\nu(C-N) + \nu(C-NH_2)$
1323	1343	1336	1336	1329	$\delta(N-H)$
1130	1126	1127	1140	1120	$\nu(C-N) + \delta(N-H)$
1081	1054	1100	1099	1092	$\nu(C-N)$
946	923	943	925	939	$\nu(C-N) + \nu(C-NH_2)$
770	781	760	780	757	$\delta(C-NH_2)$
710	646	646	634	621	$\delta(C=O)$

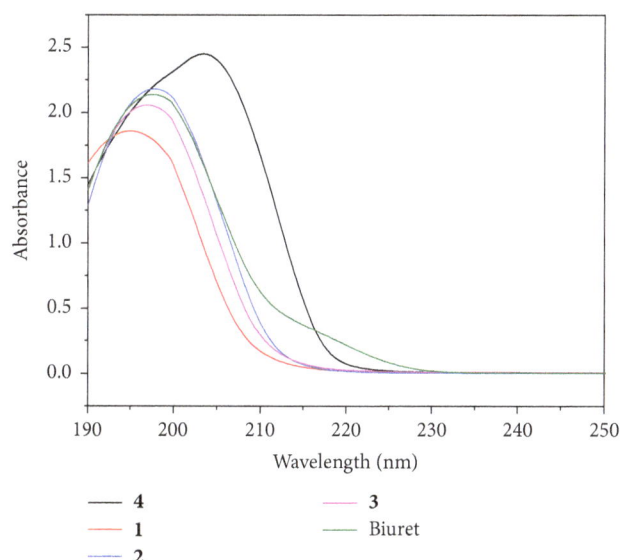

FIGURE 1: UV spectra of the complexes and biuret.

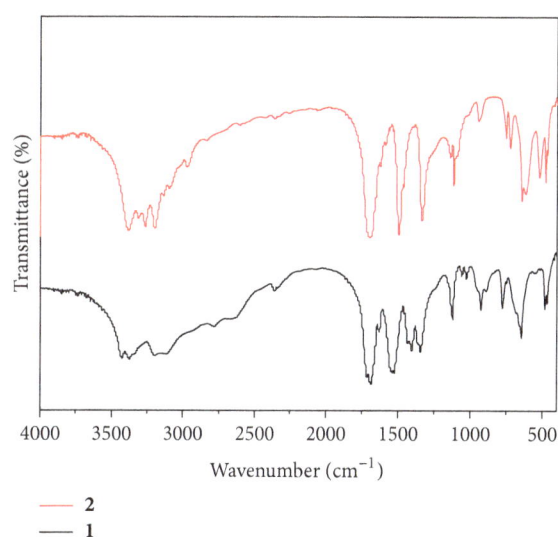

FIGURE 2: IR spectra of complexes 1 and 2.

bands (3115–3430 cm^{-1}) are observed in the N–H stretching region. The former one is alternatively assigned to bridging hydroxide whereas the water of hydration band appears near 3400 cm^{-1} in complexes 1 and 3. The N–H deformation vibrations are observed at 1530 cm^{-1} with significant intensities. The carbonyl stretching frequencies in compounds containing the CO–NH–CO group are reported to give rise to two bands [10, 11], the asymmetric stretching vibration peak appears above 1700 cm^{-1}, and the symmetric vibration peak appears near 1700 cm^{-1}. When coordination occurs it determines the amount of electron delocalization in the N–CO–N system; thus, coordination through the oxygen atom will produce a decrease in the double bond character of the C=O bonds and conduct a shift of the carbonyl stretching mode to lower frequencies [12]. The stretching vibration peaks of the C=O bonds in the complexes are detected at 1686, 1693, 1693, and 1685 cm^{-1}, respectively, which are a little lower than the frequency of biuret (1719 cm^{-1}). It is

believed that the unprotonated biuret M(II) (M = Ni, Mn) complexes should have the M–O coordinated bonds. On the other hand, the bending vibration peaks of the C=O bonds found in the region of 621–664 cm^{-1} are the evidence of the coordination between metal ions and oxygen atoms in the ligand as well. In the nickel complexes, the frequency of the stretching vibration peak of the C=O bonds in complex 1 containing acetate anions is a little lower than that of complex 2 containing chloride anions. However, in the manganese complexes, it is higher in complex 3 containing acetate anions than that of complex 4 containing chloride anions. Compared with complex 1, the stretching vibration peak of the C=O bonds in complex 3 moves to high wavenumber, which can illustrate that both the oxygen atoms in biuret molecules and acetate anions are coordinated to the manganese ions in complex 3. The absorption peaks at 476 cm^{-1} in complexes 1 and 2 are originated to the stretching vibrations of the Ni–O bonds, and the absorption peaks at 459 cm^{-1} in complexes 3 and 4 are the characteristic peaks of the Mn–O bonds [13–17].

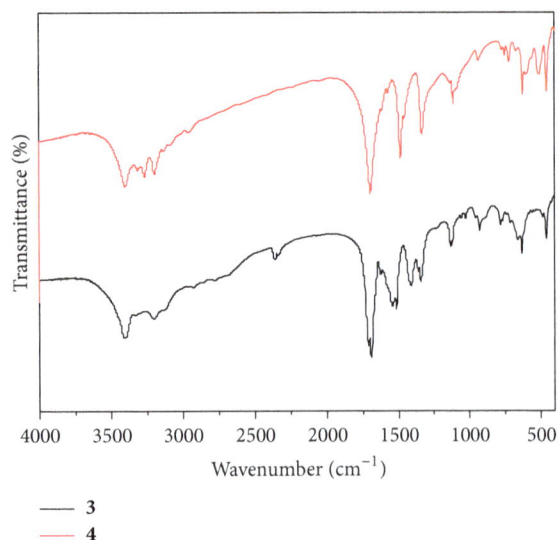

FIGURE 3: IR spectra of complexes **3** and **4**.

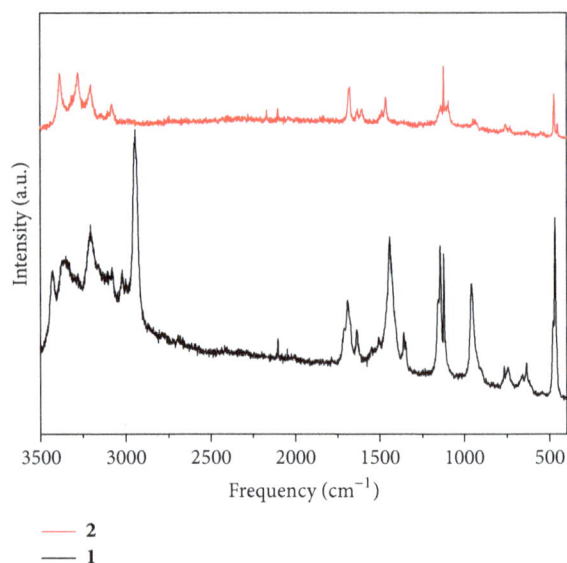

FIGURE 5: Raman spectra of complexes **1** and **2**.

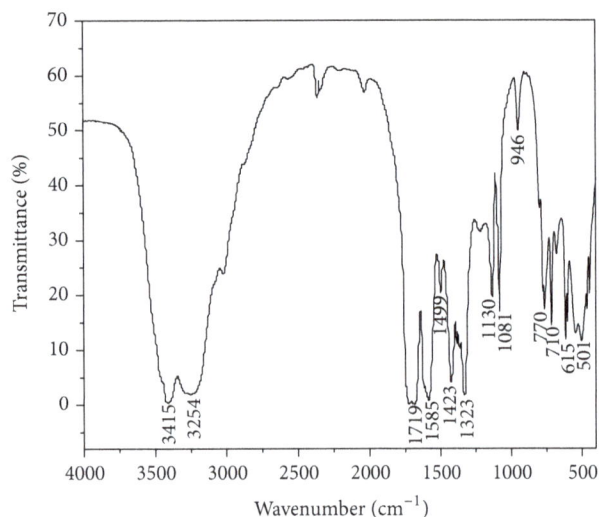

FIGURE 4: IR spectrum of biuret.

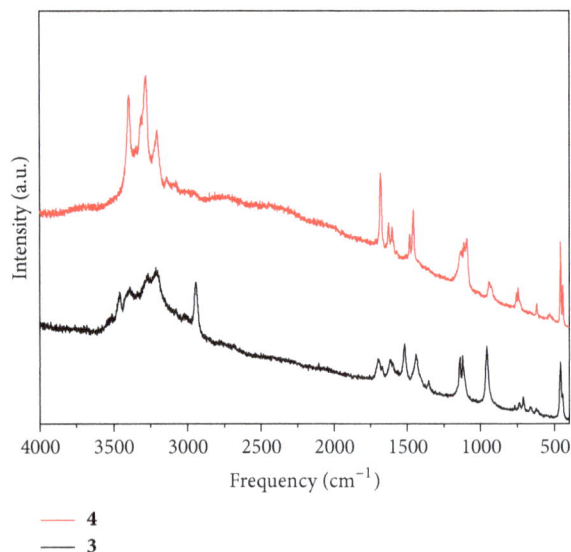

FIGURE 6: Raman spectra of complexes **3** and **4**.

3.4. Raman Spectroscopy Analysis. Figures 5 and 6 show the Raman spectra of the complexes obtained in the 400–4000 cm^{-1} range, and the frequencies data of the biuret and its complexes are listed in Table 3. Obviously, there are many correlations among peaks when comparing the Raman spectra with the IR spectra. For example, the wide absorption peaks in the four complexes from the stretching vibrations of the O–H and N–H bonds appear in the region of 3430–3200 cm^{-1} in the IR spectra as well as in the Raman spectra. But some differences can be discovered that the bending vibration peaks of the N–H bond near 1320 cm^{-1} in the Raman spectra are found to be quite weak and even cannot be found comparing with those in the IR spectra. The result may be explained that the N–H bond in biuret is a polar bond, and it is Raman negative. It is well known that when the interaction between the metal cations (M^{2+}) and the coordinated water molecules is strong enough, a Raman

band due to the symmetric stretching vibration of M–OH$_2$ is observed in the low-frequency region from 300 to 550 cm^{-1} [18–22]. The absorption peak at 467 cm^{-1} for complex **1** is assigned to the symmetric M–OH$_2$ stretching vibration.

3.5. X-Ray Powder Diffraction Analysis. X-ray powder diffraction (XRD) is measured to confirm the phase purity of the samples. The XRD patterns of the complexes and the ligand are shown in Figures 7–9. The background of the XRD patterns is small and the diffractive intensity is strong, indicating that the complex has a fine crystalline state. The strong peak locations of the complexes are shown in Table 4, which are obviously changed comparing with biuret and nickel acetate (JCPDS 26-1282) or nickel chloride (JCPDS 22-0765), manganese acetate (JCPDS 29-0879),

TABLE 3: Raman spectra of the complexes (cm^{-1}).

bi	$[Ni(bi)_2(H_2O)_2](Ac)_2 \cdot H_2O$	$[Ni(bi)_2Cl_2]$	$[Mn(bi)_2(Ac)_2] \cdot 1.5H_2O$	$[Mn(bi)_2Cl_2]$	Vibration type
3370	3433	3382	3398	3398	$\nu_{as}(NH_2) + \nu(H_2O)/\nu_{as}(NH_2)$
3230	3203	3276	3210	3282	$\nu_s(NH_2)$
1690	1691	1685	1698	1685	$\nu(C=O)$
1440	1439	1467	1441	1460	$\nu(C-N) + \nu(C-NH_2)$
1150	1134	1123	1129	1129	$\nu(C-N) + \delta(N-H)$
965	955	938	956	936	$\nu(C-N) + \nu(C-NH_2)$
775	750	750	709	748	$\delta(C-NH_2)$
647	638	627	618	619	$\delta(C=O)$
—	467	473	456	463	$\nu(M-O)$

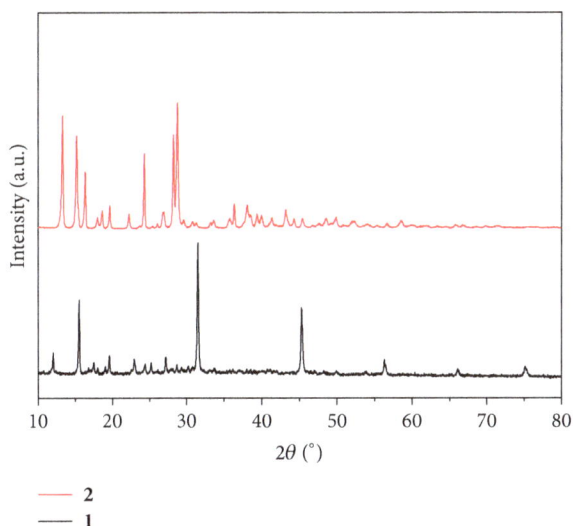

FIGURE 7: X-ray powder diffraction patterns of complexes 1 and 2.

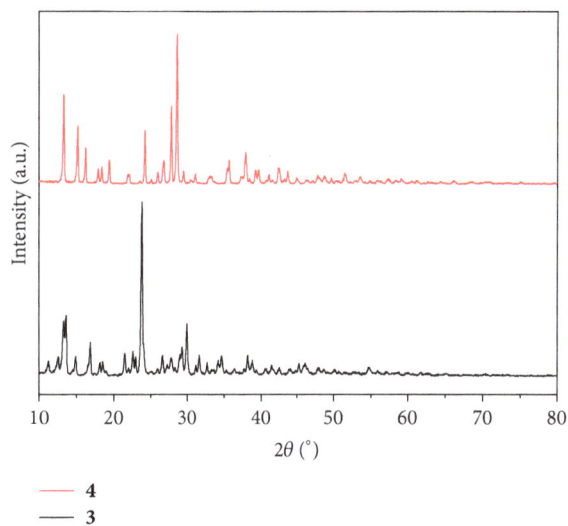

FIGURE 8: X-ray powder diffraction patterns of complexes 3 and 4.

and manganese chloride (JCPDS 22-0721). All these strong peaks of the reactants are disappeared in the X-ray powder diffraction patterns of the complexes. The diffraction angle (2θ), diffractive intensity, and spacing (d) of the products are completely different from the reactive materials, which may illuminate that the resultants are new compounds instead of the reactant mixture [23].

3.6. Thermogravimetric Analysis.

The thermal behavior of the four complexes is studied from 25°C to 800°C under air. The TG-DTG curves are shown in Figures 10–13. The TG analysis (Figure 10) reveals that complex 1 is decomposed through four major processes, namely, the loss of lattice water molecules, the coordinated water molecules, and the combustion of biuret ligand and nickel acetate. The first weight loss is approximately 4.27% (calcd. 4.12%) in the range of 64–105°C, corresponding to the weight of one lattice water molecule. There is a weight loss of 8.03% near 137°C for complex 1, which is ascribed to the loss of two coordinated water molecules, and the measured value is in agreement with the calculated one (8.25%). The third weight loss occurs between 153°C and 514°C and is characteristic of the combustion of biuret ligand (found 47.58%, calcd. 47.17%).

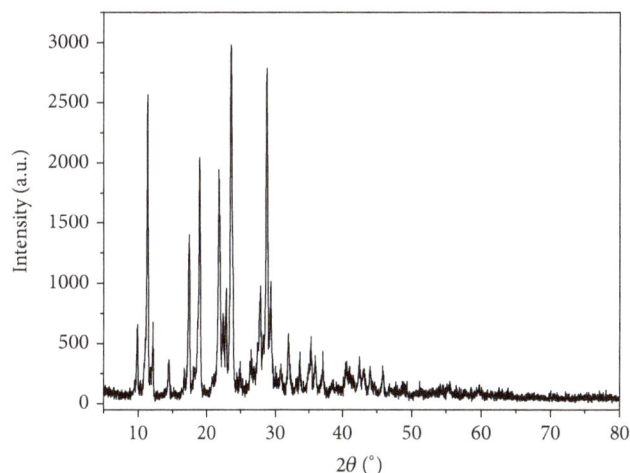

FIGURE 9: X-ray powder diffraction pattern of biuret.

The last weight loss is considered to be the decomposition of nickel acetate (found 23.25%, calcd. 23.36%). As a result, the final residue is nickel oxide. As shown in Figure 11, there are two large weight losses of 60.67% (calcd. 61.40%, loss of two

TABLE 4: The data for X-ray powder diffraction patterns of the complexes and biuret.

Compound	The $2\theta(°)$ locations of the main strong peaks (relative intensities are in brackets)				
bi	23.51 (100)	28.71 (94)	11.30 (86)	18.88 (69)	21.83 (65)
$[Ni(bi)_2(H_2O)_2](Ac)_2 \cdot H_2O$	31.50 (100)	15.45 (60)	45.18 (56)	12.06 (24)	19.51 (22)
$[Ni(bi)_2Cl_2]$	28.68 (100)	13.18 (90)	28.23 (75)	15.19 (74)	24.27 (61)
$[Mn(bi)_2(Ac)_2] \cdot 1.5H_2O$	23.78 (100)	13.52 (35)	13.13 (32)	29.92 (30)	16.86 (20)
$[Mn(bi)_2Cl_2]$	28.45 (100)	13.13 (60)	27.58 (49)	14.99 (39)	24.12 (36)
$Ni(Ac)_2 \cdot 4H_2O$	12.93 (100)	18.62 (30)	28.34 (20)	22.13 (19)	21.09 (13)
$NiCl_2 \cdot 6H_2O$	15.26 (100)	36.24 (79)	30.08 (30)	51.78 (18)	52.58 (17)
$Mn(Ac)_2 \cdot 4H_2O$	9.068 (100)	11.59 (32)	12.18 (25)	26.20 (8)	27.55 (8)
$MnCl_2 \cdot 4H_2O$	17.93 (100)	20.03 (75)	30.67 (75)	34.54 (45)	16.04 (40)

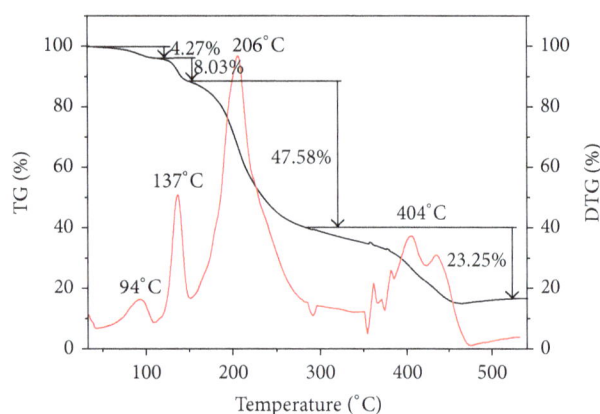

FIGURE 10: TG-DTG curves of complex **1**.

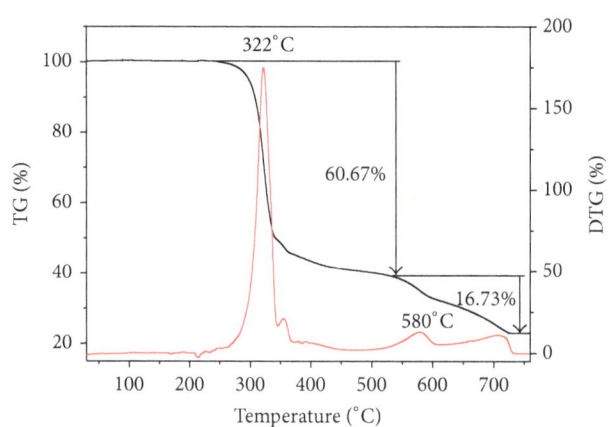

FIGURE 11: TG-DTG curves of complex **2**.

biuret molecules) and 16.73% (calcd. 16.35%, decomposition of nickel chloride) for complex **2**, and the residual weight in the TG curve is 22.60%, which agrees with the theoretical value (22.25%), and the final residue is determined as nickel oxide. Complex **4** is similar to complex **2**. There are two stages weight losses with the increasing of the temperature, namely, the combustion of biuret ligand (found 61.88%, calcd. 62.10%) and the oxidation of manganese chloride (found 16.24%, calcd. 16.53%), and the final residue is MnO (found 21.88%, calcd. 21.37%). Compared with three other complexes, complex **3** undergoes three weight losses of 6.85%, 49.65%, and 25.5% near 104°C, 215°C, and 485°C, respectively, and the first two weight losses agree with the loss of $1.5H_2O$ (calcd. 6.65%) and two biuret molecules (calcd. 49.65%). The last weight losses are considered to be the decomposition of manganese acetate (found 25.40%, calcd. 25.14%) and the final residue is MnO (found 18.10%, calcd. 17.46%). All in all, the thermal behavior of the complexes corresponds to their composition.

4. Conclusion

In summary, the complexes $[Ni(bi)_2(H_2O)_2](Ac)_2 \cdot H_2O$ (**1**), $[Ni(bi)_2Cl_2]$ (**2**), $[Mn(bi)_2(Ac)_2] \cdot 1.5H_2O$ (**3**), and $[Mn(bi)_2Cl_2]$ (**4**) were successfully synthesized with nickel acetate or nickel chloride, manganese acetate, manganese chloride, and

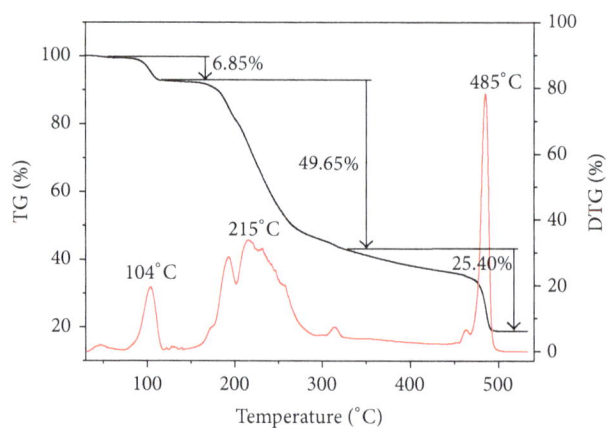

FIGURE 12: TG-DTG curves of complex **3**.

biuret as raw materials. The four complexes of Ni(II) and Mn(II) were hexacoordinated. It was special that the acetate anions in complex **3** were coordinated to the Mn(II) ion, which was not common in most complexes. The optical properties of the complexes were studied via UV, FTIR, Raman spectra, and X-ray powder diffraction. The FTIR spectra were complementary to the Raman spectra, and the structures of the complexes were further verified. The thermal analysis results showed that the decomposition of complexes **1**

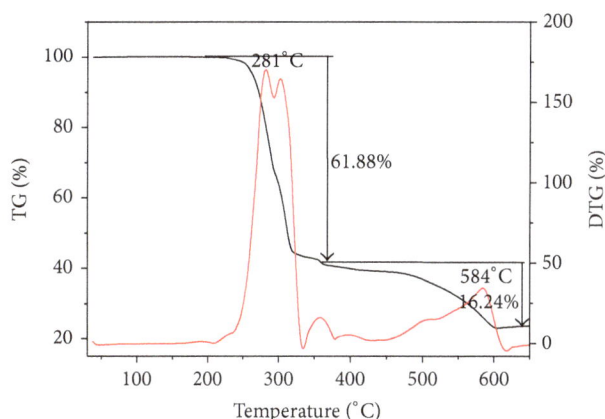

FIGURE 13: TG-DTG curves of complex **4**.

and **3** contained the loss of water molecules, the oxidation and decomposition of biuret, and the oxidation of inorganic salts. The thermal decomposition processes of complexes **2** and **4** are thought to be only the oxidation and decomposition of biuret and the oxidation of inorganic salts, in which there is no water molecule. The final residues were NiO for complexes **1** and **2** and MnO for complexes **3** and **4**.

Conflict of Interests

The authors declare that there is no conflict of interests regarding the publication of this paper.

Acknowledgments

This work was supported by the Scientific Research Funds of Education Department of Sichuan Province (10ZA016). The authors are very grateful to Analytical and Testing Center of Southwest University of Science and Technology and Engineering Research Center of Biomass Materials of Education Ministry for the testing of elemental analyses, XRD, UV, FTIR, Raman spectra, and TG-DTG.

References

[1] Y. L. Zhai, N. Tang, M. Y. Tan, G. M. Yu, and D. B. Wang, "Synthesis and properties of solid complexes between lanthanide nitrates and biuret," *Chinese Journal of Inorganic Chemistry*, vol. 2, no. 3, pp. 88–95, 1986.

[2] N. Tang, M. Y. Tan, Y. Li Zhai, and K. M. Wang, "Synthesis and characterization of the solid complex of thorium nitrate with biuret," *Journal of Lanzhou University*, vol. 22, no. 1, pp. 90–94, 1986.

[3] N. Tang, M. Y. Tan, Y. L. Zhai, and K. M. Wang, "Solid coordination compound of uranyl nitrate with biuret," *Chinese Journal of Applied Chemistry*, vol. 3, no. 4, pp. 13–16, 1986.

[4] S. Haddad and P. S. Gentile, "The crystal and molecular structure of tetrakis(biuret)strontium(II) perchlorate," *Inorganica Chimica Acta*, vol. 12, no. 1, pp. 131–138, 1975.

[5] G.-Q. Zhong and R.-Q. Zeng, "Synthesis of the complex of potassium bis(biureto) cuprate eetrahydrate by solid phase reaction," *Chemical Research and Application*, vol. 14, no. 4, pp. 461–462, 2002.

[6] G.-Q. Zhong and R.-Q. Zeng, "Syntheses of the copper(II) bomplexes of biuret by solid phase reaction with microwave radiation," *Chinese Journal of Inorganic Chemistry*, vol. 18, no. 8, pp. 849–853, 2002.

[7] A. I. Fischer, D. O. Ruzanov, M. Y. Gorlov, A. V. Shchukarev, A. N. Belyaev, and S. A. Simanova, "Synthesis, crystal and molecular structures of the octanuclear cationic mixed-valence cobalt acetate complex," *Russian Journal of Coordination Chemistry*, vol. 33, no. 11, pp. 789–794, 2007.

[8] T. Sato and F. Ambe, "An oxo-centered trinuclear cobalt(II)-diiron(III) acetate-aqua complex," *Acta Crystallographica Section C*, vol. 52, pp. 3005–3007, 1996.

[9] G. Liu, B. Chen, and D. X. Chen, "Synthesis, crystal structure and ferromagnetic properties of a novel acetate bridged dicadmium(II) complex with nitronyl nitroxide," *Russian Journal of Coordination Chemistry*, vol. 37, no. 10, pp. 738–742, 2011.

[10] T. Uno and K. Machida, "Infrared spectra of succinimide and maleimide in the crystalline state," *Bulletin of the Chemical Society of Japan*, vol. 35, no. 2, pp. 276–283, 1962.

[11] C. S. Kraihanzel and S. C. Grenda, "Acyclic imides as ligands. I. Diacetamide complexes of manganese(II), iron(II), cobalt(II), nickel(II), copper(II), and zinc(II) perchlorates," *Inorganic Chemistry*, vol. 4, no. 7, pp. 1037–1042, 1965.

[12] M. R. Udupa and V. Indira, "Biuret complexes of copper(II) and nickel(II)," *Journal of Indian Chemical Society*, vol. 52, pp. 585–588, 1975.

[13] J. Dharmaraja, P. Subbaraj, T. Esakkidurai, and S. Shobana, "Coordination behavior and bio-potent aspects of Ni(II) with 2-aminobenzamide and some amino acid mixed ligands—part II: synthesis, spectral, morphological, pharmacological and DNA interaction studies," *Spectrochimica Acta Part A: Molecular and Biomolecular Spectroscopy*, vol. 132, pp. 604–614, 2014.

[14] K. R. Sangeetha Gowda, H. S. Bhojya Naik, B. Vinay Kumar et al., "Synthesis, antimicrobial, DNA-binding and photonuclease studies of Cobalt(III) and Nickel(II) Schiff base complexes," *Spectrochimica Acta Part A*, vol. 105, pp. 229–237, 2013.

[15] M. S. Refat, "Spectroscopic and thermal degradation behavior of Cr(III), Mn(II), Fe(III), Co(II), Ni(II), Cu(II) and Zn(II) complexes with thiopental sodium anesthesia drug," *Journal of Molecular Structure*, vol. 1037, pp. 170–185, 2013.

[16] B. Cabir, B. Avar, M. Gulcan, A. Kayraldiz, and M. Kurtoglu, "Synthesis, spectroscopic characterization, and genotoxicity of a new group of azo-oxime metal chelates," *Turkish Journal of Chemistry*, vol. 37, no. 3, pp. 422–438, 2013.

[17] A. A. Soayed, H. M. Refaat, and D. A. Noor El-Din, "Metal complexes of moxifloxacin-imidazole mixed ligands: characterization and biological studies," *Inorganica Chimica Acta*, vol. 406, pp. 230–240, 2013.

[18] R. J. H. Clark and R. E. Hester, *Advances in Infrared and Raman Spectroscopy*, vol. 12, John Wiley & Sons, New York, NY, USA, 1985.

[19] R. R. Dogonadze, E. Kalman, A. A. Kornyshev, and J. Ulstrup, *The Chemical Physics of Solvation. Part B: Spectroscopy of Solvation*, Elsevier Science, Amsterdam, The Netherlands, 1986.

[20] H. Kanno, "Hydrations of metal ions in aqueous electrolyte solutions: a Raman study," *Journal of Physical Chemistry*, vol. 92, no. 14, pp. 4232–4236, 1988.

[21] H. Kanno and H. Kanno, "Correlation of the Raman ν_1 bands of aquated divalent metal ions with the cation-hydrated water

distance," *Journal of Raman Spectroscopy*, vol. 18, pp. 301–304, 1987.

[22] H. Kanno and J. Hiraishi, "A Raman study of aqueous solutions of ferric nitrate, ferrous chloride and ferric chloride in the glassy state," *Journal of Raman Spectroscopy*, vol. 12, no. 3, pp. 224–227, 1982.

[23] D. Li and G.-Q. Zhong, "Synthesis and crystal structure of the bioinorganic complex [Sb(Hedta)]·2H2O," *Bioinorganic Chemistry and Applications*, vol. 2014, Article ID 461605, 7 pages, 2014.

16

Digital Image Encryption Algorithm Design Based on Genetic Hyperchaos

Jian Wang[1,2,3]

[1]*Graduate School, Yanshan University, Qinhuangdao 066004, China*
[2]*School of Information Science and Engineering, Yanshan University, Qinhuangdao 066004, China*
[3]*The First Hospital of Qinhuangdao, Qinhuangdao 066000, China*

Correspondence should be addressed to Jian Wang; dupeng198510@163.com

Academic Editor: Chenggen Quan

In view of the present chaotic image encryption algorithm based on scrambling (diffusion is vulnerable to choosing plaintext (ciphertext) attack in the process of pixel position scrambling), we put forward a image encryption algorithm based on genetic super chaotic system. The algorithm, by introducing clear feedback to the process of scrambling, makes the scrambling effect related to the initial chaos sequence and the clear text itself; it has realized the image features and the organic fusion of encryption algorithm. By introduction in the process of diffusion to encrypt plaintext feedback mechanism, it improves sensitivity of plaintext, algorithm selection plaintext, and ciphertext attack resistance. At the same time, it also makes full use of the characteristics of image information. Finally, experimental simulation and theoretical analysis show that our proposed algorithm can not only effectively resist plaintext (ciphertext) attack, statistical attack, and information entropy attack but also effectively improve the efficiency of image encryption, which is a relatively secure and effective way of image communication.

1. Introduction

With the rapid development of Internet technology and information technology, digital communication is more and broader: people can release on the Internet all kinds of information anytime and anywhere. Digital image is the most intuitive, visual, and abundant information carrier, due to its convenience, speed, lack of geographical restrictions, low cost, high efficiency, and so forth; it has been more widely used and has become one of the main information network era expressions. However, people enjoy all sorts of convenience brought about by the digital image but also face some difficult security problems, such as personal privacy protection, business and military information protection, and electronic products illegal copying and dissemination. So how to protect digital image in the transmission process has become the focus of the industry.

In order to protect security of images which contain data and information, we use the original image that is encrypted to resolve the security hidden danger, so the image encryption research has become a hot research topic in the field of image analysis and processing. In general, the conventional image encryption mainly has image encryption in the spatial domain, transform domain of image encryption, image encryption based on neural network and based on chaotic image encryption. Spatial domain image encryption basically has the following two ways: one is scrambling by changing the position relationship of each image's pixels [1]. The other is to use certain encryption rules that change the pixel values of the original image and make the information entropy close to the maximum, namely, information entropy encryption [2]. Image scrambling encryption scheme is mainly used in digital image security process of pretreatment and posttreatment stage, so as to further guarantee the security of information contained in image; it can be used as a special digital image encryption method, but it is vulnerable to be attacked just by image scrambling encryption by statistical analysis; it cannot solve the information contained in the original image security problems. Image replacement and diffusion both change the relevance of the original image, making the information entropy change. And image diffusion is based on image correlation transformation among adjacent pixels according to certain rules, but it may cause some image information loss. Transform domain image encryption is mainly through some sort of orthogonal transformation on the image; then, it is encrypted when it is coding processing. Like image encryption based on tree structure [3] and image encryption based on SCAN language [4], these image encryption schemes involve the problem of how to generate pseudorandom sequences; now the problem has no good solution. By using neural network with the parallel distributed processing, highly nonlinear association memory [5], and other characteristics, to encrypt the image information, we call it artificial neural network image encryption. But the neural network needs a lot of neurons data to encrypt, because it cannot be adaptive to generate neural networks, then increases complexity of encryption, and reduces efficiency of encryption. In order to solve this problem, the related research scholars put forward using chaos theory to encrypt digital image, because the chaos theory [6] is sensitive to initial conditions and system parameters extremely, randomness of trajectory, pseudorandomness and ergodicity, and other special complex dynamics properties, making it very suitable for digital image encryption, forming a kind of chaotic image encryption algorithm. For example, Fridrich in 1998 for the first time introduced chaos theory as digital image encryption and put forward the two-dimensional Baker map image encryption [7] for image scrambling operation and then extended it to 3d. Due to 3d, Baker map image encryption efficiency is low, so Chen et al. put forward 3d Baker map [8] fast image encryption algorithm, which makes the security of encryption and efficiency improve to a certain extent. In [9], a standard map image encryption algorithm is proposed; then Wong et al. proposed on the basis of Baptista algorithm an improved fast image encryption algorithm [10]. At the same time, the document [11–17] was also, respectively, proposed by using Logistic mapping, Tent, Lorenz system, and one-way coupled map lattice, such as a variety of chaotic mapping algorithm for digital image encryption. Though these chaotic image encryption algorithms to a certain extent improve the security performance, but due to the varying complex degrees of the image, causing them to fail to solve the problem of encryption efficiency, it well often fail to solve the problem of security threats to a certain extent.

But genetic algorithm (GA) [18, 19] may solve this problem which provides a feasible technical way. Genetic algorithm is proposed by Holland of Michigan university in the United States [18] in the 1960s, and then in the late 80s Goldberg [19] summarized the basis of predecessors' research, finally forming the basic theoretical framework of genetic algorithm. Genetic algorithm is a new global optimization search algorithm, because it has the characteristics of group search technology; it can represent a set of solutions using the population and then through the operation of the selection, crossover and mutation of species, and so on finally gets a new generation of population, so that gradually makes the population evolution to the optimal solution or near optimal solution evolving, getting the best state of population. Because optimization constraints are less and objective function and constraint condition requirements are low, its actual operation is simple and practical, suitable for optimization. At the same time, its search is in the whole solution space and also is most likely to look for an optimal solution or approximate optimal solution; because it has such advantages, it is widely applied to aviation system [20], machine learning [21], pattern recognition [22], and so forth.

To this end, in this paper, we propose a new image encryption algorithm, which uses genetic algorithm optimization features and pseudorandomness and ergodicity of chaotic theory to solve the image features such as encryption of security threats and inefficiencies. First it goes through the chaos theory of image encryption, in the encryption process, using the genetic algorithm to carry out adaptive optimization on encryption process design parameters and then get the best encryption parameters, so that it will solve the problem of encryption security threats and inefficiencies, to achieve efficient, reliable, and secure encrypted image. Experimental analysis shows that this method not only can solve the problem of security threats, which improves the security performance, and can solve the problem of encryption efficiency but also greatly improves the efficiency of encryption, namely, to realize image encryption effect and receive a significant boost in performance.

2. The Basic Principle of Genetic Algorithm

2.1. The Basic Concept of Genetic Algorithm and Steps. Genetic algorithm is different from traditional search algorithm; it first randomly generates a set of initial solutions, namely, "population," where each individual in population, namely, a solution vector, which is called "chromosomes," begins the search process. These chromosomes evolve in the subsequent iterations and generate the next generation of chromosomes, called "offspring". The stand or fall of chromosomes in each generation through chromosome "fitness" is evaluated: a chromosome of higher fitness is more likely to be selected; instead, the possibility of fitness small chromosome which is selected is smaller, the selected chromosome by cross-generating (crossover and mutation) new chromosomes, "offspring." It passes through several generations, the algorithm converges to the best chromosome, and the chromosomes are likely to be the optimal solution of the problem or approximate optimal solution. The operation of the genetic algorithm steps is shown as follows:

(1) Randomly generate initial population pop(k).
(2) Go through fitness function to evaluate the chromosomes.
(3) Select chromosomes according to the fitness level and form a new population.
(4) Go through crossover and mutation operation which produces new chromosomes that *offspring*.
(5) Repeat steps (2)–(4), until getting the scheduled evolution algebra.

This method is shown in Figure 1; it is shown that the genetic algorithm is mainly composed of genetic operators

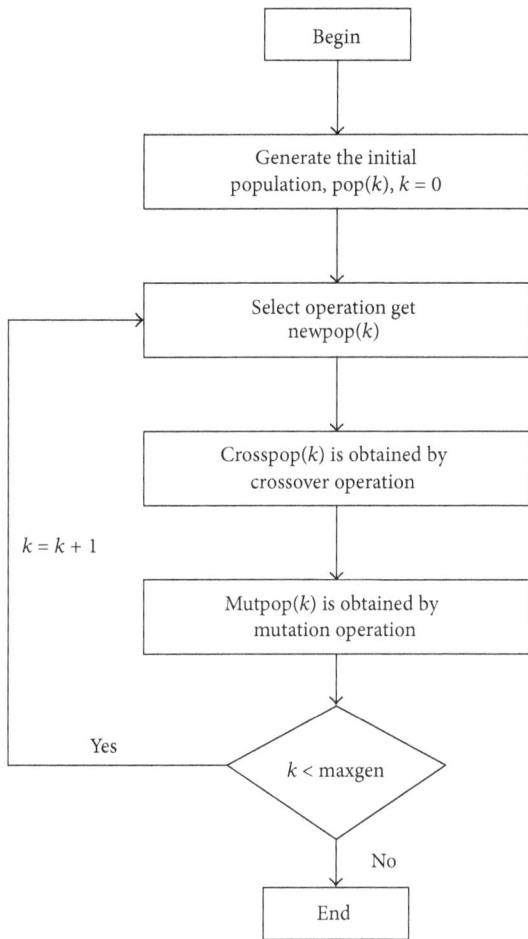

FIGURE 1: The flow chart of genetic algorithm iteration.

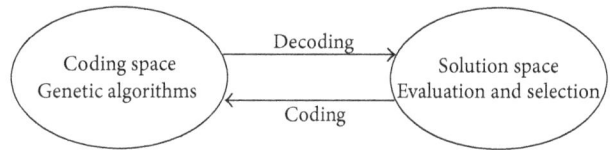

FIGURE 2: Coding space and solution space.

coding space and solution space and its genetic operation on chromosome in the coding, and it evaluates and chooses the solution in the solution space. The bridge between them is the encoding and decoding. Code is converting the solution of the problem space to variable chromosome of genetic space. On the contrary, the decoding is the operation where the chromosome coding is mapped in the problem space. The relationship between them is shown in Figure 2.

The original genetic algorithm uses the Holland coding scheme, namely, binary code. But for the application of many genetic algorithms, such as multidimensional, high accuracy algorithms, especially in the optimization of complex system, the simple encoding method shows many drawbacks: it cannot directly reflect the structure of the requirement problem To the problem of large range and high precision, the chromosome length and the length of the search space will be very big; such genetic search is very difficult. Adjacent binary code may have larger Hamming distance, so as to make the search efficiency of genetic operators reduced.

Selecting an appropriate coding method is the basis of genetic algorithm to solve practical problems. For the problem of any application, coding must consider the following aspects:

(1) *Completeness*. All the points (candidates) in the problem space can be used as genetic points (chromosome) in genetic space for performance.

(2) *Integrity*. The genetic space chromosome corresponds to all the problems of candidate solution space.

(3) *Nonredundancy*. Chromosomes and the candidate solution are in a one-to-one relationship.

One of the problems we know of is that it is difficult to design the coding scheme that could satisfy the requirement of the above three aspects at the same time, but these designs must meet the requirements of completeness.

2.2.2. Fitness Function. In the evolution process genetic algorithm, the stand or fall of chromosome is evaluated by a fitness function; the fitness function value is the basis for the selection operation. For theoretical analysis of convenience, it is best to guarantee the fitness function value is nonnegative; proper transformation must be adapted to the value of the negative. The relationship between the objective function and fitness function is different according to the optimization problem categories. Optimization problem was divided into two categories, a class of global maximum values for the objective function and another for the objective function of the global minimum. For these two kinds of optimization

(crossover and mutation) and evolutionary computing (select). Genetic algorithm simulates the natural evolution of species turnover mechanism; it generates a new species to reach the purpose of searching the global optimal solution. Its evolutionary computation is through the competition mechanism constantly updating population process.

Crossover operation is the main genetic algorithm; the performance of genetic algorithm depends largely on the performance of its adopted crossover operation. Crossover operation operates the two chromosomes at the same time, combining the two features to produce new offspring. Variation is a basic operation; it spontaneously generates random variation on chromosome. Variation can offer gene which is not contained in the initial population or find missing gene in the selection process, providing new content for population.

2.2. The Application Design of Genetic Algorithm Design

2.2.1. Encoding Problem. Genetic algorithm through genetic operators (crossover and mutation) restructures individual in the population; by selecting operation, it constantly optimizes the individual structure and searches the optimal structure of the individual and, finally, achieves the goal of becoming closer to problem of the optimal solution. It can be seen that the genetic algorithm is the process of alternating work in the problem, a certain point of the objective function in the solution space converts to the fitness function of corresponding individual in search space, shown as follows.

For the problem of the maximum, one has the following:

$$F(X) = \begin{cases} f(X) + C_{\min}, & \text{if } f(X) + C_{\min} > 0 \\ 0, & \text{if } f(X) + C_{\min} \leq 0, \end{cases} \quad (1)$$

where $C_{\min}$ is an appropriate number of relatively small value.

The problem of the minimum is shown as follows:

$$F(X) = \begin{cases} C_{\max} - f(X), & \text{if } f(X) < C_{\max} \\ 0, & \text{if } f(X) \geq C_{\min}, \end{cases} \quad (2)$$

where $C_{\max}$ is an appropriate number of relatively big value.

The mapping relationship formula between the objective function and fitness function has other forms. Objective function converting into fitness function normally needs to follow two principles: (1) the objective function in the optimization process of the optimization direction (e.g., seeking the maximum or minimum value of the objective function) and in the process of population evolution fitness function value is increasing is in the same direction; (2) fitness function value must be greater than or equal to zero. In practical problems, we adopt the kind of conversion form according to the specific circumstances.

2.2.3. Selection Problem. Select operation is the direct driving force of evolution. Selection pressure is an implicit rule, where pressure is too large; the search will be prematurely terminated; Pressure is too small as the search will be very slow. Options include three basic aspects: sample space, the sampling mechanism, and selection probability.

(1) Sample Space. The size and the constitution of the sample space constitute the sample space. The sample space is divided into two: the regulatory sample space and the expanding sample space.

Put *popsize* as the size of the population and *popsize* 1 as the offspring size after crossover and mutation; the regulatory sample space which is to keep the population size remains the same. The choice diagram based on regulatory sample space is shown in Figure 3.

The expanding sample space is *popsize* + *popsize* 1; namely, sampling space includes all parents and offspring. Figure 4 depicts the choice diagram based on expanding sampling space. The most distinguishing feature of the expanding sample space is effective limiting the random fluctuation caused by high crossover rate and mutation rate.

(2) Sampling Mechanism. Sampling mechanism is a theory on how to choose the chromosomes from sampling space theory. The selection of individual species in principle can be divided into three types: random sampling, determined sampling, and mixed sampling.

(3) Select Probability. Holland's original roulette algorithm adopted by the genetic algorithm is a kind of proportional selection method; selection probability is proportional to the chromosome's fitness.

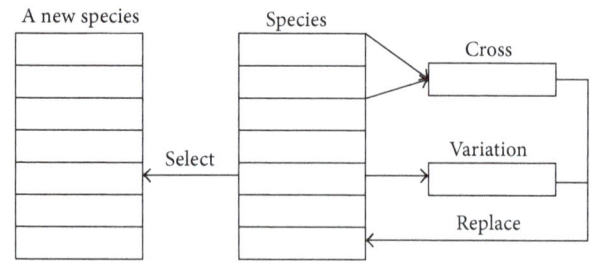

FIGURE 3: The choice diagram based on regulatory sample space.

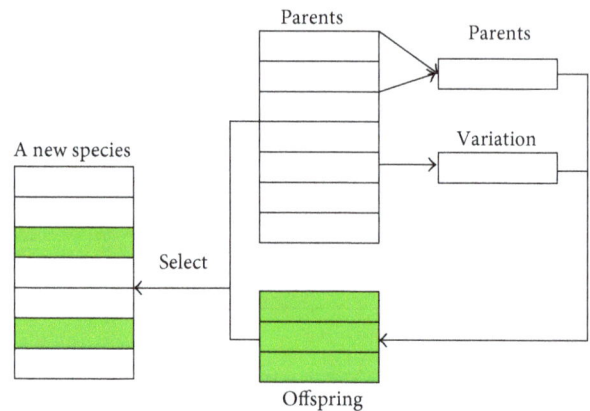

FIGURE 4: The choice diagram based on the expanding sampling space.

2.2.4. Crossover Operation. Crossover operation is the most important genetic operation; the population by cross generates new chromosomes and constantly expands the search space, finally achieving the goal of global search.

2.2.5. Mutation Operation. Mutation operation is changing some genes points of chromosome string.

2.2.6. The Selection of Main Parameters. Parameters in the genetic algorithm design mainly include population size, crossover rate, mutation rate, and evolution algebra. In addition, when choosing a specific operator, sometimes it involves the selecting of the parameters related with operator. The selection of genetic algorithm parameters is reasonable or does not directly relate to the convergence speed and accuracy of the algorithm. However, because there are many factors which can affect the parameter selection, some relate to the problem itself of connotation of objective laws and some relate to the selection operator, so it is difficult to find common rules.

(1) Population Size. Population size is the first parameter of genetic algorithm that needs to be determined; it is the main influencing factor in a local solution of the algorithm.

(2) Cross Rate. Cross rate is the main genetic algorithm; the performance of genetic algorithm depends largely on adopted crossover operator performance and the size of the cross rate.

TABLE 1: Genetic algorithm (GA) trial scope commonly used parameters.

Trial parameters	Population size	Crossover probability	Mutation probability	The biggest evolution algebra
Trial parameters	20–100	0.4–0.9	0.00001–0.1	50–1000

(3) Mutation Rate. Mutation rate refers to the number of the gene variations in a population accounting for the percentage of the total number of genes.

(4) Evolution Termination Conditions. Termination conditions can be controlled from two aspects: preset evolution algebra or control according to the evolution of the population. The evolution of population refers to the relationship between the current generation of maximum adaptation and the population average fitness. The evolution of termination conditions is decided according to specific circumstances.

The common value range of main operation parameters of genetic algorithm is shown in Table 1.

2.2.7. Handling of Illegal Individual Strategy. We usually obtain infeasible offspring when using genetic algorithm on chromosomes, so the core of the solution of nonlinear programming problem using the genetic algorithm is how to meet the problem of constraint. It can be used to reject and fix penalty policy for processing.

3. Hyperchaos Algorithm Basic Principle

3.1. The Definition of Chaos. We assume there exist continuous mappings in the interval I; if it meets the following conditions, then the mapping is called chaotic.

(1) f is the cycle without upper bound.

(2) On the closed interval I, there is uncountable subset S, and it can meet the following conditions:

(i) $\forall x, y \in S, x \neq y, \lim_{n \to \infty} \sup |f^n(x) - f^n(y)| > 0$.

(ii) $\forall x, y \in S, \lim_{n \to \infty} \inf |f^n(x) - f^n(y)| = 0$.

(iii) $\forall x \in S$ and f in any cycle point y, it get $\lim_{n \to \infty} \sup |f^n(x) - f^n(y)| > 0$.

3.2. Hyperchaotic System. Hyperchaos system is described using the following equation:

$$\begin{aligned} x_1 &= a(x_2 - x_1) + x_4 \\ x_2 &= dx_1 - x_1 x_3 + cx_2 \\ x_3 &= x_1 x_2 - bx_3 \\ x_4 &= x_2 x_3 + rx_4, \end{aligned} \quad (3)$$

where a, b, c, d, and r are the control parameters, often taking $a = 35, b = 3, c = 12, d = 7$, under the condition of r in the following range: $[0, 0.085], (0.085, 0.798], (0.798, 0.90]$, the system of chaotic motion, chaotic motion, and periodic motion [23].

4. A Digital Image Encryption Algorithm Based on Genetic Hyperchaos System

4.1. Hyperchaos Randomly Generated Initial Population. The Logistic map gives m different initial values in the model; chaotic variables $x_i, i = 1, 2, \ldots, m$, m a different trajectory, according to formula (4), will make m a chaotic variable, respectively, mapped to the optimization variables within the scope of making it into a chaotic variable $x_{n_i}^*$:

$$x_{n_i}^* = a_i + (b_i - a_i) x_{n_i}. \quad (4)$$

Fixed $n, x_k^* = [x_{k,1}^*, x_{k,2}^*, \ldots, x_{k,m}^*]$ represents a feasible solution. Each feasible solution to calculate the fitness chooses fitness of high N individual initial population. In the process of chaos generation of initial population, chaotic sequence should take enough number of iterations (average value is 400), that is, to ensure that chaotic variables can adequately be traversal.

4.2. Pixel Position Scrambling. Advantages of the characteristics of digital image with digital array are as follows: it is used to analyze the image matrix step finite elementary matrix transformation and disturb the arrangement of image pixel position into a chaotic image and it is impossible to identify the purpose of the original image, which has the effect of image encryption.

Assume that the size of $M \times N$ of the original image $P_{M \times N}$ pixels matrix is expressed as

$$P_{M \times N} = \begin{pmatrix} p_1 & p_2 & \cdots & p_N \\ \cdots & \cdots \cdots & \cdots \\ p_{N(M-1)+1} & \cdots \cdots & p_{MN} \end{pmatrix}. \quad (5)$$

Specific steps are as follows.

(1) The Line Displacement. Transform matrix $P_{M \times N}$ according to the line for the row vector form is

$$\begin{aligned} &P_{M \times N} \\ &= \left(\underbrace{p_1 p_2 \cdots p_{MN/k}}_{P_1} \underbrace{p_{MN/k+1} \cdots p_{2MN/k}}_{P_2} \underbrace{p_{2MN/k+1} \cdots p_{3MN/k}}_{P_3} \right. \\ &\quad \left. \cdots \underbrace{p_{(k-1)MN/k+1} \cdots p_{MN}}_{P_k} \right). \end{aligned} \quad (6)$$

Among them, the k in the parallel system is defined as the number of processors; number of k is defined as a group in a serial system, so each child vector, respectively, is

$$
\begin{aligned}
P_1 &= (p_1 p_2 \cdots p_{MN/k}), \\
P_2 &= (p_{MN/k+1} \cdots p_{2MN/k}), \\
P_3 &= (p_{2MN/k+1} \cdots p_{3MN/k}), \\
P_k &= (p_{(k-1)MN/k+1} \cdots p_{MN}).
\end{aligned} \tag{7}
$$

Using Runge-Kutta algorithm will cause hyperchaos of the system iterations N times, which is used to prevent the transition effect. For a given system, the number of iterations N may be associated with initial conditions and system parameters, $N = 200$ iterations before we throw away the data; then, we calculate

$$
r_k = \mathrm{mod}\left((\mathrm{abs}\,(x_k) - \mathrm{floor}\,(\mathrm{abs}\,(x_k))) \times 10^{14}, \frac{MN}{k} \right) \tag{8}
$$
$$
+ 1.
$$

Obviously, $r_k \in [1, MN/k]$. Iteration continues until this chaotic system produces MN/k completely different fairly r_k values. It is denoted by $\{r_k(i), i = 1, \ldots, MN/k\}$. Based on fairly r_k scrambling for vector P_k, we have

$$
P_k^r(i) = P_k(r_k(i)), \tag{9}
$$

after row scrambling matrix is expressed as P^r.

(2) Replacement. It converts line displacement matrix P^r according to the column to row vector form as

$$
P_{M\times N}^r
$$
$$
= \left(\overbrace{\underbrace{p_1^r p_2^r \cdots p_{MN/k}^r}_{P_1^r} \underbrace{p_{MN/k+1}^r \cdots p_{2MN/k}^r}_{P_2^r} \underbrace{p_{2MN/k+1}^r \cdots p_{3MN/k}^r}_{P_3^r}} \right.
$$
$$
\left. \cdots \underbrace{p_{(k-1)MN/k+1}^r \cdots p_{MN}^r}_{P_k^r} \right). \tag{10}
$$

Among them, each child vector, respectively, is

$$
\begin{aligned}
P_1^r &= p_1^r p_2^r \cdots p_{MN/k}^r, \\
P_2^r &= p_{MN/k+1}^r \cdots p_{2MN/k}^r, \\
P_3^r &= p_{2MN/k+1}^r \cdots p_{3MN/k}^r, \\
P_k^r &= p_{(k-1)MN/k+1}^r \cdots p_{MN}^r.
\end{aligned} \tag{11}
$$

By the same token, we calculate

$$
c_k
$$
$$
= \mathrm{mod}\left((\mathrm{abs}\,(x_k) - \mathrm{floor}\,(\mathrm{abs}\,(x_k))) \times 10^{14}, \frac{MN}{k} \right) \tag{12}
$$
$$
+ 1.
$$

Clearly, $c_k \in [1, MN/k]$. Iteration continues until this chaotic system produces MN/k completely different c_k values, as follows: $\{c_k(i), i = 1, \ldots, MN/k\}$. According to c_k, scrambling for vector P_k^r is shown as follows:

$$
P_k^{rc}(i) = P_k^r(c_k(i)). \tag{13}
$$

P^{rc} is the original matrix P after hyperchaos scrambling matrix; Section 4.3 will introduce how to encrypt the P^{rc}.

4.3. The Diffusion Process of Pixels. First of all, for P' of scrambling image preprocessing, scrambling image P' is divided into P_a' and P_b'. Then, by adaptive image processing of the two parts, the P_a' data matrix and matrix P_b' are different or are replaced matrix operation points; likewise, P_{aa}' and P_b' are exclusive or replace the corresponding matrix operation points and will get the new image matrix P_{aa}' to go up and down. Finally, a random sequence of matrix and matrix PP' processing complete diffusion process. Specific process is shown as follows:

$$
E_{i,j} = \begin{cases} qq_{ij} \oplus \mathrm{mod}\left(x_{m,n} \times 10^{10}, Q\right) & \text{if } i = 1 \text{ or } j = 1, \\ qq_{ij} \oplus \mathrm{mod}\left(x_{m,n} \times 10^{10}, Q\right) \oplus E_{i-1,j} \oplus E_{i,j-1} & \text{else.} \end{cases} \tag{14}
$$

In formula (14), the radius is binary x or operations, $i = m = 1, 2, \ldots, M$ and $j = m = 1, 2, \ldots, N$ are image pixels and the space coordinates of chaotic state value, $E_{i,j}$ is the final image encryption, $qq_{i,j}$ is the adaptive processing of the image pixels, $x_{m,n}$ is the status value of chaos, and Q is digital image grayscale. Formula (14) corresponding to the inverse operation is as follows:

$$
E_{i,j} = qq_{ij} \oplus \mathrm{mod}\left(x_{m,n} \times 10^{10}, Q\right)
$$
$$
\text{if } i = 1 \text{ or } j = 1,
$$

$$
qq_{ij} = E_{i,j} \oplus \mathrm{mod}\left(x_{m,n} \times 10^{10}, Q\right) \oplus E_{i-1,j} \oplus E_{i,j-1}
$$
$$
\text{otherwise.} \tag{15}
$$

Decryption is the inverse of the encryption process, namely, counter proliferation first and then the scrambling processing.

4.4. Pixel Values to Replace. Produced by Logistic mapping and hyperchaos system of random sequence, it is only related to the initial value and system parameters and does not rely on what seems to lead to clear bytes that can only affect the bytes of an encrypted cryptograph, bringing it to choose plaintext attack and chosen-ciphertext attack. In order to overcome this defect and improve the efficiency of encryption, we put forward new encryption schemes; the specific steps are shown as follows.

(1) Random sequences generated from chaotic system are calculated:

$$x_j = \mathrm{mod}\left(\left(\mathrm{abs}\left(x_k\right) - \mathrm{floor}\left(\mathrm{abs}\left(x_k\right)\right)\right) \times 10^{14}, 256\right)$$
$$(J = 1, 2, 3, 4),$$

$$(16)$$

where $x_j \in [0, 255]$.

(2) P^{rc} ranks of displacement of matrix according to the line are converted to vector, respectively, for each child:

$$P_1^r = p_1^r p_2^r \cdots p_{MN/k}^r,$$
$$P_2^r = p_{MN/k+1}^r \cdots p_{2MN/k}^r,$$
$$P_3^r = p_{2MN/k+1}^r \cdots p_{3MN/k}^r,$$
$$P_k^r = p_{(k-1)MN/k+1}^r \cdots p_{MN}^r.$$

$$(17)$$

And then we calculate the type:

$$E_1^r(i) = x_1(i) \oplus P_1^{rc}(i) E_1^r(i-1)$$
$$E_2^r(i) = x_2(i) \oplus P_2^{rc}(i) E_2^r(i-1)$$
$$\vdots$$
$$E_k^r(i) = x_k(i) \oplus P_k^{rc}(i) E_k^r(i-1),$$
$$i \in \left[1, \frac{MN}{k}\right].$$

$$(18)$$

By the same token, the transform matrix E^r in columns as the row vector, and each child vector is obtained encryption cipher to:

$$E_1 = E_1^{rc}(j) = x_1(j) \oplus E_1^R(j) E_1^{rc}(j-1)$$
$$E_2 = E_2^{rc}(j) = x_2(j) \oplus E_2^r(j) E_2^{rc}(j-1)$$
$$\vdots$$
$$E_k = E_k^{rc}(j) = x_k(j) \oplus E_k^r(j) E_k^{rc}(j-1),$$
$$j \in \left[1, \frac{MN}{k}\right].$$

$$(19)$$

(3) If all plaintext is encrypted, the encryption process is ended. Otherwise, go back to step (1). Decryption process is similar to the encryption process.

4.5. To Improve Species Diversity Index. In the process of genetic algorithm in the late fitness, some of the biggest individuals within the population repeat or converge. At this point, they have bigger probability to participate in the choice of the next generation of copy; offspring of crossover between them will not have the too big change with the father generation; it may lead to genetic algorithm to search the optimization process which is very slow, and it may also reduce the search efficiency. Therefore, we should correctly judge whether a species occurs at premature convergence and mainly should look at the population of the current fitness value which is larger in the individual, whether to repeat or mutually converge. Local optimization genetic algorithm is introduced in this paper; we first determine the degree of species diversity in Section 4.1. It is defined as follows.

Set the first t generation population by individual $X(t)_1, X(t)_2, \ldots, X(t)_N$, fitness value of $f(t)_1, f(t)_2, \ldots, f(t)_N$, which is the best individual fitness value of $f_{\max}$, individual $\overline{f} = (1/N) \sum_{t=1}^{N} f(t)$ total average fitness; fitness value is greater than individual fitness value of $\overline{f}$ which will do an average of $\overline{f}^*$, and using the defined $\overline{f}^*$ and the difference in value between $\overline{f}^*$ as $\Delta^* = f_{\max} - \overline{f}^*$, the $\Delta^* = f_{\max} - \overline{f}^*$ indicators are used to characterize the population of premature convergence.

4.6. Chaos Genetic Algorithm Basic Steps. Using chaos genetic algorithm in Sections 4.1–4.4 of the image encryption process for optimization operation, in order to obtain the best encryption to decrypt steps and performance plan, the optimization process is shown in Figure 5. Concrete steps are as follows.

(1) Coding and the parameters setting are still using real number coding; the method of natural intuitive can save the time and space overhead, the advantage of high computation efficiency. To the population size, chaos optimization iterations, such as fitness function parameters, are set.

(2) Chaos is used to generate the initial population.

(3) The expected value method and the choice of the optimal preservation strategy are used to implement the replication.

(4) The improved genetic algorithm is used in the adaptive adjusting the crossover probability of pc, pm mutation probability, crossover, and mutation operator.

(5) Calculate degree of species diversity, its outstanding individual to decide whether to use contemporary chaos optimization; if the degree of species diversity is less than a random number, use chaos optimization, and if they can get better individuals, it will be used as the optimal solution in the algebra.

(6) Repeat steps (3) to (5), until meeting the termination conditions of evolution. Termination conditions may be evolution algebra or best individual fitness function.

4.7. Theoretical Analysis and Experimental Simulation

4.7.1. Statistical Analysis. Simulating the process, in order to assess the performance of the proposed algorithm, this

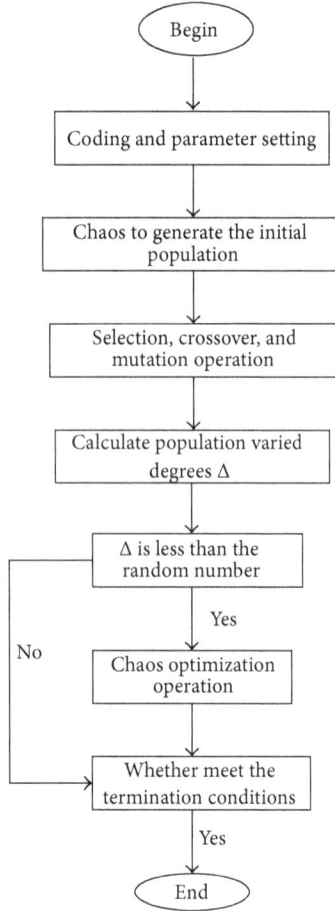

FIGURE 5: Chaos genetic algorithm flow chart.

TABLE 2: Two kinds of algorithm ciphertext image information entropy.

$H(x)$	7.1.10	7.2.01	Boat	Elaine	Gray21	Numbers
Gao	7.8673	7.6298	7.8967	7.5678	7.7645	7.7694
Ours	7.9378	7.9679	7.9786	7.9897	7.9903	7.9936

point ratio is 3.21‰; it can obtain good scrambling effect. But after scrambling, image can only be destroyed by the original correlation between adjacent pixels, without changing the pixel value at every point, so the image of gray distribution histogram will not change and must be of scrambling image pixel values for further encryption. Eventually, the encrypted results are shown in Figure 6(c); it has completely hidden the original image, and it cannot see the outline of the original image. Figures 6(d) and 6(e) show these problems; we can compare with the uneven distribution of the original histogram; the encrypted flat histogram and gray value are evenly distributed. This shows that the ciphertext pixel values in [0, 255] are within the scope of the equal probability values, namely, uniform for the whole ciphertext space distribution characteristics. Thus, this algorithm can effectively prevent the statistical attack.

4.7.2. The Key Space and the Key Sensitivity Analysis. The key of the encryption algorithm for $K = \{\mu, \omega, \lambda, x_{0,0}, x_{m+1,0} = \lambda x_{m,0}(1 - x_{m,0}), x_{0,n+1} = \lambda x_{0,n}(1 - x_{0,n}, T)\}$; if the computer precision is effective for 10^{15}, key space size is about $10^{15 \times 5 \times M \times N}$; the algorithm has a large enough key space and it can effectively resist brute force attack. $x_{0,0} = 0.6167$ Barbara encryption image, respectively, in $x_{0,0} = 0.6167 + 0.1 \times 10^{-9}$ and $x_{0,0} = 0.6167 + 0.1 \times 10^{-14}$ under the condition of the same (other decryption key and encryption key) decrypted image is shown in Figure 7. We can see from Figure 7 that it is extremely sensitive to key encryption algorithm.

4.7.3. Information Entropy Analysis. A digital image information entropy can show the distribution of the gray value. If the gray value distribution is uniform, the amount of information of the image is greater. The definition of information entropy is the average amount of information; it is as follows:

$$H(x) = -\sum_i^n p(x_i) \log_2 p(x_i),$$

$$(21)$$

$$0 \le p(x_i) \le 1, \quad (i = 1, 2, \ldots, n) \sum_i^n p(x_i) = 1.$$

In formula (21), $p(x_i)$ is concentrated symbol probability x_i in the message. In order to have a good contrast, Table 2 shows the 6 different images using this algorithm and Gao and Chen [24] encryption algorithm using the entropy value comparison. As shown, Gao algorithm of ciphertext image information entropy is smaller than the information entropy of the proposed algorithm, so its gray value distribution is not uniform. The algorithm of ciphertext image information entropy is close to a maximum of 8, of image gray value

chapter chose 256 KB image file for simulation, according to the theory of Shannon; statistical analysis is often used to analyze and decipher the algorithm. Therefore, a password system should have good performance in terms of statistical attack resistance. The simulation results are shown in Figure 6: Figure 6(a) is 256 KB of original image; Figure 6(b) for image scrambling after visible, scrambling image pixel shading distribution, while great changes have taken place in picture similar to white noise, showed good scrambling effect. In the image scrambling, the fewer the numbers of the fixed points, the higher the secrecy, and then scrambling effect is better. The fixed point of statistical formula is as follows:

$$\delta = \left(\sum_{n=1}^{M \times N} \frac{\nabla_n}{M \times N}\right) \times 100\%,$$

$$(20)$$

$$\nabla_n = \begin{cases} 0 & \left(\text{find}(p_n)|_{p_n \in P} \ne \text{find}(p_n)|_{p_n \in P^{rc}}\right) \\ 1 & \left(\text{find}(p_n)|_{p_n \in P} = \text{find}(p_n)|_{p_n \in P^{rc}}\right). \end{cases}$$

Among them, *find* () used for the position of the Matlab command and p_n corresponding pixel values, M and N, are used to clear the size of the matrix. According to formula (20). The statistical Figure 6(b) is the scrambling of the fixed point. It is got from Figure 6(b) that the scrambling of the fixed

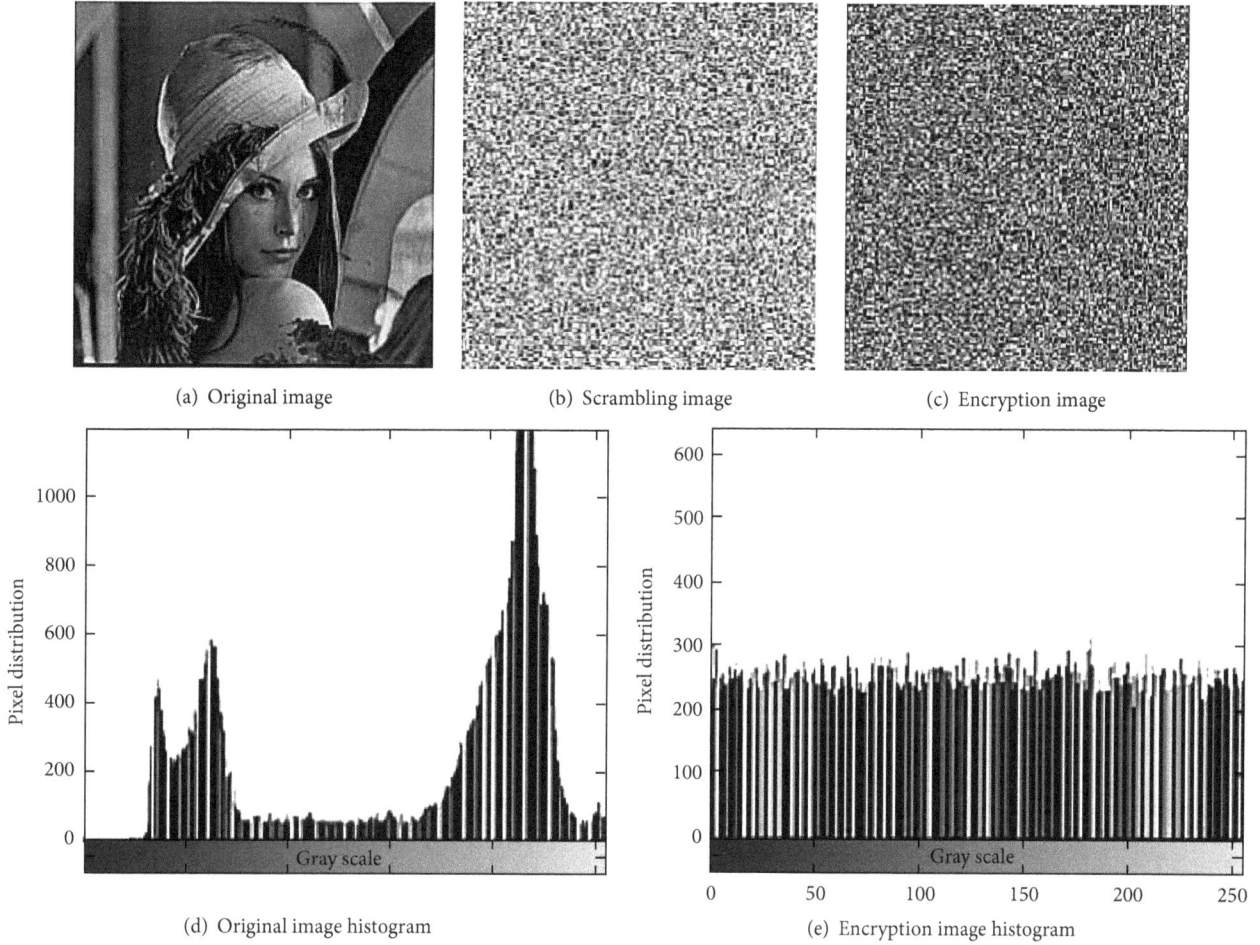

(a) Original image   (b) Scrambling image   (c) Encryption image

(d) Original image histogram   (e) Encryption image histogram

FIGURE 6: The original image and the encryption image histogram.

distribution, and uniform. Therefore, from the perspective of information entropy attack, the algorithm is secure.

4.7.4. Correlation Analysis. Digital image of each pixel is not independent; its correlation is very large. This suggests the large area of gray value. In one of the digital TV images, for example, the same line of two adjacent pixels or adjacent two rows of pixels, the correlation coefficient is 0.9, and the correlation between the adjacent two television images is larger than frame correlation, so the image information redundancy is very big. For image encryption, one of the goals is to reduce the correlation between adjacent pixels, mainly including horizontal pixel, vertical pixels, and the correlation between diagonal pixels. Obviously correlation is smaller, and the better the image encryption, the higher the security.

Figure 8 is expressed as the vertical direction and horizontal direction, the original image, and the algorithm of encryption image correlation of adjacent pixels. The correlation between the original image pixel rendering obvious linear relationship and encryption image pixels of correlation among random corresponding relation are visible.

Table 3 is for the original image and the proposed algorithm of ciphertext image between adjacent pixels according to the horizontal, vertical, and diagonal direction calculated

TABLE 3: The correlation coefficient of the original image and the cipher image pixel.

Correlation index	Original image	Encryption image
Level	0.8766893	0.0003875
Opposite angles	0.9047526	0.0098546
Vertical	0.9537689	0.0209381

using the correlation coefficient. For pixel correlation coefficient ρ_{xy} calculation method is shown as follows:

$$E(x) = \frac{1}{N} \sum_{i=1}^{N} x_i,$$

$$D(x) = \frac{1}{N} \sum_{i=1}^{N} (x_i - E(x))^2,$$

$$\text{cov}(x, y) = \frac{1}{N} \sum_{i=1}^{N} (x_i - E(x))(y_i - E(y)),$$

$$\rho_{xy} = \frac{\text{cov}(x, y)}{\sqrt{D(x) D(y)}}.$$

(22)

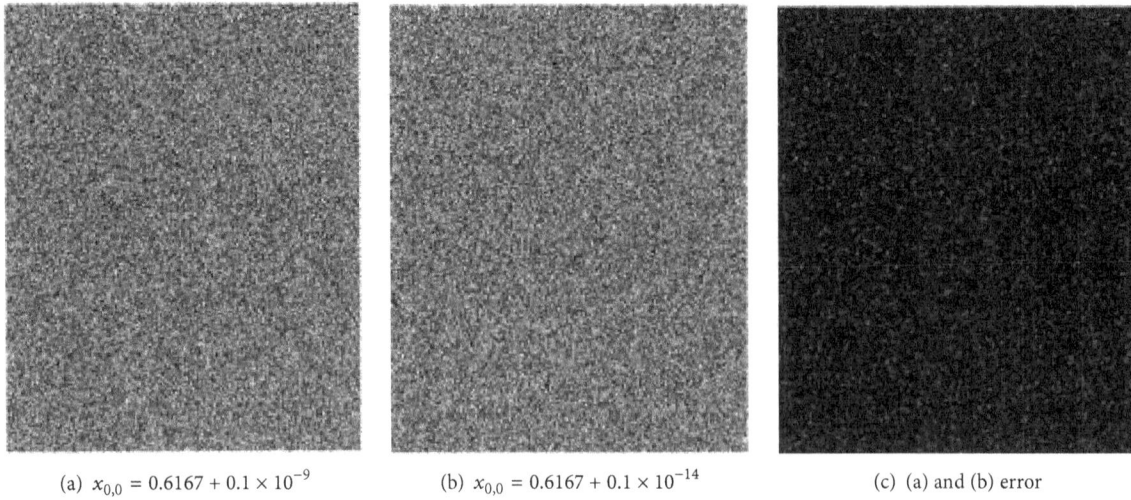

(a) $x_{0,0} = 0.6167 + 0.1 \times 10^{-9}$ · · · · (b) $x_{0,0} = 0.6167 + 0.1 \times 10^{-14}$ · · · · (c) (a) and (b) error

FIGURE 7: Wrong decryption keys under the image and the difference between them.

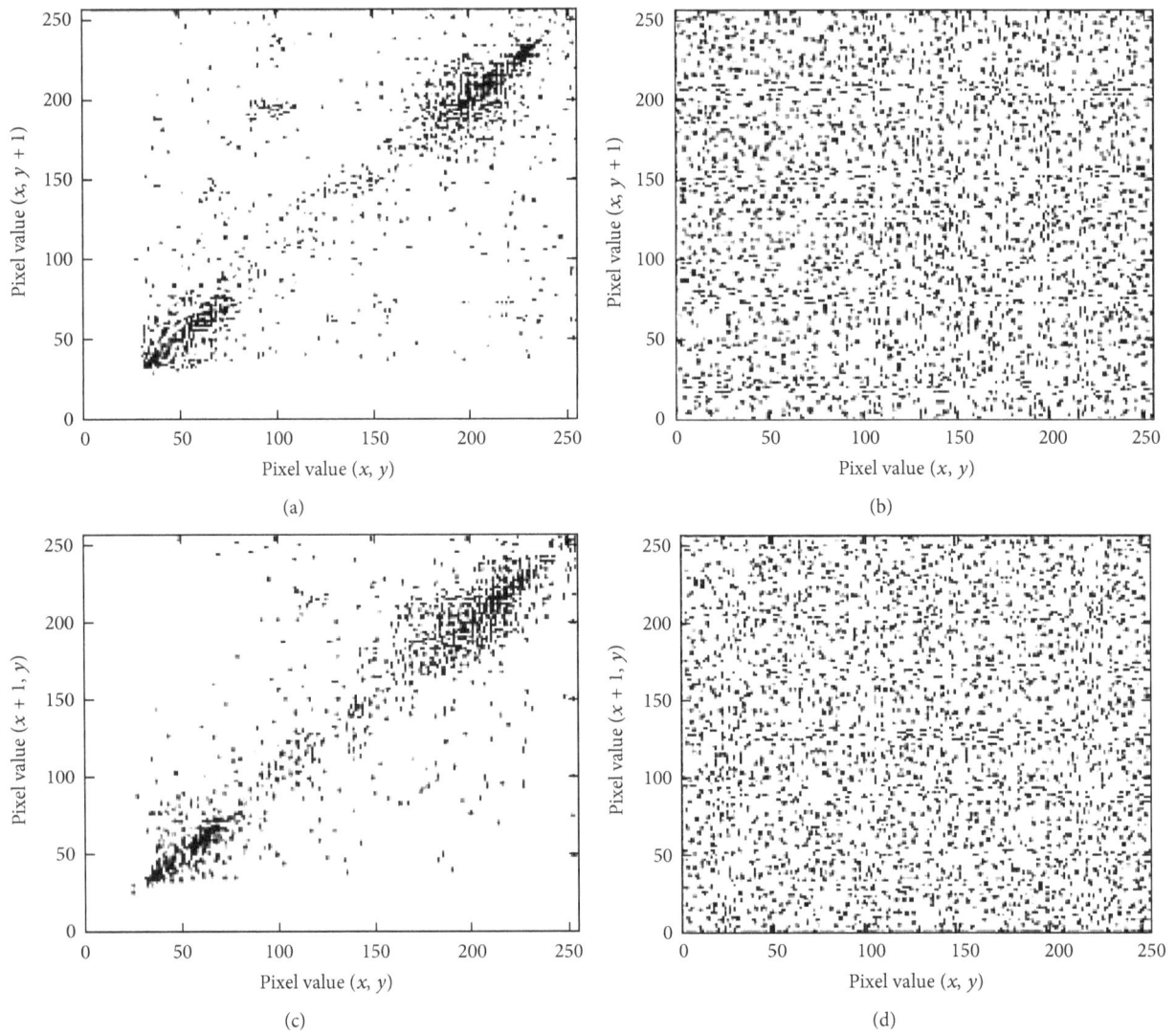

(a)

(b)

(c)

(d)

FIGURE 8: Adjacent pixels correlation.

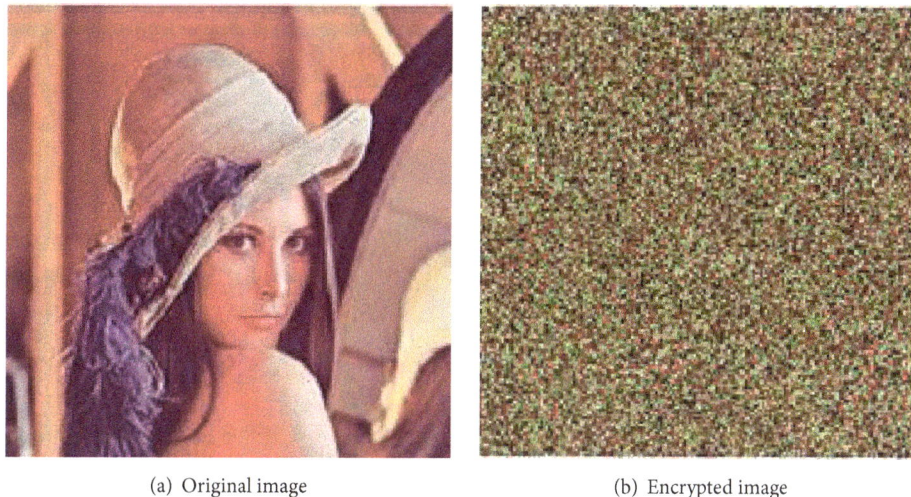

(a) Original image

(b) Encrypted image

FIGURE 9: Original image and encrypted image.

In formula (22), x and y, respectively, are the image pixel values of two adjacent pixels and ρ_{xy} is the correlation coefficient of two adjacent pixels. As shown in Table 3, the adjacent pixels of the original image are highly correlated; their correlation coefficient is close to 1. And we concluded that encrypted ciphertext correlation coefficient of adjacent pixels of the image is very small; they are close to zero; the adjacent pixels have been largely irrelevant, which shows that the statistical characteristic of the original image has been spread to random ciphertext image.

4.7.5. Color Image. In order to demonstrate the effectiveness of the proposed encryption algorithm, the following will use the algorithm to analyze color image.

(1) Image Encryption and Statistical Analysis. Select 256 × 256 color Lena standard drawing as experimental object; make use of MATLAB 7.6 programming to do the experiment simulation; the experimental simulation results are shown in Figure 1. Figures 9(a) and 9(b) are the original image and the encryption image, respectively. We can see intuitively that the encrypted image cannot see any effective information of the original image. First of all, on the vision the encryption algorithm achieves good encryption effect.

Here, we calculate, respectively, scrambling degree of the three primary colors R, G, and B of encrypted image; after several test calculations we get the average:

R component image scrambling degree: $SM_R = 0.942$.

G component image scrambling degree: $SM_G = 0.975$.

B component image scrambling degree: $SM_B = 0.914$.

The result shows that scrambling degree of the algorithm is similar with magic square transformation and has good scrambling effect.

(2) The Key Space and the Key Sensitivity Analysis. Key of the encryption algorithm $K = \{\mu, \omega, \lambda, x_{0,0}, x_{m+1,0} = \lambda x_{m,0}(1 - x_{m,0}), x_{0,n+1} = \lambda x_{0,n}(1 - x_{0,n}, T)\}$; if the computer precision effectively is 10^{15}, key space size is about $10^{15*5*M*N}$, and due to the extreme sensitivity to the initial value of the chaos system, the key sequence generated by a key generator is complex and unpredictable, so the key sequence space is large enough; this indicates that the algorithm can resist brute force attack.

Through the tiny change of authenticated key influencing encryption result to measure sensitivity of the algorithm, as shown in Figure 10(b), the encrypted image uses right key and wrong key to decrypt the algorithm, respectively; the simulation results are shown in Figure 10, of which Figure 10(a) is the decrypted image which uses the correct key to decrypt, and Figure 10(b) is the error decrypted image obtained by keeping the other key parameters unchanged, taking x_0 values differently.

By observing the decrypted image, one can know that, with the key orders of magnitude difference error of only 10^{-10} keys to decrypt, the image is similar to the noise of the image, on the vision which fully does not see any effective information of the original image; it shows that the algorithm has a good sensitivity to key; that is, the initial key of tiny change will lead to the result of the encryption algorithm being completely different; this makes the attacker to not only attack from the ciphertext fragments obtained to judge the information of the original image.

(3) Ability to Resist Noise Analysis. Shown in Figure 9(b) are the images after adding noise using the correct secret key to decrypt them; the decrypted image is shown in Figure 11, and Figures 11(a) and 11(b), respectively, show the decryption result after the encryption image is joined with the interference intensity of 20% salt and pepper noise and interference intensity of 15% gaussian noise, respectively, as shown in Figures 11(a) and 11(b). As you can see, to decrypt right the encryption image after adding noise, it can still be

(a) Correct decryption image

(b) Error decrypting image

FIGURE 10: Key sensitivity testing.

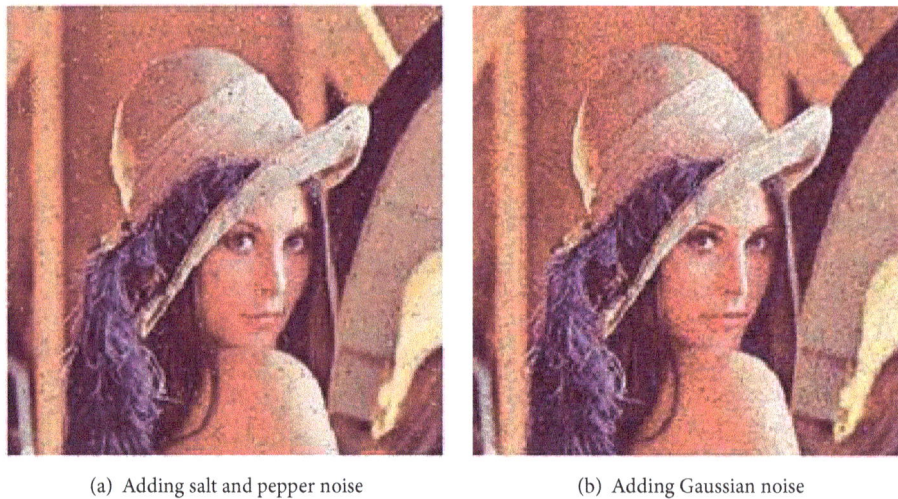

(a) Adding salt and pepper noise

(b) Adding Gaussian noise

FIGURE 11: Figure 9(b), the decrypted image, after different noise.

good to restore the original image information and to retain the original image information effectively, using decryption without distortion, and this shows that the algorithm has good anti-interference ability.

(4) The Correlation Analysis. The nature of the scrambling effect of encryption algorithm lies in breaking the correlation between pixel values of adjacent pixels, to resist the attack by any clear way. Therefore, the comparison of neighboring pixels correlation can be a good measure of a scrambling effect of the encryption algorithm; here, we compare the correlation of adjacent pixels of the three color components R, G, and B of the original image and encrypted image, respectively; pixel point, respectively, on the horizontal, vertical, and diagonal direction are analyzed, and the pixel correlation coefficient calculation method is shown as in formulae (22).

Do statistical tests on the correlation of adjacent pixels in the original image and the encrypted image, and randomly

TABLE 4: The correlation among adjacent pixels of R component.

Related systems	Original image	Encryption image
Horizontal	0.9372451	−0.0058392
Diagonal	0.8965389	0.0031983
Vertical	0.9209471	−0.0089638

draw multiple pixels and calculate separately the correlation of adjacent pixels of the R, G, and B components on horizontal, vertical, and diagonal directions; the averaged results after multiple computations are shown in Tables 4, 5, and 6 [25].

Observe that the correlation coefficient of the adjacent pixels no matter in horizontal, vertical, or diagonal line direction of the encrypted image is far less than the correlation coefficient of adjacent pixels of original image; this shows that the encryption algorithm has good scrambling effect, destroys the correlation between image pixels, and effectively conceals the statistical characteristics of the image.

TABLE 5: The correlation among adjacent pixels of G component.

Related systems	Original image	Encryption image
Horizontal	0.9689230	−0.00329842
Diagonal	0.8276401	0.00214621
Vertical	0.9737210	−0.00231832

TABLE 6: The correlation among adjacent pixels of B component.

Related systems	Original image	Encryption image
Horizontal	0.9012730	0.0005832
Diagonal	0.9103892	−0.0065932
Vertical	0.9101020	0.0090382

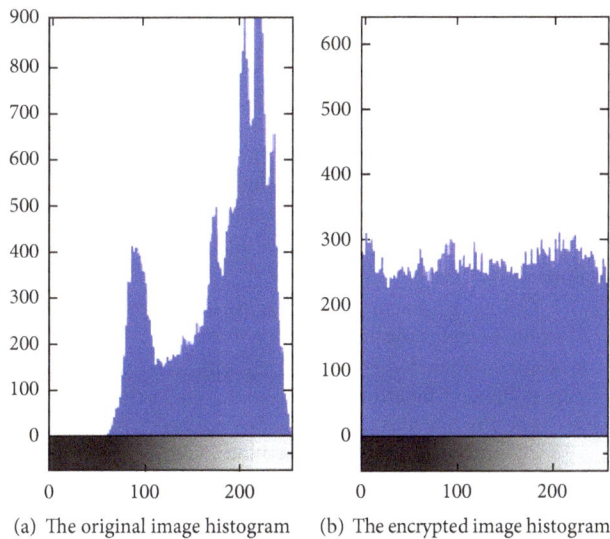

(a) The original image histogram (b) The encrypted image histogram

FIGURE 12: The histogram comparison of R component.

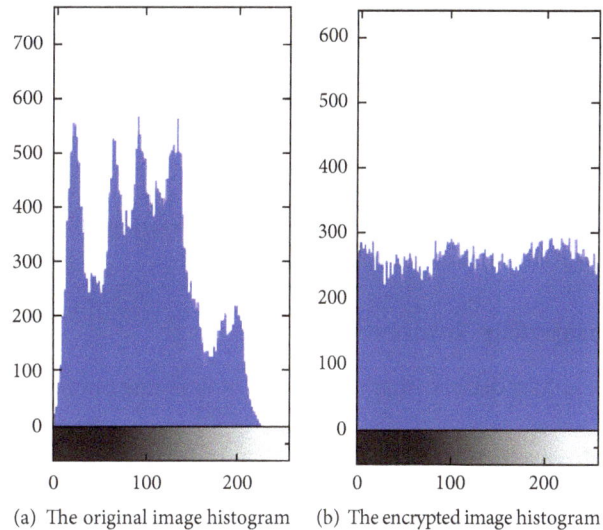

(a) The original image histogram (b) The encrypted image histogram

FIGURE 13: The histogram comparison of G component.

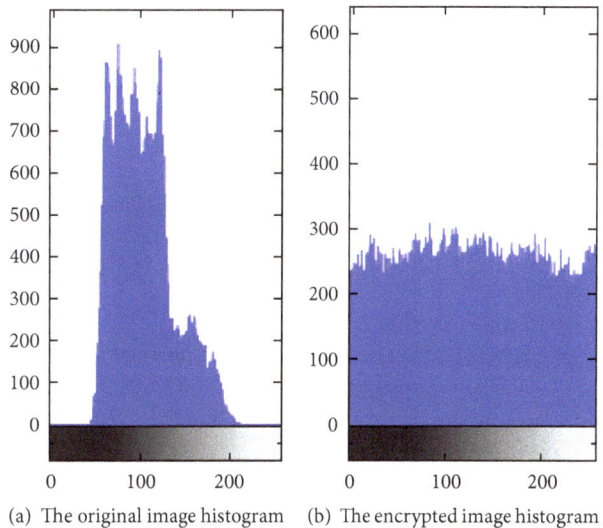

(a) The original image histogram (b) The encrypted image histogram

FIGURE 14: The histogram comparison of B component.

(5) Histogram Analysis. Histogram of the image is one of the important statistical characteristics of image; it directly reflects the relationship between the digital image of each grayscale and the occurrence of the grayscale. Extract the three colors of R, G, and B components from the original image of Figure 9(a) and the encrypted image of Figure 9(b); histograms are calculated, respectively, as shown in Figures 12, 13, and 14.

Looking at these figures, it can be seen that the encrypted image histogram is uniform; there is a world of difference with the original image; it shows that the encrypted image will not give the attacker any clear information; the algorithm has a strong ability to resist statistical attack.

5. Conclusion

This paper analyzes the existing "scrambling to replace" type of some shortages which are chaotic image encryption algorithms; namely a plaintext byte can affect a cipher byte; it can be made use of to select plaintext attack and chosen-ciphertext attack is easy to decipher. That is to say, it will not be able to resist choosing plaintext attack and

chosen-ciphertext attack. Aiming at the existence of some high dimensional chaotic encryption algorithm in the weak resistance to choose plaintext attack and chosen-ciphertext attack problems, we put forward a new kind of digital image encryption algorithm based on genetic chaos. The new algorithm is to make the following improvements: (1) in the process of scrambling encryption algorithm, the hyperchaos system is also introduced; it makes the image scrambling result not only depend on the chaotic sequence generated by the initial key but also depend on the characteristics of the image itself, making it have adaptive features to a certain extent; (2) for the introduction of the genetic algorithm for image encryption-decryption process parameters of dynamic optimization, it gets the best encryption-decryption process and steps; (3) for the diffusion in the encryption process, clear feedback mechanism is introduced in the algorithm and improves the sensitivity to the plaintext, objectively improving the algorithm of plaintext and ciphertext attack resistance. Through the relevant experiment and safety analysis, we show that our proposed digital image encryption algorithm based on genetic hyperchaos not only can effectively resolve the current problems of weak resistance to aggressive encryption algorithm but also strengthened the sensitivity to the plaintext, enhanced ability to resist differential attacks, and improved the efficiency of encryption.

Competing Interests

The author declares that they have no competing interests.

Acknowledgments

This work is supported in part by the National Natural Science Foundation of China under Grants 61272466 and 61303233 and the Natural Science Foundation of Hebei Province under Grant F2014203062.

References

[1] Y. Liu, X. Tong, and S. Hu, "A family of new complex number chaotic maps based image encryption algorithm," *Signal Processing: Image Communication*, vol. 28, no. 10, pp. 1548–1559, 2013.

[2] Z.-H. Guan, F. Huang, and W. Guan, "Chaos-based image encryption algorithm," *Physics Letters, Section A: General, Atomic and Solid State Physics*, vol. 346, no. 1–3, pp. 153–157, 2005.

[3] W.-T. Wong, F. Y. Shih, and T.-F. Su, "Thinning algorithms based on quadtree and octree representations," *Information Sciences*, vol. 176, no. 10, pp. 1379–1394, 2006.

[4] R.-J. Chen and S.-J. Horng, "Novel SCAN-CA-based image security system using SCAN and 2-D von Neumann cellular automata," *Signal Processing: Image Communication*, vol. 25, no. 6, pp. 413–426, 2010.

[5] X. Xu, Z. Tang, and J. Wang, "A method to improve the transiently chaotic neural network," *Neurocomputing*, vol. 67, no. 1–4, pp. 456–463, 2005.

[6] S.-D. Liu, Shi-Shi, and Z.-W. Yan, *The Essence of Chaos*, Meteorological Press, Beijing, China, 1997.

[7] J. Fridrich, "Symmetric ciphers based on two-dimensional chaotic maps," *International Journal of Bifurcation and Chaos in Applied Sciences and Engineering*, vol. 8, no. 6, pp. 1259–1284, 1998.

[8] G. Chen, Y. Mao, and C. K. Chui, "A symmetric image encryption scheme based on 3D chaotic cat maps," *Chaos, Solitons and Fractals*, vol. 21, no. 3, pp. 749–761, 2004.

[9] S. Lian, J. Sun, and Z. Wang, "A block cipher based on a suitable use of the chaotic standard map," *Chaos, Solitons & Fractals*, vol. 26, no. 1, pp. 117–129, 2005.

[10] K.-W. Wong, B. S.-H. Kwok, and W.-S. Law, "A fast image encryption scheme based on chaotic standard map," *Physics Letters, Section A: General, Atomic and Solid State Physics*, vol. 372, no. 15, pp. 2645–2652, 2008.

[11] H. Zhu, C. Zhao, and X. Zhang, "A novel image encryption-compression scheme using hyper-chaos and Chinese remainder theorem," *Signal Processing: Image Communication*, vol. 28, no. 6, pp. 670–680, 2013.

[12] G. Alvarez and S. Li, "Cryptanalyzing a nonlinear chaotic algorithm (NCA) for image encryption," *Communications in Nonlinear Science and Numerical Simulation*, vol. 14, no. 11, pp. 3743–3749, 2009.

[13] M. Ghebleh, A. Kanso, and H. Noura, "An image encryption scheme based on irregularly decimated chaotic maps," *Signal Processing: Image Communication*, vol. 29, no. 5, pp. 618–627, 2014.

[14] S. Banerjee, L. Rondoni, S. Mukhopadhyay, and A. P. Misra, "Synchronization of spatiotemporal semiconductor lasers and its application in color image encryption," *Optics Communications*, vol. 284, no. 9, pp. 2278–2291, 2011.

[15] N. K. Pareek, V. Patidar, and K. K. Sud, "Image encryption using chaotic logistic map," *Image and Vision Computing*, vol. 24, no. 9, pp. 926–934, 2006.

[16] C. K. Volos, I. M. Kyprianidis, and I. N. Stouboulos, "Image encryption process based on chaotic synchronization phenomena," *Signal Processing*, vol. 93, no. 5, pp. 1328–1340, 2013.

[17] T. Xiang, X. F. Liao, G. Tang, Y. Chen, and K.-W. Wong, "A novel block cryptosystem based on iterating a chaotic map," *Physics Letters A*, vol. 349, no. 1–4, pp. 109–115, 2006.

[18] J. H. Holland, *Adaptation in Nature and Artificial Systems*, MIT Press, 1992.

[19] D. E. Goldberg, *Genetic Algorithms in Search, Optimization and Machine Learning*, Addison-Wesley, 1989.

[20] K. Krishnakumar, "Micro-genetic algorithms for stationary and non-stationary function optimization," in *Intelligent Control and Adaptive Systems*, vol. 1196 of *Proceedings of SPIE*, pp. 289–296, Philadelphia, Pa, USA, November 1989.

[21] D. Maclay and R. Dorey, "Applying genetic search techniques to drivetrain modeling," *IEEE Control Systems Magazine*, vol. 13, no. 3, pp. 50–55, 1993.

[22] Y. Davidor, *Genetic Algorithms and Robotics*, World Scientific, Singapore, 1991.

[23] J. H. Park, "Adaptive synchronization of hyperchaotic Chen system with uncertain parameters," *Chaos, Solitons and Fractals*, vol. 26, no. 3, pp. 959–964, 2005.

[24] T. G. Gao and Z. Q. Chen, "A new image encryption algorithm based on hyper-chaos," *Physics Letters A*, vol. 372, no. 4, pp. 394–400, 2008.

[25] Y.-Q. Zhang and X.-Y. Wang, "A new image encryption algorithm based on non-adjacent coupled map lattices," *Applied Soft Computing*, vol. 26, pp. 10–20, 2015.

17

Multiple-Beams Splitter Based on Graphene

Xiao Bing Li, Hong Ju Xu, Wei Bing Lu, and Jian Wang

State Key Laboratory of Millimeter Waves, School of Information Science and Engineering, Southeast University, Nanjing 210096, China

Correspondence should be addressed to Wei Bing Lu; wblu@seu.edu.cn

Academic Editor: Zhihao Jiang

Due to its tunability of conductivity, graphene can be considered as a novel epsilon-near-zero (ENZ) material. Based on this property, we propose a wave splitter using graphene. Simulation results show that the circular surface plasmon polariton waves excited by a point source can be transferred to narrow beams through a graphene-based wave splitter, which is formed by a polygonal contour of the ENZ graphene layer. The number of beams can be easily controlled by adjusting the shape of the polygonal ENZ graphene layer, and the operation frequency can also be chosen.

1. Introduction

In recent years, there has been a great deal of interest in physics and engineering of artificially constructed metamaterials, due to their exciting properties caused by unconventional values of permittivity (ε) or permeability (μ) [1–6]. In particular, much attention has been focused on structures for which the real part of one or both of the constitutive parameters approaches zero [2]. These structures have been used to form interesting devices such as highly directive antennas [7] and compact resonators [8]. It is proved that materials with epsilon-near-zero (ENZ) may be directly found in nature in infrared and optical frequency band [3]. A well-known example is electron gas in which the current created by the drift of free electrons effectively interacts with the radiation as continuous medium characterized by a Drude-type dispersion model that has near-zero ε around its plasma frequency [1]. In infrared and optical frequencies, some low loss noble metals like silver and gold, semiconductors (e.g., indium antimonide) [9], and polar dielectrics like silicon carbide (SiC) [10] may behave as ENZ materials near their plasma frequencies. However, due to their dispersive properties, such ENZ materials only work well near their plasma frequencies. So, finding new ENZ materials at desired frequencies is important to design functional devices.

Graphene [11–17], due to its various intriguing properties, has been considered as one of the promising optical materials in the future [16, 18]. The one-atom-thick graphene is characterized by a surface conductivity ($\sigma_g = \sigma_{g,r} + i\sigma_{g,i}$), which can be computed by the Kubo formula [10, 19]:

$$
\sigma\left(\omega, \mu_c, \Gamma, T\right) = \frac{je^2\left(\omega - j2\Gamma\right)}{\pi\hbar^2} \left[\frac{1}{\left(\omega - j2\Gamma\right)^2} \right.
$$
$$
\cdot \int_0^\infty \varepsilon\left(\frac{\partial f_d\left(\varepsilon\right)}{\partial \varepsilon} - \frac{\partial f_d\left(-\varepsilon\right)}{\partial \varepsilon}\right) d\varepsilon \tag{1}
$$
$$
\left. - \int_0^\infty \frac{f_d\left(-\varepsilon\right) - f_d\left(\varepsilon\right)}{\left(\omega - j2\Gamma\right)^2 - 4\left(\varepsilon/\hbar\right)^2} d\varepsilon \right],
$$

where ω is the radian frequency, Γ is the phenomenological scattering rate, T is the temperature, μ_c is the chemical potential, e is the electron charge, $\hbar = h/2\pi$ is the reduced Planck constant, $f_d(\varepsilon) = (e^{(\varepsilon-\mu_c)/k_BT} + 1)^{-1}$ is the Fermi-Dirac distribution, and k_B is Boltzmann's constant. Importantly, the conductivity of graphene can be changed via tuning the chemical potential, which is controlled by gate voltage, electric field, magnetic field, and/or chemical doping [20, 21]. According to the relationship between equivalent complex permittivity and conductivity of the Δ-thick graphene layer (graphene has a very small thickness Δ), given in [22], we can tune the equivalent complex permittivity of the one-atom-thick graphene layer. In addition, graphene can support the highly confined transverse-magnetic (TM) surface plasmon polariton (SPP) wave if and only if $\sigma_{g,i} > 0$ (in other words, $\text{Re}(\varepsilon_{g,\text{eq}}) < 0$) [16, 23]. Based on the above conclusions, graphene shows the potential to construct transformational metamaterials on a one-atom-thick surface.

In graphene, the chemical potential depends on the carrier density (n) which can be modified dynamically by applying a gate voltage [24]. The relation between μ_c and n can be described as the following formula: $|\mu_c| = \hbar v_F (\pi n)^{1/2}$, where $v_F \approx 1 \times 10^6 \text{ m s}^{-1}$ is the Fermi velocity, n is given by $n = (n_0^2 + \alpha^2 |\Delta V|^2)^{1/2}$, n_0 is the residual carrier concentration, α is the gate capacitance effected by the specific electrode configuration, $|\Delta V| = |V_{\text{CNP}} - V_g|$, with $|V_{\text{CNP}}|$ being the charge neutral gate voltage, and V_g is the applied gate voltage. The chemical potential of graphene can be controlled using an ion-gel top gate [25], which allows a large doping range through low electrostatic voltage. Through varying the chemical potential, we can find an equivalent complex graphene permittivity value at the working frequency, the real part of which is close to zero, either positive or negative. Here we choose negative ENZ values in order to support the propagation of SPP waves on the graphene. In this paper, we propose a tunable wave coupler based on a single sheet of graphene. The low wave number of propagation characteristics of such materials implies that the phase variation of the electromagnetic fields is negligible over a physically long distance, providing the possibility of manipulating the phase fronts into a desired pattern by controlling the shape of the interfaces of the ENZ material.

In the infrared frequency band, the near-zero epsilon values can be obtained via changing the gate voltage, which is related to the chemical potential [26]. Figure 1 shows the relation between the chemical potential and the real part of graphene's equivalent dielectric permittivity at different frequencies with $T = 3 \text{ K}$ and $\Gamma = 0.43 \text{ meV}$. As can be seen, when the working frequency increases, the chemical potential corresponding to zero permittivity rises, implying that the wave coupler we have designed can also operate at other frequencies through tuning the gate voltage.

2. Simulation Model

Figure 2 shows the simulation model of the wave splitter based on graphene. The radius of the inner cylinder is denoted as $R1$, the apothem of the polygon is denoted as $R2$, and the diameter of proposed wave splitter is D. The red zone corresponds to the ENZ graphene layer. In this area, the real part of graphene's dielectric permittivity approaches zero. In other areas, it can be defined as a negative value to support the propagation of SPP waves. According to [17], uneven ground plane can be used to create inhomogeneous conductivity or permittivity pattern along graphene layer. This is schematically shown in Figure 3(a). Highly doped silicon substrate with uneven height profile serves as the ground plane. The distance between the ground plane and graphene can be filled up with a regular dielectric spacer, for example, silicon oxide. In the numerical simulation, a point source on top of the graphene layer is utilized for

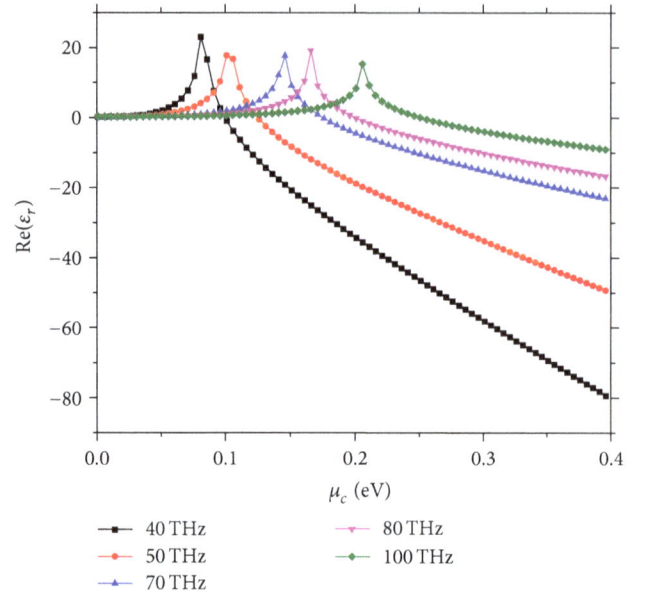

FIGURE 1: Relation between chemical potential and real part of graphene's equivalent dielectric permittivity at different frequencies with $T = 3 \text{ K}$ and $\Gamma = 0.43 \text{ meV}$.

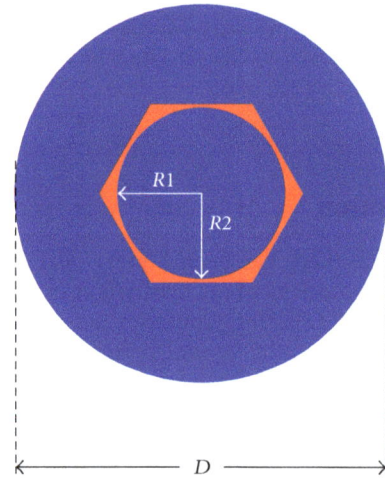

FIGURE 2: Simulation model of the wave splitter based on graphene. The red zone is the ENZ graphene layer, and the blue zone is the graphene with negative dielectric permittivity to support the propagation of SPP waves.

exciting the circular SPP waves. The distance between the source and the wave splitter is $H = 10 \text{ nm}$, as shown in Figure 3(b). To satisfy the wave vector matching condition, the evanescent fields emitted from the infinitesimal dipole are used to excite the circular SPP waves on the graphene. So, the distance between the dipole and the graphene is much smaller than the working wavelength. In our design, the distance is chosen to be less than a thousand of the working wavelength.

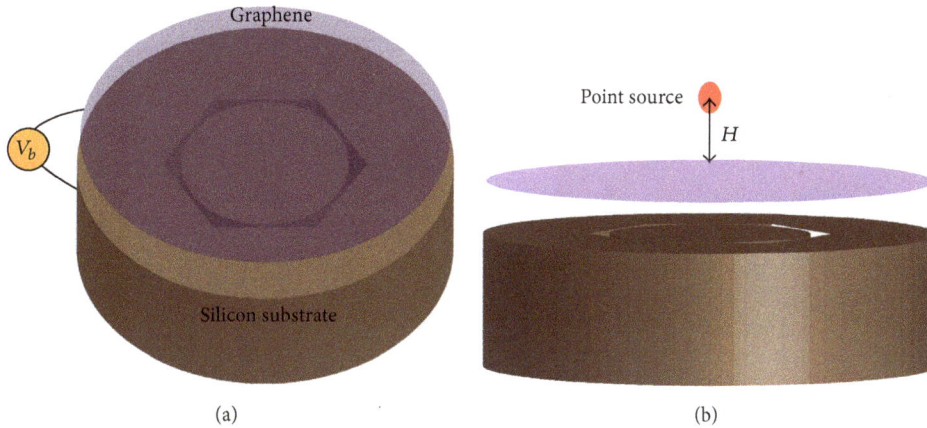

FIGURE 3: (a) Sketch of our proposed one-atom-thick wave splitter, which consists of a single sheet of graphene with inhomogeneous conductivity or permittivity distributions. (b) A point source on top of the graphene layer is used to excite circular SPP waves. The distance between the source and the wave splitter is $H = 10$ nm.

3. Numerical Results and Discussions

In this paper, we use a full-wave electromagnetic simulator software, CST Microwave Studio [27], to obtain three-dimensional numerical results. The numerical calculations use frequency-domain solver, performed with adaptive tetrahedral meshing and open boundary conditions for a free-standing graphene in vacuum without ground plane. An infinitesimal dipole is used as a point source to generate the circular SPP waves. To confirm the effectiveness of the proposed wave splitter based on graphene, we take three-, four-, five-, and six-beam splitter as specific examples. In all simulations, the working frequency is chosen as 50 THz and we set $R1 = 90$ nm, $R2 = 95$ nm, and $D = 400$ nm. The thickness of graphene is taken as 1 nm. According to Figure 1, we can obtain the equivalent complex permittivity $\varepsilon_{r,eq} = -2.28 \times 10^{-4} + i0.2178$ with $T = 3$ K, $\Gamma = 0.43$ meV, and $\mu_c = 0.12155$ eV, which acts as ENZ graphene layer. However, the loss of the ENZ graphene area is large. As a consequence, when the graphene layer is used as ENZ material, the geometrical size of ENZ region should be set very small in order to reduce the propagation loss.

The chemical potential of the other background graphene layer is chosen as $\mu_c = 0.1799$ eV, corresponding to the complex permittivity $\varepsilon_{r,eq} = -15.05 + i0.1511$. It should be emphasized that, for the background graphene layer, any chemical potential larger than 0.129 eV (when $\mu_c > 0.129$ eV, $\mathrm{Re}(\varepsilon_r) < 0$ and $|\mathrm{Re}(\varepsilon_r)| > 1$) is capable of supporting the SPP waves propagation. Figure 4(a) shows the snapshot of the z-component electric field E_z distribution of the TM SPP wave of the graphene-based wave splitter, of which the outer contour of the ENZ graphene is triangle. When the circular SPP waves propagate through the ENZ graphene region, three beams of the almost linear SPP waves emerge. When the outer contour of the ENZ graphene layer is a square with length of $R2$, a four-beam wave splitter is realized, as

shown in Figure 4(b). In the same way, five- and six-beam wave splitters have been designed and demonstrated. The simulation results of electric field distribution are shown in Figures 4(c) and 4(d). It should be noted that the device only operates effectively when the infinitesimal dipole source is placed right above the geometric center of the device. When the source is moved away from the center, the excited circular wave front is no longer a series of concentric circles centered at the device's center [28]. In this case, the excited circular wave front is unparallel to the ENZ material interface, leading to the spatial phase difference along the arc interface. Thus, the wave fronts of the output wave after propagating through the ENZ region are not plane waves perpendicular to the plane interface, due to the spatial phase difference. Therefore, the source should be placed above the device concentrically to ensure that the input wave front is conformal with the interface and thus the output waves are perpendicular to the output interface.

4. Conclusion

In this paper, we theoretically demonstrate that graphene can behave as a novel ENZ material. As one of the applications of ENZ graphene, multiple-beams wave splitter is presented. The beams of the wave splitter are controlled by designing the polygon contours of the ENZ graphene layer. The operation frequency can be designed through changing the gate voltage. Simulation results verify the effectiveness of our design. The ENZ graphene may have important applications in transformational plasmon optics and the proposed wave splitter can be used for generating multiple-beams of SPP waves.

Competing Interests

The authors declare that they have no competing interests.

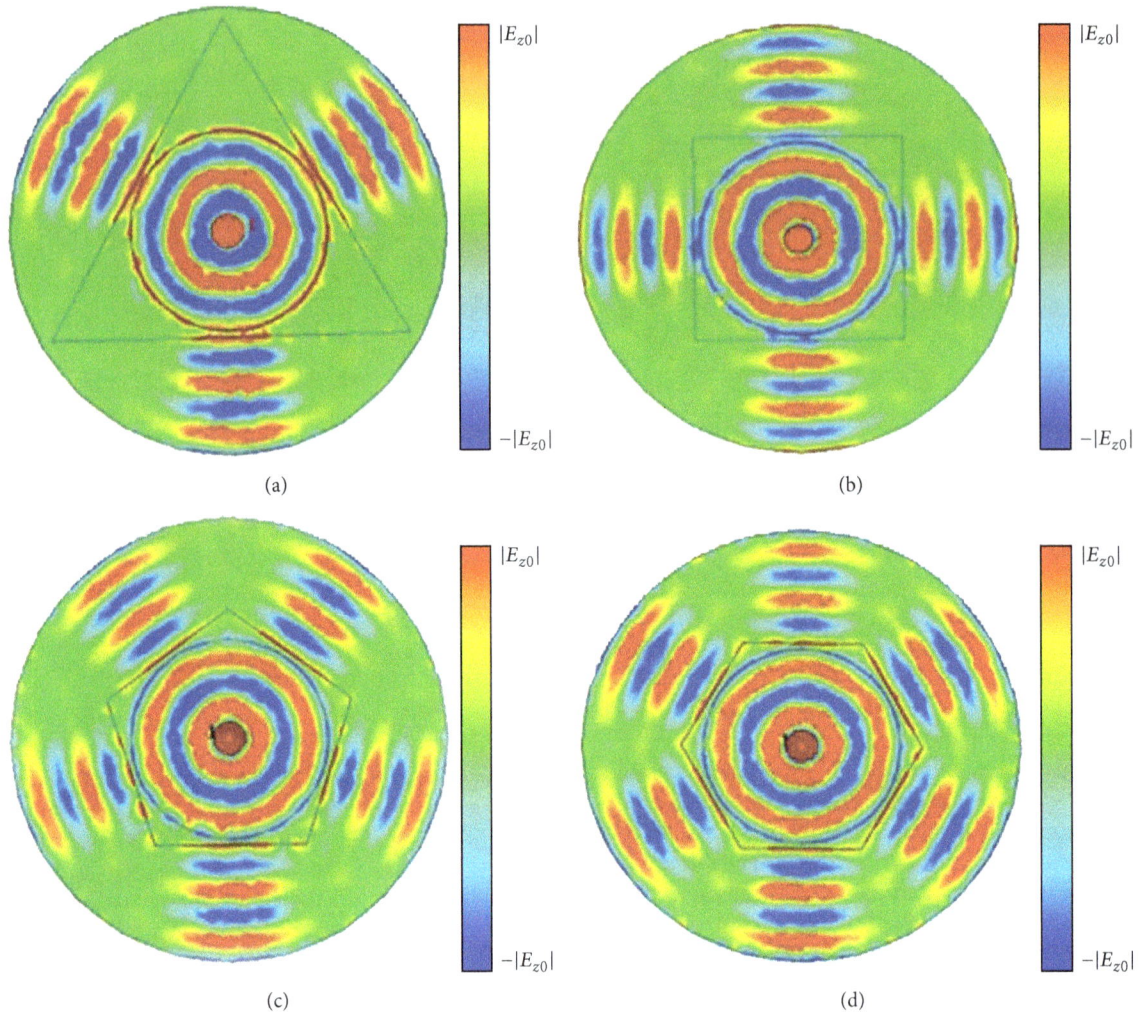

FIGURE 4: Snapshots of the z-component electric field distribution of the TM SPP wave on graphene-based wave splitters with three, four, five, and six beams from (a) to (d), respectively.

Acknowledgments

This work was partially supported by the National Natural Science Foundation of China [Grant no. 61271057]; the Scientific Research Foundation of Graduate School of Southeast University; the Fundamental Research Funds for the Central Universities; and the Innovation Program for Graduate Education of Jiangsu Province [Grant no. CXLX13_092].

References

[1] A. Alù, M. G. Silveirinha, A. Salandrino, and N. Engheta, "Epsilon-near-zero metamaterials and electromagnetic sources: tailoring the radiation phase pattern," *Physical Review B*, vol. 75, no. 15, Article ID 155410, 2007.

[2] R. Liu, Q. Cheng, T. Hand et al., "Experimental demonstration of electromagnetic tunneling through an epsilon-near-zero metamaterial at microwave frequencies," *Physical Review Letters*, vol. 100, no. 2, Article ID 023903, 2008.

[3] M. Silveirinha and N. Engheta, "Tunneling of electromagnetic energy through subwavelength channels and bends using ε-near-zero materials," *Physical Review Letters*, vol. 97, no. 15, Article ID 157403, 2006.

[4] T. Y. Kim, M. A. Badsha, J. Yoon, S. Y. Lee, Y. C. Jun, and C. K. Hwangbo, "General strategy for broadband coherent perfect absorption and multi-wavelength all-optical switching based on epsilon-near-zero multilayer films," *Scientific Reports*, vol. 6, Article ID 22941, 2016.

[5] S. Campione, J. R. Wendt, G. A. Keeler, and T. S. Luk, "Near-infrared strong coupling between metamaterials and epsilon-near-zero modes in degenerately doped semiconductor nanolayers," *ACS Photonics*, vol. 3, no. 2, pp. 293–297, 2016.

[6] S. Lee, T. Q. Tran, M. Kim, H. Heo, J. Heo, and S. Kim, "Angle- and position-insensitive electrically tunable absorption in graphene by epsilon-near-zero effect," *Optics Express*, vol. 23, no. 26, pp. 33350–33358, 2015.

[7] S. Enoch, G. Tayeb, P. Sabouroux, N. Guérin, and P. Vincent, "A metamaterial for directive emission," *Physical Review Letters*, vol. 89, no. 21, Article ID 213902, 2002.

[8] A. Lai, C. Caloz, and T. Itoh, "Composite right/left-handed transmission line metamaterials," *IEEE Microwave Magazine*, vol. 5, no. 3, pp. 34–50, 2004.

[9] J. G. Rivas, C. Janke, P. H. Bolivar, and H. Kurz, "Transmission of THz radiation through InSb gratings of subwavelength apertures," *Optics Express*, vol. 13, no. 3, pp. 847–859, 2005.

[10] W. G. Spitzer, D. Kleinman, and D. Walsh, "Infrared properties of hexagonal silicon carbide," *Physical Review*, vol. 113, no. 1, pp. 127–132, 1959.

[11] F. J. G. de Abajo, "Graphene plasmonics: challenges and opportunities," *ACS Photonics*, vol. 1, no. 3, pp. 135–152, 2014.

[12] Y. Fan, N.-H. Shen, T. Koschny, and C. M. Soukoulis, "Tunable terahertz meta-surface with graphene cut-wires," *ACS Photonics*, vol. 2, no. 1, pp. 151–156, 2015.

[13] A. K. Geim and K. S. Novoselov, "The rise of graphene," *Nature Materials*, vol. 6, no. 3, pp. 183–191, 2007.

[14] A. H. Castro Neto, F. Guinea, N. M. R. Peres, K. S. Novoselov, and A. K. Geim, "The electronic properties of graphene," *Reviews of Modern Physics*, vol. 81, no. 1, pp. 109–162, 2009.

[15] A. K. Geim, "Graphene: status and prospects," *Science*, vol. 324, no. 5934, pp. 1530–1534, 2009.

[16] A. Vakil and N. Engheta, "Transformation optics using graphene," *Science*, vol. 332, no. 6035, pp. 1291–1294, 2011.

[17] Y. Fan, Z. Liu, F. Zhang et al., "Tunable mid-infrared coherent perfect absorption in a graphene meta-surface," *Scientific Reports*, vol. 5, Article ID 13956, 2015.

[18] M. Jablan, H. Buljan, and M. Soljačić, "Plasmonics in graphene at infrared frequencies," *Physical Review B*, vol. 80, no. 24, Article ID 245435, 7 pages, 2009.

[19] L. A. Falkovsky and A. A. Varlamov, "Space-time dispersion of graphene conductivity," *The European Physical Journal B*, vol. 56, no. 4, pp. 281–284, 2007.

[20] C. R. Dean, A. F. Young, I. Meric et al., "Boron nitride substrates for high-quality graphene electronics," *Nature Nanotechnology*, vol. 5, no. 10, pp. 722–726, 2010.

[21] K. S. Novoselov, A. K. Geim, S. V. Morozov et al., "Electric field effect in atomically thin carbon films," *Science*, vol. 306, no. 5696, pp. 666–669, 2004.

[22] L. A. Falkovsky and S. S. Pershoguba, "Optical far-infrared properties of a graphene monolayer and multilayer," *Physical Review B*, vol. 76, no. 15, Article ID 153410, 2007.

[23] A. Vakil and N. Engheta, "Fourier optics on graphene," *Physical Review B*, vol. 85, no. 7, Article ID 075434, 4 pages, 2012.

[24] S. H. Lee, M. Choi, T.-T. Kim et al., "Switching terahertz waves with gate-controlled active graphene metamaterials," *Nature Materials*, vol. 11, no. 11, pp. 936–941, 2012.

[25] L. Ju, B. Geng, J. Horng et al., "Graphene plasmonics for tunable terahertz metamaterials," *Nature Nanotechnology*, vol. 6, no. 10, pp. 630–634, 2011.

[26] G. W. Hanson, "Dyadic green's functions for an anisotropic, non-local model of biased graphene," *IEEE Transactions on Antennas and Propagation*, vol. 56, no. 3, pp. 747–757, 2008.

[27] CST Microwave Studio, CST Computer Simulation Technology, http://www.cst.com/.

[28] Q. Zhao, Z. Xiao, F. Zhang et al., "Tailorable zero-phase delay of subwavelength particles toward miniaturized wave manipulation devices," *Advanced Materials*, vol. 27, no. 40, pp. 6187–6194, 2015.

18

Disorder Improves Light Absorption in Thin Film Silicon Solar Cells with Hybrid Light Trapping Structure

Yanpeng Shi,[1] Xiaodong Wang,[2] and Fuhua Yang[2]

[1]*School of Physics, Shandong University, Jinan 250100, China*
[2]*Engineering Research Center for Semiconductor Integrated Technology, Institute of Semiconductors, Chinese Academy of Sciences, Beijing 100083, China*

Correspondence should be addressed to Xiaodong Wang; xdwang@semi.ac.cn

Academic Editor: Marek Samoc

We present a systematic simulation study on the impact of disorder in thin film silicon solar cells with hybrid light trapping structure. For the periodical structures introducing certain randomness in some parameters, the nanophotonic light trapping effect is demonstrated to be superior to their periodic counterparts. The nanophotonic light trapping effect can be associated with the increased modes induced by the structural disorders. Our study is a systematic proof that certain disorder is conceptually an advantage for nanophotonic light trapping concepts in thin film solar cells. The result is relevant to the large field of research on nanophotonic light trapping which currently investigates and prototypes a number of new concepts including disordered periodic and quasiperiodic textures. The random effect on the shape of the pattern (position, height, and radius) investigated in this paper could be a good approach to estimate the influence of experimental inaccuracies for periodic or quasi-periodic structures.

1. Introduction

Thin film crystalline silicon (TF c-Si) solar cells show a great potential in the worldwide application of photovoltaic technologies. It has an active layer thickness of a few micrometers which decreases the material cost greatly and it could be fabricated in a low-cost, feasible way. However, the photocurrent conversion efficiency is largely constrained by the deteriorated light absorption due to the ultrathin active layer. For this reason, advanced light trapping strategies in solar cells are essential since they increase the absorption of incident sunlight. In the last decade, a variety of photonic nanostructure designs have opened unprecedented opportunities for boosting the light absorption, such as photonic crystal [1, 2], light gratings [3–8], plasmonic metal nanoparticles [9–12], and other nanostructures [13, 14].

The question whether random or periodic photonic nanostructures lead to better light trapping in solar cells is currently hotly debated and remains controversial. Lots of works have been done to find which structure behaves better in enhancing light absorption in solar cells, the random structure or the periodic structure. Battaglia et al. [15] reported that periodic structures rival random textures unambiguously based on their work. Han and Chen [16] found that randomness that destroyed the periodic arrangement could bring further light absorption enhancement, as the mirror symmetries are broken. Ferry et al. [17] demonstrated that certain randomness could give broadband and isotropic photocurrent enhancement in TF c-Si solar cells. Recently, Hong et al. [18] have conducted a symmetric study comparison between solar cells with periodic and random nanohole, and they found that random nanohole behaves better in the light absorption enhancement performance. However, little effort has been done trying to combine the advantages of both the periodic and the random structures. Theory predicts that periodic structures should outperform random textures, as they avoid scattering into lossy radiation modes; meanwhile random structures perform better in some realistic experiments and they can be obtained by low-cost, feasible ways [19]. Actually, controlling the optical modes of two-dimensional disordered structures constitutes a new approach for photon management. The disordered

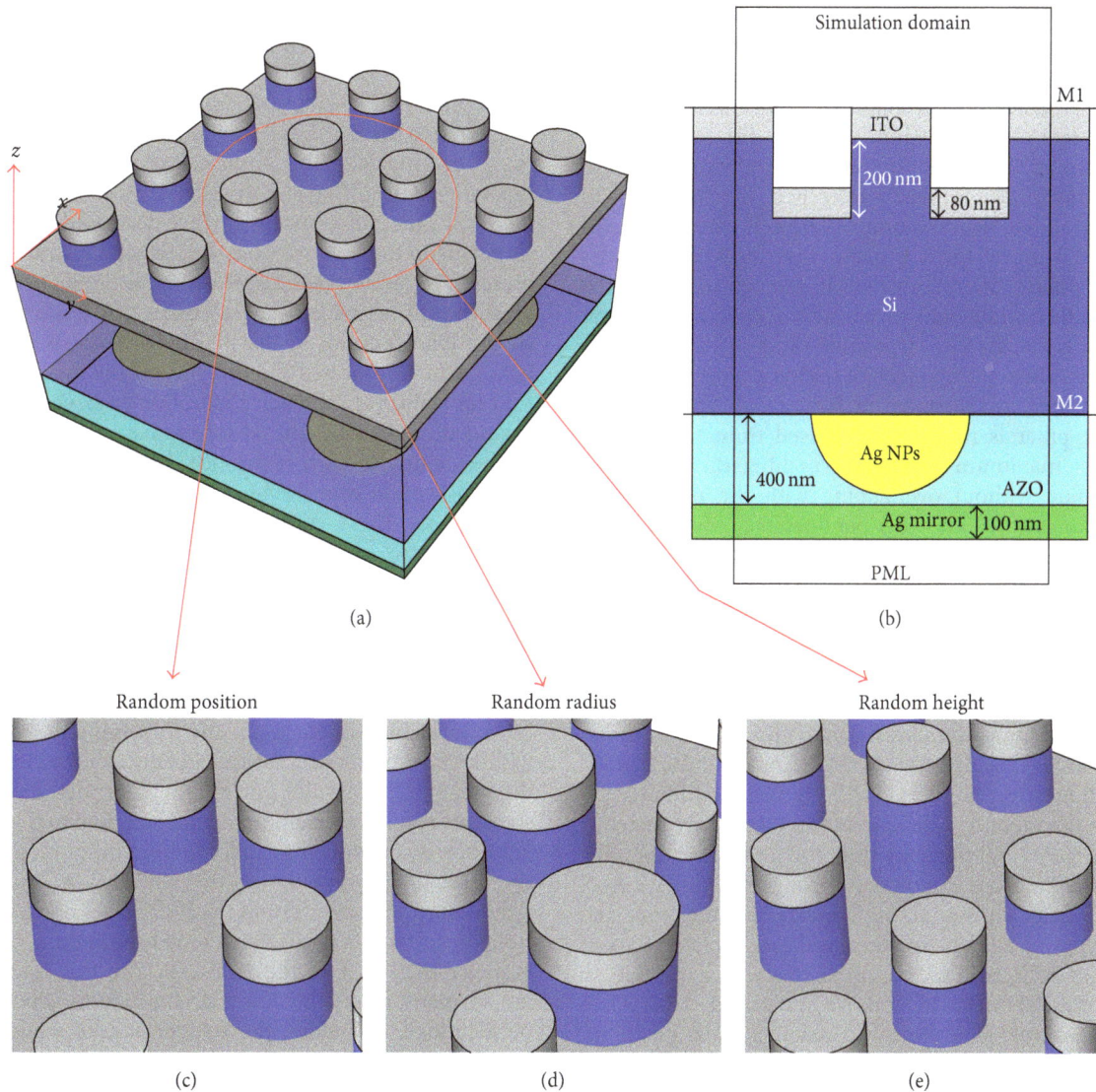

FIGURE 1: The solar cell with hybrid light trapping structure (a) and the solar cells with random grating pillar position (PRSC) (b), with random grating pillar radius (RRSC) (c), and with random grating pillar height (HRSC) (d) used in the simulations.

nanopatterning which provides quasi-guided modes formed by an engineered multiple-scattering process can improve the light coupling efficiency between free space and the films [20]. In our previous work [3, 21, 22], we promoted a hybrid light trapping structure combined of front grating and rear-located silver nanoparticles in TF c-Si solar cell and obtained a short circuit current density (J_{sc}) as high as 29.7 mA/cm^2 within 1 μm c-Si film. In this paper, we attempt to further improve the performance of TF c-Si cells with hybrid light trapping structures through introducing disorder. We study systematically the impact of disorder of initially periodic arrangements of nanostructures as part of an already advanced nanophotonic light trapping concept of prototype thin film solar cells in our study before [3, 22]. We mainly aim to demonstrate the mechanism of the randomness enhancing light absorption and supply a new way of improving thin film solar cell

performance. In addition, the random effect on the shape of the pattern investigated in this paper may also find practical use for estimating the influence of experimental inaccuracies of periodic and quasi-periodic structures.

2. Numerical Method

Figures 1(a) and 1(b) show the schematic of the hybrid light trapping structure we have investigated in our previous work [22], which is combined of front grating and rear-located silver (Ag) nanoparticles. Figures 1(c), 1(d), and 1(e) display the solar cells with randomness in the position, radius, and height of the grating pillar labeled as PRSC, RRSC, and HRSC, respectively. We have simulated the effects of the structural parameters on the optical absorption of the solar cells and identified that the optimum structure with the radius is equal to 160 nm and the height is 200 nm, which gives rise to the highest short circuit current density of 29.7 mA/cm^2. In terms of the optimum silver nanoparticles, the light absorption enhancement is almost constant when the radius is in the range of 350–400 nm, so we fix the radius of Ag nanoparticles as 350 nm. The optimum periodic structure will serve as a starting point and a reference (the REF cell). In the simulation, we define a unit square cell with an area of 2.8 μm by 2.8 μm on the x-y plane. Each unit cell includes 49 grating pillars. The parameters in the cell including the position, radius, and height were randomly generated within certain range. For the PRSC (position random solar cell), the randomness is introduced through setting a certain deviation from the center, as illustrated in Figure 1(c). The center of the grating pillar is randomly displaced from the original center by a maximum of 40 nm, reaching the maximum possible value without overlapping with the neighboring grating pillar; meanwhile other parameters are fixed at their optimum values. For the RRSC, the radius varies from 120 nm to 200 nm, and, in the HRSC, the height varies from 50 nm to 200 nm, while other parameters are fixed at their optimum values.

A numerical analysis of the electromagnetic field features is performed utilizing finite-difference time-domain (FDTD) method (https://www.lumerical.com/). The dielectric functions are modeled using a Drude model for Ag and a Drude–Lorentz model for Si. Periodic boundary conditions are applied in the x- and y-directions while perfectly matched layer boundary conditions are used for the z-direction. The calculated absorption is obtained through the difference of power flux through the monitors located on the front and back surfaces of silicon. In the simulations, plane wave is incident normally onto the nanostructure with a wavelength ranging from 300 nm to 1100 nm. We firstly used MATLAB to randomly generate 49 uniformly distributed pillar positions, which are then fed into the FDTD software to simulate the optical absorption. Note that since the unit simulation area includes 49 grating pillar periods, it can be described as quasi-periodicity more exactly. The randomness is introduced on the basis of the periodic structure.

In addition, a quantitative measure is needed for evaluating the light absorption over the entire solar spectrum. A suitable measure of the performance of the structure is the short circuit current density, that is, J_{sc}, excited by the AM1.5 solar spectrum, which is calculated according to the following equation assuming unit internal quantum efficiency:

$$J_{sc} = e \int \frac{\lambda}{hc} \frac{P_{abs}(\lambda)}{P_{in}(\lambda)} I_{AM1.5}(\lambda)\,d\lambda, \tag{1}$$

where $I_{AM1.5}$ is reference solar spectral irradiance, P_{abs} is the power absorbed by silicon, P_{in} is the incident power, h is Planck's constant, e is the charge of the electron, and c is the speed of light.

3. Simulation Results

3.1. The Simulation Result of the PRSC with Random Position of the Grating Pillar.
Figure 2 shows the optical characteristics of the PRSC with random position of the grating pillar and compared with those of the BARE cell (without any light trapping structure) and the REF cell. It can be seen in Figure 2(a) that both the PRSC and REF cell have lower reflectance compared to the BARE cell over almost the entire wavelength range investigated. This indicates the great antireflection capability of the diffraction grating. Between the REF cell and the PRSC, the reflectance is further decreased in the PRSC, especially in the long wavelengths, which illustrates that the presence of randomness improves the antireflection capability of the diffraction grating. In Figure 2(b), it is seen that the absorption spectrum of the BARE cell consists of multiple peaks, which indicated the presence of the guided mode in thin film silicon [23]. Each absorption peak corresponds to a guided resonance. The absorption is strongly enhanced in the vicinity of each resonance. Hence, one can enhance light absorption over the broad spectrum by a collection of these peaks. These guided modes can be guided into the nanostructures, and this is accomplished by light trapping. This can be seen from the absorption spectrum of the REF cell. Much more peaks and higher peak value appear in the spectrum compared to that of the BARE cell, which indicates the strong light trapping effect of the hybrid light trapping structure [22]. However, there are resonances that cannot be coupled to incident light due to symmetric constraints within a symmetric structure. Generally, symmetries result in degeneracy of the modes, and mirror planes in the symmetric structure result in certain mode not coupling to the incident light [16]. By introducing randomness in the grating position, the structural symmetry in the REF cell is broken. The light absorption spectrum of the PRSC does not show many isolated diffraction peaks, but a broad continuum, flatter and stronger than that of the REF cell. The fact is that there are so many peaks that form a continuum, which cannot be identified. Therefore, the structural randomness results in additional resonances and broadening of the existing resonance. Consequently more light coupled to the guided modes of the PRSC and enhanced the light absorption [24]. J_{sc} of these cells is shown in Figure 2(c), which gives 17.7, 29.7, and 30.4 mA/cm^2, respectively. With the assistance of hybrid light trapping structure, J_{sc} is greatly increased up to 29.7 mA/cm^2, while Yablonovitch [25] gives a theoretical limit of 33.0 mA/cm^2 within 1 μm c-Si. J_{sc} of the REF cell is so high that it is very difficult to be further increased. However, J_{sc} of the PRSC is still boosted up to 30.4 mA/cm^2 resulting from the disordered structures. It is noticed that no structural optimization has been made on the disordered structures, indicating that the disorder structure is a powerful approach for photon management in energy efficiency technologies and may form a new generation of high-efficiency thin film photovoltaic devices.

3.2. The Simulation Result of the RRSC with Random Radius of the Grating Pillar.
Figure 3 shows the optical characteristics of the RRSC. It is found that their light reflection and absorption spectra reveal a similar trend as in the case of the PRSC. This radius distribution has impacts on the incident light of both the long wavelengths and the short wavelengths.

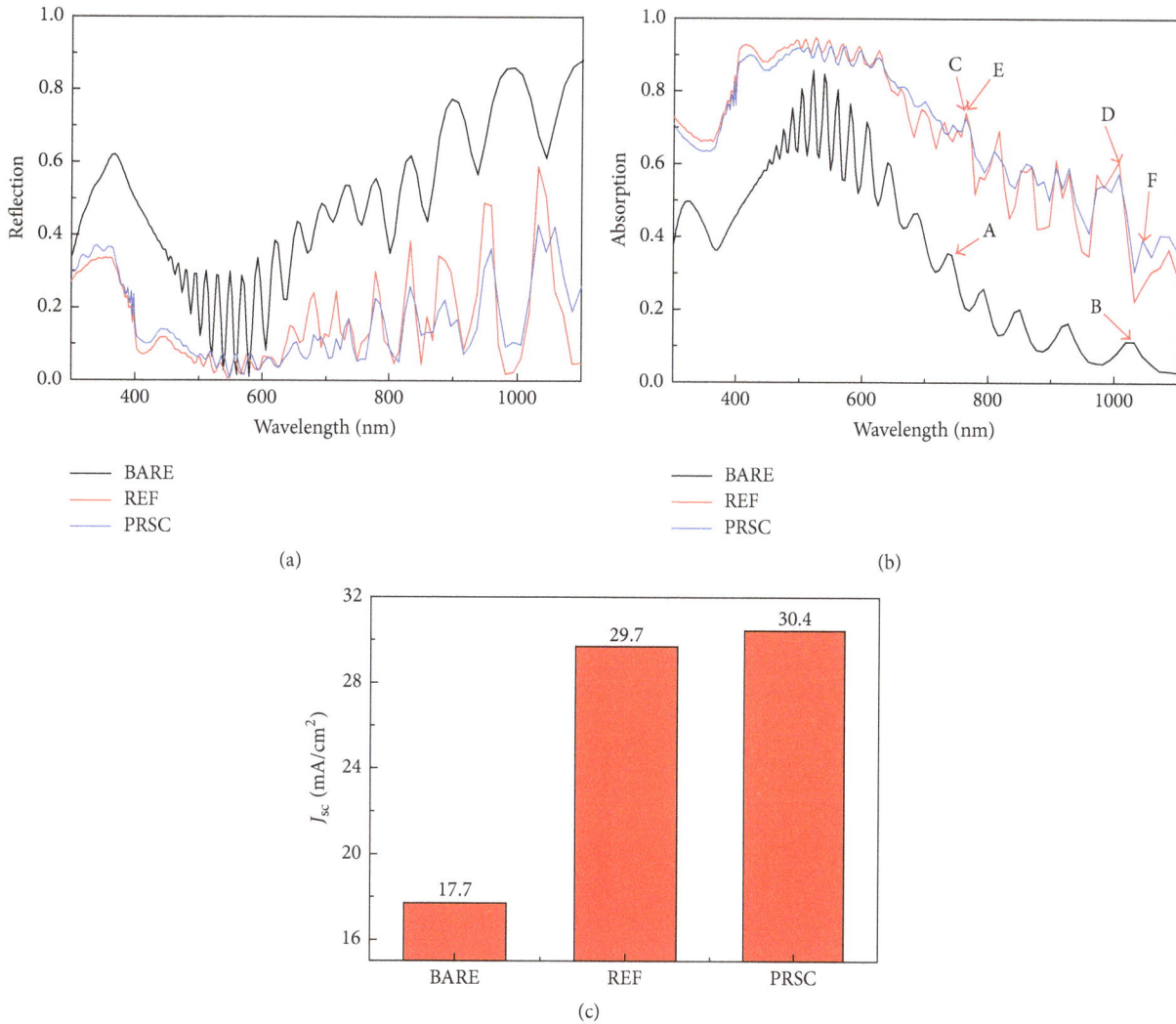

FIGURE 2: The reflection (a), the absorption (b), and the short circuit current density (c) of the BARE cell, the REF cell, and the PRSC with random position of the grating pillar.

Generally, for the asymmetric case, the number of resonances that can contribute to the absorption doubles compared to the symmetric case [5]. It is seen that the reflection of the long wavelength light is reduced dramatically as more guided modes are excited by the randomness introduced into the light trapping structure. These broadening modes are ascribed to the variations in the structural parameters of the RRSC. In the absorption spectrum of Figure 3(b), the higher frequency region has denser peaks, since the density of resonances in the film in general increases with frequency. As for the long wavelengths, a new mechanism for photon absorption is introduced by Oskooi et al. [26] based on quasi resonances, which combine the large absorption of impedance-matched resonances with the broadband and robust characteristics of disordered systems. This makes the absorption spectrum of less sharp peaks but broadening, higher absorption values. J_{sc} of these cells with all these configures are 17.7 mA/cm^2, 29.7 mA/cm^2, and 29.8 mA/cm^2.

3.3. The Simulation Result of the HRSC with Random Height of the Grating Pillar.
For the case of the HRSC, due to the height randomness, the incident light experiences a more gradual change in the effective refractive index, which is also known as the "impedance matching." Consequently, with better index matching between air and the HRSC, light reflection is substantially reduced as seen in Figure 4(a). For the absorption spectrum shown in Figure 4(b), the absorption enhancement can be explained by the richer Fourier spectrum of the disordered structures [7], which increases the number of accessible diffraction orders. However, there is also certain decrease in the absorption of some wavelengths, especially in the short wavelength range. The optimal HRSC is thus determined by a trade-off between the increased coupling efficiency to the guided modes, which improves light trapping, and the coupling to the radiative components, which leads to diffraction in air and degrades light trapping when the disorder becomes too large. Meanwhile, similar

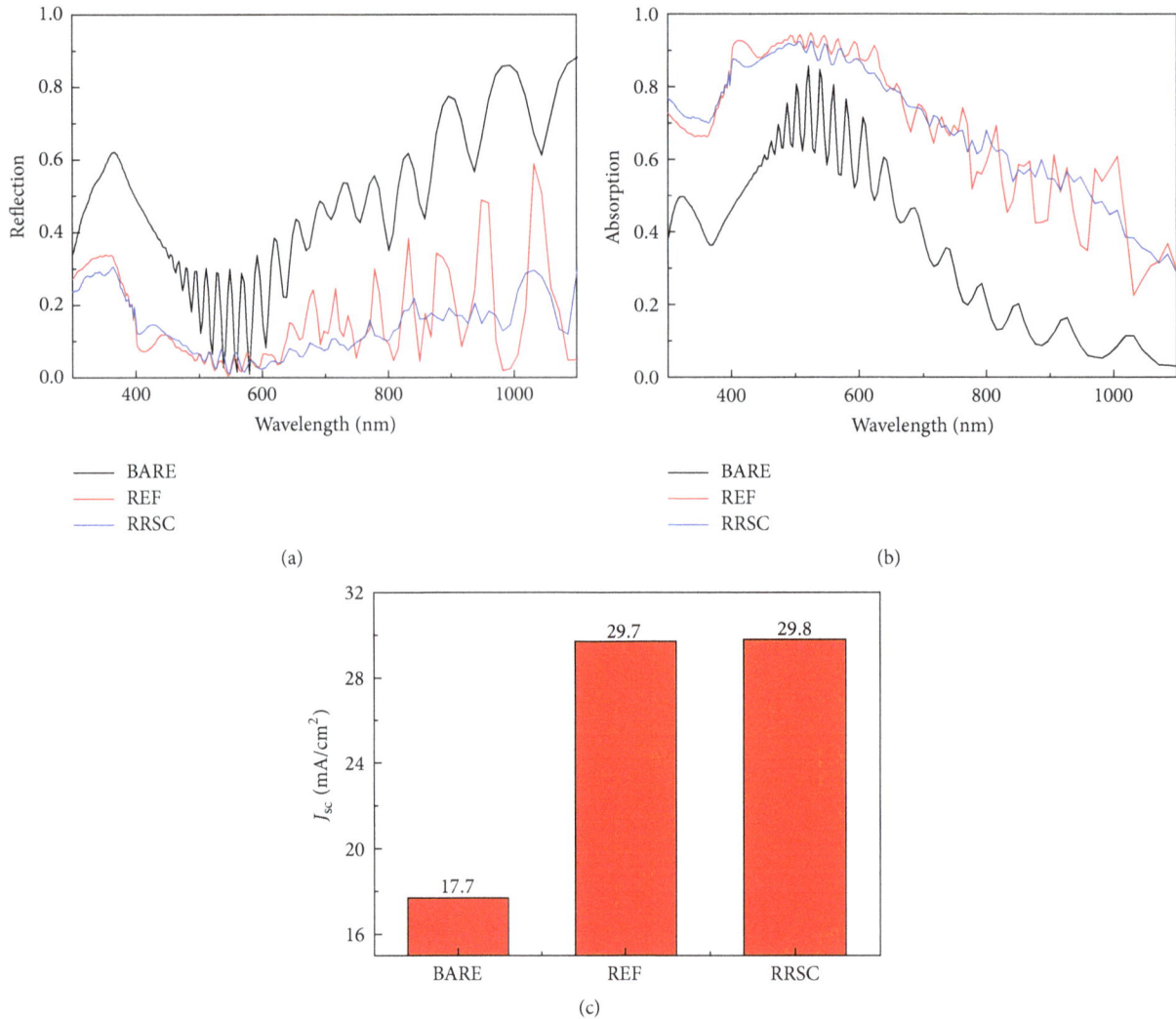

FIGURE 3: The reflection (a), the absorption (b), and the short circuit current density (c) of the BARE cell, the REF cell, and the RRSC with random radius of the grating pillar.

peak-broadening behavior also happens due to height disorder. This phenomenon also exists in the PRSC and RRSC. The final J_{sc} of the HRSC is 30.3 mA/cm^2.

4. Discussion

To gain insight into the mechanism behind this strong absorption enhancement, it is instructive to observe how the electromagnetic energy density is distributed in the cells. The repartition of the energy density between a dielectric material and free space allows the determination of the maximal enhancement of absorption in the material [23, 27]. Figure 5 illustrates the electromagnetic field maps at six representative wavelengths (i.e., 736 and 1019 nm for the BARE cell, 763 and 1006 nm for the REF cell, 763 and 1045 nm for the PRSC). These several wavelengths correspond to the absorption peaks labeled in Figure 2(b). For the BARE cell, the electromagnetic field maps are typical Fabry-Perot

resonances. In planar system, the Fabry-Perot resonance can only induce few absorption peaks at typical wavelength where the normal incident light strongly interferes with the reflected light by the rear flat surface. On the contrary, the REF cell is composed of much richer interfaces and cavity modes rather than the planar system. Therefore, much more guided modes could be excited in the REF cell. Its electromagnetic field maps exhibit a Bloch mode-like diffraction pattern, with a periodicity in the horizontal x-direction. However, strict periodicity brings degeneracy of the modes, which prevents the light absorption from further increasing. In the PRSC, strict symmetry is broken by the randomness of the grating pillar position which could also be clearly distinguished through comparing the electromagnetic field maps between C and E. Once the strict symmetry disappears, more modes could be guided in the cell. The complex field pattern is due to the superposition of the different optical modes in the active layer and reflects the underlying energy transport process.

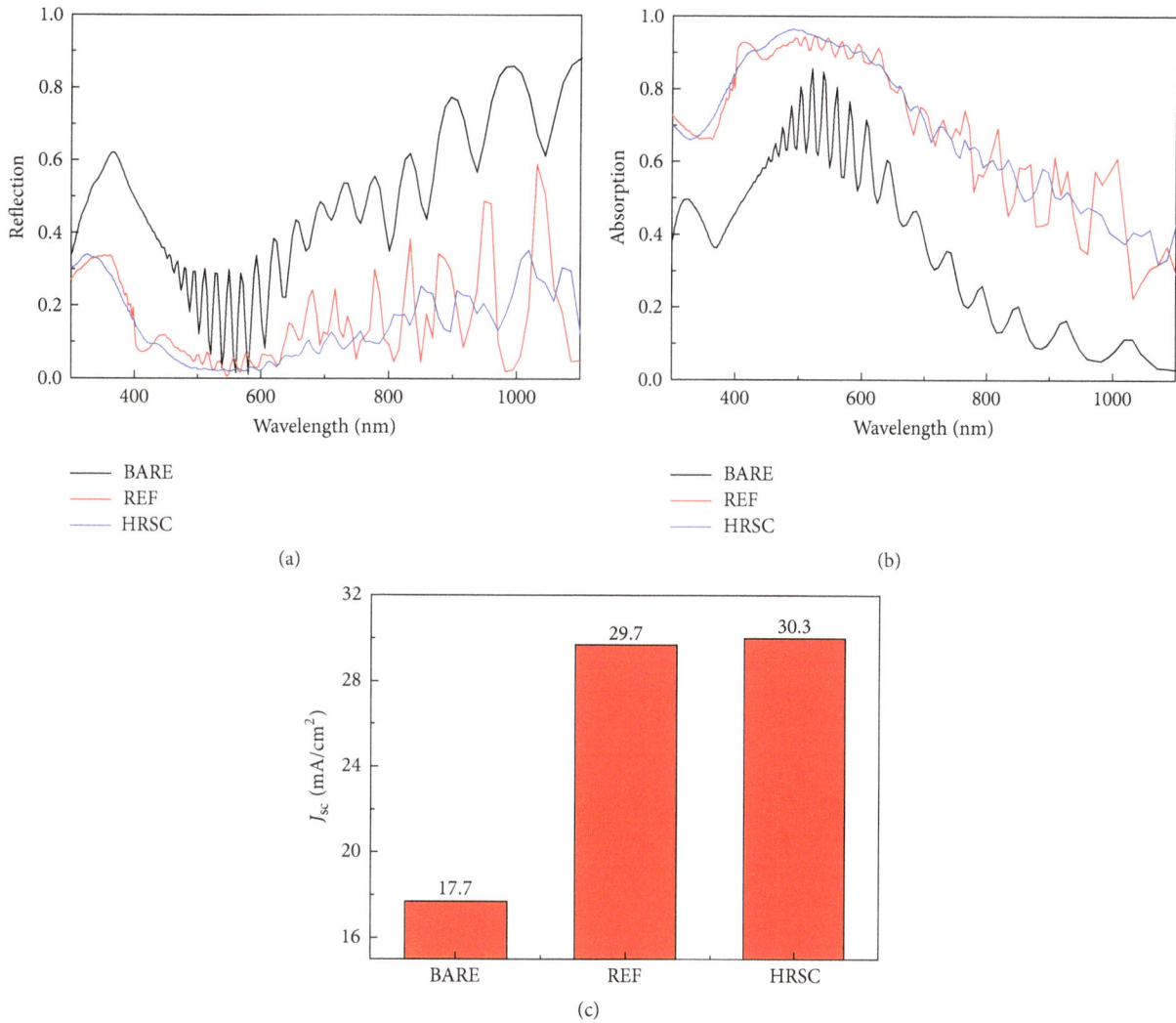

FIGURE 4: The reflection (a), the absorption (b), and the short circuit current density (c) of the BARE cell, the REF cell, and the PRSC with random height of the grating pillar.

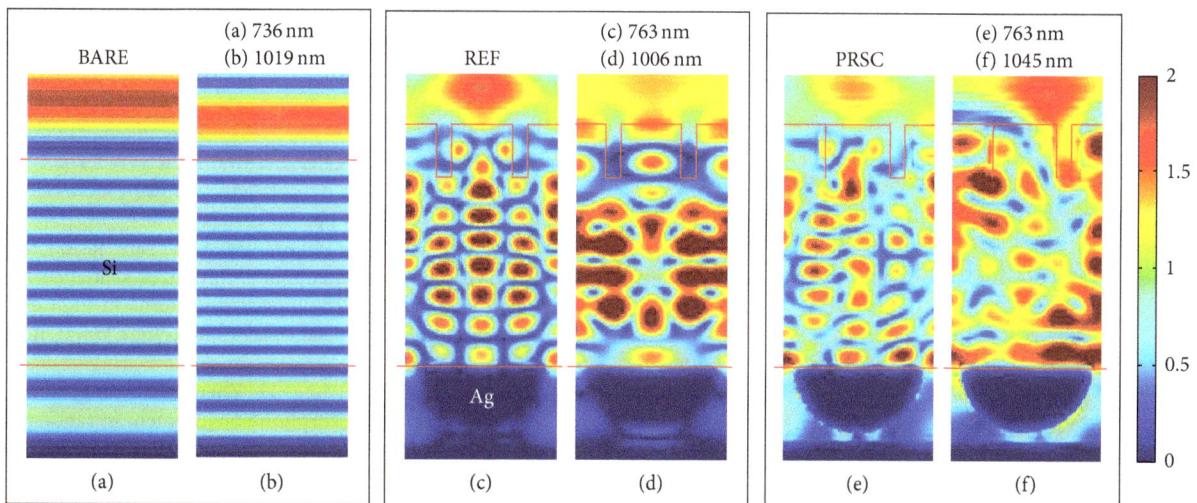

FIGURE 5: The electromagnetic field maps of the BARE cell (a, b), the REF cell (c, d), and the PRSC (e, f), respectively. The wavelength selected for the BARE cell is of (a) 736 nm and (b) 1019 nm and for the REF cell is of (c) 763 nm and (d) 1006 nm, while for the PRSC it is of (e) 763 nm and (f) 1045 nm.

5. Conclusion

In summary, we calculated the optical characteristics of a hybrid light trapping structure with randomness introduced into the structural parameters which include the radius, depth, and position. The light absorption of solar cells with these quasi-periodic nanostructures is improved compared to the reference structure due to reduced reflection, additional resonances induced, and broadening of the existing resonance. Therefore, the structural randomness is beneficial for light absorption enhancement in thin film silicon solar cells. Due to the loss of the strict symmetry resulting from the disorder, many dark modes are introduced into the solar cell, resulting in the increase of the short circuit current density. We conclude that periodic structures contributed a lot in the light absorption enhancement, and on the basis of periodic structures, random structure could further improve the light absorption by breaking the mirror symmetry. The most prominent strategy of employing a periodic nanostructure is to diffract light into several light beams with different directions, resulting in increased light path length in solar cells. Through introducing disorder into the periodic structures, both the periodic and random structures can be utilized in improving the performance of thin film silicon solar cells. The optimal configuration from the point of view of light trapping is neither perfectly ordered nor totally random, but rather an engineered combination of both order and disorder. In conclusion, utilizing certain disorder on the basis of the periodic structures, the performance of TF solar cells can be further improved, and the random effect on the shape of the pattern (position, height and radius) investigated in this literature could be a good approach to estimating the influence of experimental inaccuracies for periodic and quasi-periodic structures.

Conflict of Interests

The authors declare that there is no conflict of interests regarding the publication of this paper.

Acknowledgments

The authors greatly acknowledge the support from the National Basic Research Program of China (973 Program) under Grant no. 2012CB934204 and the National Natural Science Foundation of China under Grants nos. 61076077 and 61274066.

References

[1] G. Gomard, E. Drouard, X. Letartre et al., "Two-dimensional photonic crystal for absorption enhancement in hydrogenated amorphous silicon thin film solar cells," *Journal of Applied Physics*, vol. 108, no. 12, Article ID 123102, 2010.

[2] S. B. Mallick, M. Agrawal, and P. Peumans, "Optimal light trapping in ultra-thin photonic crystal crystalline silicon solar cells," *Optics Express*, vol. 18, no. 6, pp. 5691–5706, 2010.

[3] Y. Shi, X. Wang, W. Liu, T. Yang, and F. Yang, "Hybrid light trapping structures in thin-film silicon solar cells," *Journal of Optics*, vol. 16, no. 7, Article ID 075706, 2014.

[4] X. Meng, E. Drouard, G. Gomard, R. Peretti, A. Fave, and C. Seassal, "Combined front and back diffraction gratings for broad band light trapping in thin film solar cell," *Optics Express*, vol. 20, no. 19, pp. A560–A571, 2012.

[5] Z. Yu, A. Raman, and S. Fan, "Fundamental limit of light trapping in grating structures," *Optics Express*, vol. 18, supplement 3, pp. A366–A380, 2010.

[6] C. Heine and R. H. Morf, "Submicrometer gratings for solar energy applications," *Applied Optics*, vol. 34, no. 14, pp. 2476–2482, 1995.

[7] A. Bozzola, M. Liscidini, and L. C. Andreani, "Broadband light trapping with disordered photonic structures in thin-film silicon solar cells," *Progress in Photovoltaics: Research and Applications*, vol. 22, no. 12, pp. 1237–1245, 2014.

[8] S. Mokkapati, F. J. Beck, and K. R. Catchpole, "Analytical approach for design of blazed dielectric gratings for light trapping in solar cells," *Journal of Physics D: Applied Physics*, vol. 44, no. 5, Article ID 055103, 2011.

[9] Y. Shi, X. Wang, W. Liu, T. Yang, R. Xu, and F. Yang, "Multilayer silver nanoparticles for light trapping in thin film solar cells," *Journal of Applied Physics*, vol. 113, no. 17, Article ID 176101, 2013.

[10] K. R. Catchpole and A. Polman, "Design principles for particle plasmon enhanced solar cells," *Applied Physics Letters*, vol. 93, no. 19, Article ID 191113, 2008.

[11] H. A. Atwater and A. Polman, "Plasmonics for improved photovoltaic devices," *Nature Materials*, vol. 9, no. 3, pp. 205–213, 2010.

[12] A. Ji, Sangita, and R. P. Sharma, "A study of nanoellipsoids for thin-film plasmonic solar cell applications," *Journal of Physics D: Applied Physics*, vol. 45, no. 27, Article ID 275101, 2012.

[13] A. Chutinan, C. W. W. Li, N. P. Kherani, and S. Zukotynski, "Wave-optical studies of light trapping in submicrometre-textured ultra-thin crystalline silicon solar cells," *Journal of Physics D: Applied Physics*, vol. 44, no. 26, Article ID 262001, 2011.

[14] J. P. Mailoa, Y. S. Lee, T. Buonassisi, and I. Kozinsky, "Textured conducting glass by nanosphere lithography for increased light absorption in thin-film solar cells," *Journal of Physics D: Applied Physics*, vol. 47, no. 8, Article ID 085105, 2014.

[15] C. Battaglia, C.-M. Hsu, K. Söderström et al., "Light trapping in solar cells: can periodic beat random?" *ACS Nano*, vol. 6, no. 3, pp. 2790–2797, 2012.

[16] S. E. Han and G. Chen, "Toward the lambertian limit of light trapping in thin nanostructured silicon solar cells," *Nano Letters*, vol. 10, no. 11, pp. 4692–4696, 2010.

[17] V. E. Ferry, M. A. Verschuuren, M. C. V. Lare, R. E. I. Schropp, H. A. Atwater, and A. Polman, "Optimized spatial correlations for broadband light trapping nanopatterns in high efficiency ultrathin film a-Si:H solar cells," *Nano Letters*, vol. 11, no. 10, pp. 4239–4245, 2011.

[18] L. Hong, Rusli, X. Wang, H. Zheng, H. Wang, and H. Yu, "Simulated optical absorption enhancement in random silicon nanohole structure for solar cell application," *Journal of Applied Physics*, vol. 116, no. 19, Article ID 194302, 2014.

[19] V. Depauw, Y. Qiu, K. Van Nieuwenhuysen, I. Gordon, and J. Poortmans, "Epitaxy-free monocrystalline silicon thin film: first steps beyond proof-of-concept solar cells," *Progress in Photovoltaics: Research and Applications*, vol. 19, no. 7, pp. 844–850, 2011.

[20] K. Vynck, M. Burresi, F. Riboli, and D. S. Wiersma, "Photon management in two-dimensional disordered media," *Nature Materials*, vol. 11, no. 12, pp. 1017–1022, 2012.

[21] Y. Shi, X. Wang, W. Liu, T. Yang, J. Ma, and F. Yang, "Extraordinary optical absorption based on diffraction grating and rear-located bilayer silver nanoparticles," *Applied Physics Express*, vol. 7, no. 6, Article ID 062301, 2014.

[22] Y. Shi, X. Wang, W. Liu, T. Yang, and F. Yang, "Light-absorption enhancement in thin-film silicon solar cells with front grating and rear-located nanoparticle grating," *Physica Status Solidi (A)*, vol. 212, no. 2, pp. 312–316, 2015.

[23] Z. Yu, A. Raman, and S. Fan, "Fundamental limit of nanophotonic light trapping in solar cells," *Proceedings of the National Academy of Sciences of the United States of America*, vol. 107, no. 41, pp. 17491–17496, 2010.

[24] C. Lin, L. J. Martínez, and M. L. Povinelli, "Experimental broadband absorption enhancement in silicon nanohole structures with optimized complex unit cells," *Optics Express*, vol. 21, supplement 5, pp. A872–A882, 2013.

[25] E. Yablonovitch, "Statistical ray optics," *Journal of the Optical Society of America*, vol. 72, no. 7, pp. 899–907, 1982.

[26] A. Oskooi, M. De Zoysa, K. Ishizaki, and S. Noda, "Experimental demonstration of quasi-resonant absorption in silicon thin films for enhanced solar light trapping," *ACS Photonics*, vol. 1, no. 4, pp. 304–309, 2014.

[27] D. M. Callahan, J. N. Munday, and H. A. Atwater, "Solar cell light trapping beyond the ray optic limit," *Nano Letters*, vol. 12, no. 1, pp. 214–218, 2012.

19

Components of Lens Power That Regulate Surface Principal Powers and Relative Meridians Independently

H. Abelman[1] and S. Abelman[2]

[1]School of Electrical and Information Engineering, University of the Witwatersrand, Johannesburg,
 Private Bag 3, Wits 2050, South Africa
[2]School of Computer Science and Applied Mathematics, University of the Witwatersrand,
 Johannesburg, Private Bag 3, Wits 2050, South Africa

Correspondence should be addressed to H. Abelman; herven.abelman@wits.ac.za

Academic Editor: Nicusor Iftimia

Paraxial light rays incident in air on alternate refracting surfaces of a thick lens can yield complementary powers. This paper aims to test when these powers are invariant as surface refractive powers interchange in the expression. We solve for relevant surface powers. Potential anticommutators yield the nature of surface principal refractions along obliquely crossing perpendicular meridians; commutators yield meridians that align with those on the next surface. An invariant power component orients relative meridians or the nature of the matrix power on each noncylindrical surface demands that the other component varies. Another component of lens power aligns relative meridian positions for distinct principal powers. Interchanging surface power matrices affects this component. A symmetric lens power results if perpendicular principal meridians are associated with meridians on an opposite rotationally symmetric surface. For thin lenses, meridian alignment may be waived. An astigmatic contact lens can be specified by symmetric power despite having separated surfaces.

1. Introduction

Curved lens surface elements close to an optical axis relative to the radii of the surfaces receive paraxial rays. Such elements and the pole of the surface are in sensibly equivalent planes transverse to the axis. Each refracting surface of a lens approximates a plane transverse to the axis, and the thickness of a lens is the separation of these planes, constant everywhere for rays, and equal to the axial thickness t of the lens. We assume a uniform refractive index n in air. Light rays encounter elements on separated surfaces say 1 and 2, with rotational symmetry for which meridians chosen for their distinct powers do not exist. Such a lens has the scalar power given by

$$G = F_1 + F_2 - \frac{tF_1F_2}{n} \qquad (1)$$

known as Gullstrand's equation [1]. Rays from air may enter on either face without affecting this lens power. Meridians (none is preferred) on one surface necessarily align with those on the other surface. Power F_1 of surface 1 is conjugate to power F_2 of surface 2 and powers are called conjugate variables since when these scalars interchange, the expression for G is left invariant by the transposition. Physical conjugates are ubiquitous in techniques and principles of optics and eye care [2].

In the next sections, preferred meridians of lesser symmetry introduced to each lens surface moderate rotational symmetry. Both refracting surfaces may have principal powers along meridians that may be perpendicular or oblique, aligning or crossing obliquely. Let light from air be refracted by a "back" surface of a stationary nonflipped lens. Then in expressions like (1) transposed powers denote that rays meet surfaces in a new order. Matrix surface powers that yield a symmetric or asymmetric invariant power G are found. Previous work considered systems with asymmetric powers [3]. Independent coefficients of lens thickness measuring the effect of transposition may be commutators or anticommutators of lens surface powers and are made explicit in the expression for lens power. We also show that principal meridians on the respective lens surfaces are aligned as real surface powers commute [4]. Further, if surface powers anticommute, possible principal powers on lens surfaces are equal-and-opposite powers that cross obliquely at 45°. Matrices in commutators have coincident eigenvectors and coincident eigenvectors are those of commuting matrices. Surface principal powers along oblique meridians [5] or perpendicular meridians that are not aligned are reasons for lenses to have antisymmetric dioptric power matrices. Symmetry of matrices serves as a frame of reference and leads to knowledge about the problem that can be identified with the eigenvectors often measured by instruments in the consulting area.

2. Method

Let surface 1 of a lens have matrix power $\mathbf{F}_1$ and let the opposite surface 2 have power $\mathbf{F}_2$. Suppose principal meridians on lens surfaces cross obliquely. From surfaces 1 to 2, a power matrix of the lens may be [6]

$$\mathbf{G} = \mathbf{F}_1 + \mathbf{F}_2 - \frac{t\mathbf{F}_2\mathbf{F}_1}{n} \qquad (2)$$

called the Gullstrand equation, generalized in that it follows from rays traced through toric surfaces with matrix powers $\mathbf{F}_1$ and $\mathbf{F}_2$ [4]. A lens is stationary and as surface powers in expressions are interchanged, this transposition represents light first incident on a "back" refracting surface. If the first two components of an asymmetric power $\mathbf{G}$ are symmetric, this work shows that the component $t\mathbf{F}_2\mathbf{F}_1/n$ may be asymmetric with the principal powers of the lens generally along oblique meridians. Explicit answers are available for when the principal meridians for separate lens surfaces are aligned and the nature then of the power of the lens. Does anticommutation of surface powers (see (4)) confirm this? How are powers $\mathbf{F}_1$ and $\mathbf{F}_2$ related for them to interchange in components of (2) such that $\mathbf{G}$ is invariant or not (the lens has not been flipped)?

If the thickness of the lens is neglected, simultaneous equations $t\mathbf{F}_2\mathbf{F}_1/n = \mathbf{O}$ and

$$\mathbf{F}_1 + \mathbf{F}_2 = \mathbf{G}_0 \qquad (3)$$

are satisfied in (2). The square matrices $\mathbf{F}_1$ and $\mathbf{F}_2$ with the same dimensions are conjugate surface powers that are added in any order (associative) leaving the expression for power $\mathbf{G}_0$ of the thin lens invariant. A surface with oblique principal meridians contributes an antisymmetric component in (3) so that the matrix $\mathbf{G}_0$ is asymmetric [5] and can be expressed as four components: sphere, cylinder, axis, and asymmetry [7].

In lens power equation (2) the coefficient of t/n benefits from the decomposition of the lens surface power products into bracketed terms with noteworthy distinct clinical meanings seen as likely commutators and anticommutators in the identity

$$2\mathbf{F}_2\mathbf{F}_1 = (\mathbf{F}_2\mathbf{F}_1 - \mathbf{F}_1\mathbf{F}_2) + (\mathbf{F}_2\mathbf{F}_1 + \mathbf{F}_1\mathbf{F}_2) \qquad (4)$$

that may each contribute to the symmetry of lens power. If the lens thickness is neglected, these independent meaningful potential symmetry components in (2) and (4) play no role in the lens power as in (3).

We select the left bracket in (4) to write another independent component of lens power $\mathbf{G}$ as

$$\mathbf{G}_B = \frac{t\left(\mathbf{F}_1\mathbf{F}_2 - \mathbf{F}_2\mathbf{F}_1\right)}{2n}. \qquad (5)$$

The power in (5) becomes $-\mathbf{G}_B$ when $\mathbf{F}_2$ and $\mathbf{F}_1$ interchange so that surface power $\mathbf{F}_2$ is not conjugate to $\mathbf{F}_1$. This is the only power component of the lens with this property. The reader can confirm that $\mathbf{G}_B$ is the nonzero antisymmetric component of the power $\mathbf{G}$ of a lens whose surfaces are toric but only if surface meridians are obliquely crossed. In (5) commuting matrices $(\mathbf{F}_1\mathbf{F}_2 = \mathbf{F}_2\mathbf{F}_1)$ imply alignment of principal meridians of surfaces. We return to this point after (8).

The remaining simultaneous contrasting matrix component of $\mathbf{G}$ follows from the term in the right bracket in (4):

$$\mathbf{G}_A = \mathbf{F}_1 + \mathbf{F}_2 - \frac{t\left(\mathbf{F}_1\mathbf{F}_2 + \mathbf{F}_2\mathbf{F}_1\right)}{2n}, \qquad (6)$$

where $\mathbf{G}_0$ in (3) is included in $\mathbf{G}_A$ in which surface power $\mathbf{F}_2$ is seen to be a conjugate of $\mathbf{F}_1$ since $\mathbf{G}_A$ is invariant when interchanging powers in (6) and it is immaterial whether rays first encounter the lens on surface 1 or 2. In addition $\mathbf{G}_A$ is the symmetric matrix component of the power of a thick lens whose surfaces are toric and $\mathbf{G}_A$ is closest to antisymmetric $\mathbf{G}$ in that the Frobenius norm $\|\mathbf{G} - \mathbf{G}_A\|$ is a minimum [8]. With reference to (2), (5), and (6) two components of the matrix power of the lens are

$$\mathbf{G} = \mathbf{G}_A + \mathbf{G}_B. \qquad (7)$$

As $\mathbf{F}_1$ and $\mathbf{F}_2$ trade places in (7), $\mathbf{G}_B$ changes sign and power of the lens becomes [6]

$$\mathbf{G} = \mathbf{G}_A - \mathbf{G}_B. \qquad (8)$$

Powers of surfaces with preferred meridians have been interchanged for matrices $\mathbf{G}$ from (7) to (8). For lens powers $\mathbf{G}$ not to change, surface matrix power $\mathbf{F}_1$ becomes the conjugate matrix power of $\mathbf{F}_2$ and invariant $\mathbf{G}_A$ of (6) is the power for a thick lens. For this $\mathbf{G}_B = \mathbf{O}$ or

$$\mathbf{F}_2\mathbf{F}_1 - \mathbf{F}_1\mathbf{F}_2 = \mathbf{O} \qquad (9)$$

or the powers of the refracting surfaces commute. Equation (9) can be shown to be four dependent scalar equations. The solution for the surface powers $\mathbf{F}_1$ and $\mathbf{F}_2$ of a thick lens requires arbitrary real constants p and q and $\mathbf{I}$ the 2×2 identity matrix. Matrices $\mathbf{F}_1$ and $p\mathbf{F}_1 + q\mathbf{I}$ can be shown to each have distinct eigenvalues. They represent distinct principal powers that can differ on respective lens surfaces. Different matrices $\mathbf{F}_1$ and $\mathbf{F}_2 = p\mathbf{F}_1 + q\mathbf{I}$ have a common set of eigenvectors if and only if matrices commute as in (9) [9]. Eigenvectors represent aligning principal meridian directions on respective lens surfaces. Surface powers that commute have principal values along meridians that align from surface to surface and conversely. Equation (9) for thick lenses is valid for the following reason [10]. Pre- and postmultiply $\mathbf{F}_2$ by $\mathbf{F}_1$:

$$\mathbf{F}_1\mathbf{F}_2 = \mathbf{F}_1 p\mathbf{F}_1 + \mathbf{F}_1 q\mathbf{I},$$
$$\mathbf{F}_2\mathbf{F}_1 = p\mathbf{F}_1\mathbf{F}_1 + q\mathbf{I}\mathbf{F}_1 \qquad (10)$$

TABLE 1: Aligned meridians and commuting surface power matrices. The reader can match the symmetry on respective lens surfaces of meridians in the first column with those in the first row. Meridians have been aligned on surfaces where possible. From (9) power matrices commute for aligned meridians since their power matrices have equal eigenvectors: $\mathbf{F}_2 = p\mathbf{F}_1 + q\mathbf{I}$. At not-equal signs, the meridians from opposite lens surfaces cannot physically align.

	Rotational symmetry	Oblique meridians	Toric symmetry
Rotational symmetry	$\mathbf{F}_1\mathbf{F}_2 - \mathbf{F}_2\mathbf{F}_1 = \mathbf{O}$	$\mathbf{F}_1\mathbf{F}_2 - \mathbf{F}_2\mathbf{F}_1 = \mathbf{O}$	$\mathbf{F}_1\mathbf{F}_2 - \mathbf{F}_2\mathbf{F}_1 = \mathbf{O}$
Oblique meridians	$\mathbf{F}_1\mathbf{F}_2 - \mathbf{F}_2\mathbf{F}_1 = \mathbf{O}$	$\mathbf{F}_1\mathbf{F}_2 - \mathbf{F}_2\mathbf{F}_1 = \mathbf{O}$	$\mathbf{F}_1\mathbf{F}_2 \neq \mathbf{F}_2\mathbf{F}_1$
Toric symmetry	$\mathbf{F}_1\mathbf{F}_2 - \mathbf{F}_2\mathbf{F}_1 = \mathbf{O}$	$\mathbf{F}_1\mathbf{F}_2 \neq \mathbf{F}_2\mathbf{F}_1$	$\mathbf{F}_1\mathbf{F}_2 - \mathbf{F}_2\mathbf{F}_1 = \mathbf{O}$

so that these matrices commute as in (9). Thus surface principal powers along arbitrary aligning meridians of a thick lens satisfy $\mathbf{F}_2 = p\mathbf{F}_1 + q\mathbf{I}$ and have a power matrix $\mathbf{G}_A$ that remains unchanged. A lensometer measures back and front surface vertex powers and may not be the general detector for lens power $\mathbf{G}_A$. Since $\mathbf{F}_1$ and $\mathbf{F}_2 = p\mathbf{F}_1 + q\mathbf{I}$ have a common set of linearly independent eigenvectors [9], (9) is valid and $\mathbf{F}_1$ and $\mathbf{F}_2$ commute. Only one of $\mathbf{F}_1$ and $\mathbf{F}_2$ may represent the power of a surface without preferred meridians (such matrices always commute, (1) and Table 1). For a thick lens, commuting powers $\mathbf{F}_1$ and $\mathbf{F}_2$ in $\mathbf{G}_A$ are conjugates that may be multiplied in any order in their product in (2) and (6).

Another matrix component can have zero power irrespective of lens thickness when

$$\mathbf{O} = \mathbf{F}_1\mathbf{F}_2 + \mathbf{F}_2\mathbf{F}_1 \qquad (11)$$

or the matrices $\mathbf{F}_1$ and $\mathbf{F}_2$ anticommute. General surface principal powers for which this is the case are independent of those for which (9) holds. Suppose $\mathbf{F}_1$ is invertible. This excludes pure cylinder-shaped surfaces 1. Then, (11) can be $\mathbf{F}_2 = \mathbf{F}_1^{-1}(-\mathbf{F}_2)\mathbf{F}_1$ which makes $\mathbf{F}_2$ have the same trace as $-\mathbf{F}_2$. But, generally, trace $(-\mathbf{F}_2) = -\text{trace}(\mathbf{F}_2)$. Thus the trace of $\mathbf{F}_2$ must be zero so that surface 2 has equal-and-opposite principal powers. This procedure is equally valid for $\mathbf{F}_1$ ($\mathbf{F}_2$ is invertible) and each surface is cross-cylinder-like. Anticommuting matrices $\mathbf{F}_1$ and $\mathbf{F}_2$ in (11) yield equal-and-opposite powers on the surfaces. This is not sufficient for (11) to be valid. Further investigation shows that principal meridians must cross at 45° from one surface to the next. $\mathbf{F}_1 + \mathbf{F}_2$ and $t\mathbf{F}_1\mathbf{F}_2/n$ are the simultaneous power components in (2) for a lens where neither surface is a pure cylinder. Then the equal-and-opposite powers on separate surfaces can have principal meridians crossing maximally at 45°. Equation (9) cannot be valid with (11) and associated conjugate powers cannot be present on such a thick lens simultaneously. Information on surface powers disappears if the thickness of the lens is neglected as (9) and (11) effectively become valid simultaneously.

Conjecture 1. *Energy illuminating the retina should be a significant fraction of the energy illuminating a similar area on the correcting lens. Any oblique astigmatism or asymmetry present in the power of the compensating device makes rays focus as caustic curves. Stray beams illuminate the eyeball. Contrast is thus lowered and less energy concentrates at the fovea. Spectacle or contact lenses manufacturers are trained to machine surfaces that combine and nullify the asymmetry. For efficient use of illumination energy, spherocylindrical lenses have powers that are symmetric, and perpendicular meridians align. The validity of this seems to be supported by common practice motivated in part by this work.*

General obliquely crossing principal meridians on surfaces are progressively aligned in what follows, to finally be rectangular. Corresponding lens powers are seen to have the nature that the theory predicts.

Surfaces 1 and 2 of a lens have a common vertical principal meridian β whose powers $F_{\beta 1}$ and $F_{\beta 2}$ are different and nonzero in Figure 1. Near-horizontal principal meridians of plano power on the first surface are not aligned with those on the second surface. Angles are $\pi - \alpha_1$ on the left and $\pi - \alpha_2$ on the right and α_1 and α_2 in radians are small. Eigenvector matrix of surface 1 in (12) that follows is $\left(\begin{smallmatrix} \cos\alpha_1 & \cos\beta_1 \\ \sin\alpha_1 & \sin\beta_1 \end{smallmatrix}\right) \approx \left(\begin{smallmatrix} -1 & 0 \\ \alpha_1 & 1 \end{smallmatrix}\right)$ and for surface 2 we have $\left(\begin{smallmatrix} \cos\alpha_2 & \cos\beta_2 \\ \sin\alpha_2 & \sin\beta_2 \end{smallmatrix}\right) \approx \left(\begin{smallmatrix} -1 & 0 \\ \alpha_2 & 1 \end{smallmatrix}\right)$ with corresponding singular eigenvalue matrices $\left(\begin{smallmatrix} 0 & 0 \\ 0 & F_{\beta 1} \end{smallmatrix}\right)$ and $\left(\begin{smallmatrix} 0 & 0 \\ 0 & F_{\beta 2} \end{smallmatrix}\right)$. Surface power matrices $\mathbf{F}_1$ and $\mathbf{F}_2$ are [5]

$$\mathbf{F}_1 = \begin{pmatrix} -1 & 0 \\ \alpha_1 & 1 \end{pmatrix}\begin{pmatrix} 0 & 0 \\ 0 & F_{\beta 1} \end{pmatrix}\begin{pmatrix} -1 & 0 \\ \alpha_1 & 1 \end{pmatrix},$$

$$\mathbf{F}_2 = \begin{pmatrix} -1 & 0 \\ \alpha_2 & 1 \end{pmatrix}\begin{pmatrix} 0 & 0 \\ 0 & F_{\beta 2} \end{pmatrix}\begin{pmatrix} -1 & 0 \\ \alpha_2 & 1 \end{pmatrix}. \qquad (12)$$

Slightly different principal meridians and powers on the surfaces in Figure 1 are represented in the distinct modal and spectral matrices in (12) that yield asymmetric powers $\mathbf{F}_1$ and $\mathbf{F}_2$.

Meridians of plano power on both refracting surfaces in Figure 1 are not perpendicular to meridian β so that matrices $\mathbf{F}_1 + \mathbf{F}_2$ and $\mathbf{F}_1\mathbf{F}_2$ are antisymmetric and $\mathbf{G}$ in (2) is an antisymmetric power matrix. Thus surface powers $\mathbf{F}_1$ and $\mathbf{F}_2$ in the power $\mathbf{G}$ of the lens that transmits light from surface to surface are not conjugates. For $\mathbf{F}_1$ and $\mathbf{F}_2$ to commute ($\mathbf{G}_B = \mathbf{O}$ in (5)), surfaces need to have aligning meridians (equal eigenvectors) and for this the matrix products identified by the square brackets in

$$\mathbf{F}_1\mathbf{F}_2$$

$$= \begin{pmatrix} -1 & 0 \\ \alpha_1 & 1 \end{pmatrix}\begin{pmatrix} 0 & 0 \\ 0 & F_{\beta 1} \end{pmatrix}\begin{bmatrix} -1 & 0 \\ \alpha_1 & 1 \end{bmatrix}\begin{bmatrix} -1 & 0 \\ \alpha_2 & 1 \end{bmatrix}\begin{pmatrix} 0 & 0 \\ 0 & F_{\beta 2} \end{pmatrix}\begin{pmatrix} -1 & 0 \\ \alpha_2 & 1 \end{pmatrix},$$

$$\mathbf{F}_2\mathbf{F}_1$$

$$= \begin{pmatrix} -1 & 0 \\ \alpha_2 & 1 \end{pmatrix}\begin{pmatrix} 0 & 0 \\ 0 & F_{\beta 2} \end{pmatrix}\begin{bmatrix} -1 & 0 \\ \alpha_2 & 1 \end{bmatrix}\begin{bmatrix} -1 & 0 \\ \alpha_1 & 1 \end{bmatrix}\begin{pmatrix} 0 & 0 \\ 0 & F_{\beta 1} \end{pmatrix}\begin{pmatrix} -1 & 0 \\ \alpha_1 & 1 \end{pmatrix} \qquad (13)$$

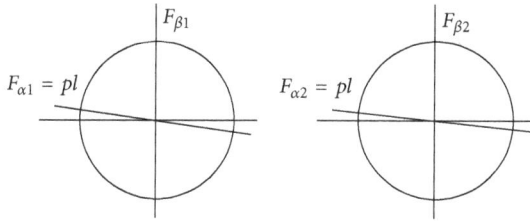

FIGURE 1: Coincident vertical principal meridians whose powers are different are found on the surfaces of a thick lens. Near-horizontal noncoincident meridians have plano power.

equal the identity matrix. This, or equally $\mathbf{F}_2 = p\mathbf{F}_1 + q\mathbf{I}$, implies that the meridians of plano power must coincide on the two surfaces, that is, $\alpha_1 = \alpha_2$ while the β meridians coincide. These conditions cause antisymmetric $\mathbf{F}_1$ to be a conjugate of $\mathbf{F}_2$ or $\mathbf{F}_1$ and $\mathbf{F}_2$ commute as in (9) and $\mathbf{F}_1\mathbf{F}_2$ is antisymmetric. Equation (2) yields the thick lens of asymmetric power

$$\mathbf{G} = \left(F_{\beta 1} + F_{\beta 2} - \frac{t F_{\beta 1} F_{\beta 2}}{n} \right) \begin{pmatrix} 0 & 0 \\ \alpha_1 & 1 \end{pmatrix} \qquad (14)$$

from aligned meridians of surfaces with conjugate matrix powers. The meridian of plano power will be horizontal on the respective surfaces when $\alpha_1 = 0$ in Figure 1. Aligned meridians on the surfaces from the remaining $\mathbf{G}_A$ will be perpendicular which yields a thick lens of symmetric power. Equations (7) and (8) show that surface powers that interchange produce a different power for lens power $\mathbf{G}$. Cylindrical surfaces whose principal meridians each are aligned, vertical and horizontal, form the lens. Power $\mathbf{G}$ is determined with respect to a principal plane of the lens. This power may be expressed as sphere, cylinder, and axis but we may not assume that $\mathbf{G}$ can be measured with respect to a back surface of the lens. All modal matrices in (12) have become $\begin{pmatrix} -1 & 0 \\ 0 & 1 \end{pmatrix}$. The columns (eigenvectors) of this symmetric matrix are along the horizontal and vertical principal meridians of a thick lens that are aligned on the respective refracting surfaces of Figure 1 and whose powers $F_{\beta 1}$ and $F_{\beta 2}$ differ.

3. Discussion

In a general lens, the ray path stipulated appropriate independent components of a power matrix. Interchanging surface powers in an expression affected some components. A demand for invariant lens power made surface powers conjugate. The resultant matrix commutation was exploited advantageously to justify the shape and power of surfaces of spherocylinder lenses. Mostly the lens was treated as thick.

The symmetries of powers along lens surface meridians in (1) are substituted by principal meridians along surfaces in the generalized Gullstrand equation (2) in which the power of a compensating lens depends on the refracting surface that rays first encounter. The thickness-dependent second-order term with a generally asymmetric power was written as autonomous components using (4) to facilitate the

recognition of conjugates. A component of lens power in (6) is invariant when matrix surface powers, now mutually conjugate, change position in the expression. For toric surfaces, in particular, this is the invariant component of lens power. Principal powers along oblique meridians may be those of conjugate surface power components of invariant lens power. A second power component changed sign when matrix surface powers change position in the expression for lens power and controlled how the meridians cross on respective surfaces. For toric surfaces, in particular, this component of asymmetric lens power peters out as principal meridians, represented by eigenvectors of surface matrix powers, align as in (9). Principal powers are distinct on a surface but differ among surfaces. We showed why meridians on pairs of toric refracting surfaces or surfaces with oblique meridians can align or cross obliquely to determine the natures of thick lens powers. All pairs of surfaces, one of which may not have preferred meridians, have distinct matrix powers that commuted in the expression for lens power when surface eigenvectors aligned as illustrated. For thin lenses, meridian alignment essential for a symmetric thick lens power component may be waived. Power component $\mathbf{G}_A$ that is independent of the order of surface powers in its expression provides a norm or reference as paraxial rays do in ray tracing. Component $\mathbf{G}_B$ measures the full deviation of power from containing conjugate variables and symmetry and contributes zero to lens power for an appropriate selection of surface variables. Spherocylindrical lenses need to have thickness to maintain adequate mechanical strength and safety. We have shown that lenses with adequate thickness can have toric power allocated to one surface and spherical power to the other or any of the other options as shown in Table 1 where the surface powers commute. Principal planes in rotationally symmetric lenses are scalar conjugates that may have a matrix analogue and we believe that conjugate variables open a possibility for matrices to shape some principal surfaces in spherocylinder lenses in future. The separation of refracting surfaces may not contribute to antisymmetry in the power matrix. Symmetric or asymmetric lens power can thus be invariant when surface powers are switched for this thick lens.

First-order image positions and sizes that paraxial optics predicts in a system provide a convenient reference from which to measure departures from perfection including aberrations. Linear paraxial expressions are simpler than trigonometric equations [11].

Parabolic mirror surfaces that cast telescope images, retinoscopy [12, 13], and automated optometers provide experimental answers or are calibrated based on principles present in conjugate variables. A common thread in thermodynamics, statistical physics, Hamiltonian mechanics, and other disciplines is the Legendre transform that contains conjugate variables. Well-known conjugate variables are present in the formats for Heisenberg's uncertainty principle. Standard methods of linear algebra [14] can be applied to fields that include medical imaging, engineering, and ophthalmic literature, where eigenvalues and eigenvectors often represent observable and measureable quantities.

Competing Interests

The authors declare that there is no competing interests regarding the publication of this paper.

Acknowledgments

The University of the Witwatersrand, Johannesburg is thanked for hospitality. S. Abelman gratefully acknowledges support from the University of the Witwatersrand, Johannesburg and the National Research Foundation, Pretoria, South Africa.

References

[1] L. Wang, A. M. Mahmoud, B. L. Anderson, D. D. Koch, and C. J. Roberts, "Total corneal power estimation: ray tracing method versus Gaussian optics formula," *Investigative Ophthalmology and Visual Science*, vol. 52, no. 3, pp. 1716–1722, 2011.

[2] A. G. Bennett, "An historical review of optometric principles and techniques," *Ophthalmic and Physiological Optics*, vol. 6, no. 1, pp. 3–21, 1986.

[3] W. F. Harris and R. D. van Gool, "Thin lenses of asymmetric power," *South African Optometrist*, vol. 68, no. 2, pp. 52–60, 2009.

[4] W. F. Harris, "Back- and front-vertex powers of astigmatic systems," *Optometry and Vision Science*, vol. 87, no. 1, pp. 70–72, 2010.

[5] H. Abelman and S. Abelman, "Paraxial ocular measurements and entries in spectral and modal matrices: analogy and application," *Computational and Mathematical Methods in Medicine*, vol. 2014, Article ID 950290, 8 pages, 2014.

[6] R. Blendowske, "Hans-Heinrich Fick. Early contributions to the theory of astigmatic systems," *South African Optometrist*, vol. 62, no. 3, pp. 105–110, 2003.

[7] W. F. Harris, "Keating's asymmetric dioptric power matrices expressed in terms of sphere, cylinder, axis, and asymmetry," *Optometry and Vision Science*, vol. 70, no. 8, pp. 666–667, 1993.

[8] H. Abelman and S. Abelman, "Modification of readings along oblique principal meridians to fit regular corneal surfaces," *Journal of Modern Optics*, vol. 62, no. 14, pp. 1187–1192, 2015.

[9] K. F. Riley, M. P. Hobson, and S. J. Bence, *Mathematical Methods for Physics and Engineering*, Cambridge University Press, Cambridge, UK, 2006.

[10] C. J. Eliezer, "A note on group commutators of 2 × 2 matrices," *American Mathematical Monthly*, vol. 75, no. 10, pp. 1090–1091, 1968.

[11] W. J. Smith, *Modern Optical Engineering*, McGraw-Hill, New York, NY, USA, 2000.

[12] M. Rosenfield, N. Logan, and K. H. Edwards, *Optometry: Science, Techniques and Clinical Management*, Butterworth-Heinemann, Edinburgh, UK, 2009.

[13] M. P. Keating, *Geometric, Physical, and Visual Optics*, Butterworth-Heinemann, Boston, Mass, USA, 1988.

[14] H. Anton and C. Rorres, *Elementary Linear Algebra: Applications Version*, JohnWiley & Sons, Philadelphia, Pa, USA, 2014.

20

Effect of Discrete Levels Width Error on the Optical Performance of the Diffractive Binary Lens

Manal Alshami, Mohamed Fawaz Mousselly, and Anas Wabby

Higher Institute for Applied Sciences and Technology, Damascus, Syria

Correspondence should be addressed to Manal Alshami; loly_394@hotmail.com

Academic Editor: Chenggen Quan

The effects of discrete levels width error developed by thin film deposition on the optical performance of diffractive binary germanium lens with four discrete levels are investigated using nonsequential mode in the optical design code ZEMAX. The thin film deposition technique errors considered are metallic mask fabrication errors. The peak value of the Point Spread Function (PSF) was used as criterion to show the effect of discrete levels width error on the optical performance of the four-level binary germanium lens.

1. Introduction

To enhance optical resolution and reduce aberrations, refractive and diffractive lenses are often combined as a hybrid lens [1]. Diffractive lenses are essentially gratings with variable groove spacing across the optical surface, which impart a change in phase of the wavefront passing through it [2].

A diffractive optical element (DOE), with continuous surface profile, is often referred to as kinoform, and the theoretical ideal profile of the diffractive surface can be approximated in a discrete fashion (the sag is approximated by discrete steps), similar to the digital representation of an analog function [3]. This discrete representation is called a multilevel or binary profile [4]. The design techniques used for binary optics were initially developed by integrated circuit (IC) manufacturers, by using the CAD software [4].

Diffractive surfaces in most of the optical design codes, such as Oslo [5] and ZEMAX [2], are closer approximation to kinoforms than true binary optics, since the phase is continuous everywhere, so the evaluation of the optical performance of that elements will be done for continuous phase profile case [6].

Swanson [3] has developed a technique using Optical Research Associates' (ORA's) Code V lens design software [7] to design diffractive optical elements. This is possible because the code uses direct ray trace with a subcode for holographic optical elements (HOEs), which can be used to partially simulate binary elements. The finite-difference time-domain method was used to simulate subwavelength diffractive lenses [8, 9]. But, for a few of the optical designers who use the optical design code ZEMAX, it will be preferable to design diffractive binary lens by it, but the optical design code ZEMAX does not model the wavelength-scale grooves directly. Instead, ZEMAX uses the phase advance or delay represented by the surface locally to change the direction of propagation of the ray [2].

The fabrication of the single-level and multilevel diffractive lenses involves the generation of a set of masks that are used sequentially for the transfer of their pattern to a substrate in conjunction with photoresist deposition, exposure, and development, as well as an etching procedure such as reactive-ion etching [10], or with thin film deposition [11, 12].

For example, K masks are needed for a lens of 2^K phase levels [3]. The development of these masks (photoresist [11] or metallic [12]) is not usually error-free. These errors are usually known as mask fabrication errors and cause significant deformations of the resulting diffractive binary lens surface and corresponding deterioration of the lens performance. Therefore, the analysis of the effects of these errors on the performance of diffractive optical elements and the determination of acceptable fabrication tolerances for each design is of central importance.

Choi and others had used geometrical and Fourier optics theory to simulate the decrease of MTF due to diffractive

TABLE 1: Specifications of the refractive lens (mm).

Surface	Type	Radius	Thickness	Glass	Diameter
OBJ	Standard	Infinity	Infinity		0.000
STO	Standard	225.371	5.000	Germanium	33.097
2	Standard	Infinity	72.849		32.787
IMA	Standard	Infinity			13.435

optical element fabrication error [1]; Glytsis and others [10] had used the Boundary-Element Method (BEM) as basic modeling tool for analyzing diffractive lenses with fabrication errors. The effects of fabrication errors on the predicted performance of surface-relief phase gratings are analyzed with a rigorous vector diffraction technique by Pommet et al. [13]. Jabbour had used the method of generalized projection to study the effect of experimental errors on the diffractive optical element performance [14].

Alshami et al. [12] had used metallic masks in the development of binary diffractive germanium lens by thin film deposition; hopefully, this paper shows the effect of discrete levels width error due to metallic mask fabrication errors on the optical performance using nonsequential mode in ZEMAX to design the four-level binary surface of a diffractive germanium lens, where the first part presents the design of 4-step binary surface of a diffractive germanium lens [12] with nonsequential mode in ZEMAX and the second part presents the effect of discrete levels width error due to the mask fabrication on the optical performance using the peak value of PSF as criterion.

2. Design of Four-Step Binary Surface in ZEMAX

The design of four-step binary surface of the diffractive binary germanium lens [12] by nonsequential mode in ZEMAX will be presented via the following procedure.

2.1. Refractive Lens. Table 1 shows the optical design specifications of refractive germanium (planoconvex) lens, as shown in Figure 1, for the wavelength band (8)–(12) μm, effective focal length of 75 mm with a 9.09-degree field of view, and diameter 33 mm.

2.2. Diffractive Lens. Table 2 shows the optical design specifications of diffractive germanium lens, with the same conditions as refractive one, and the plane surface chosen as the binary 2 surface (1), as shown in Figure 2:

$$\emptyset = -0.65554\rho^2 + 8.97589\rho^4. \tag{1}$$

2.3. Switch from Kinoform to Binary Surface. In the optical design of the considered lens [12], the diffractive surface contained one diffractive zone, and the ideal diffractive phase profile to be approximated in a binary fashion (4 steps or 4 phase levels) is (1). The diameter of each discrete phase level or binary step (equivalent to phase value $\pi/2$, π, $3\pi/2$, 2π) and the sag's thickness equivalent to each phase value are shown in Table 3 and Figure 3 [12].

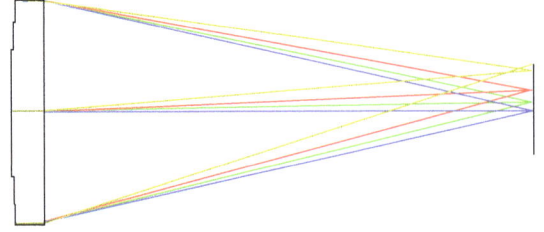

FIGURE 1: Layout of the refractive lens.

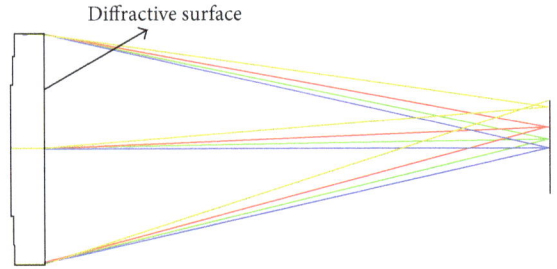

FIGURE 2: Layout of the diffractive lens.

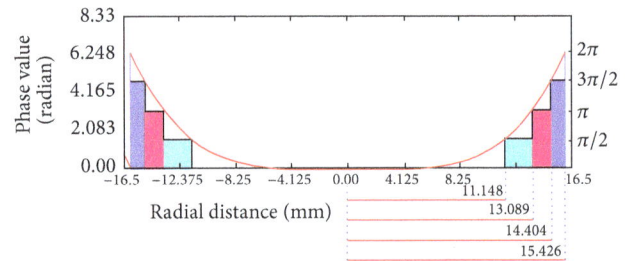

FIGURE 3: Phase curve versus aperture of the diffractive surface slicing into 2π layers and the discrete phase levels.

2.4. Design of Four-Step Binary Surface of a Diffractive Germanium Lens. The design of four-step binary surface of diffractive germanium lens by nonsequential mode in ZEMAX is presented in Tables 4 and 5, and it was done by using the *object cylinder volume* which is a rotationally symmetric volume to design each step of germanium, where the diameter of the front and rear face of each cylinder is the same as the equivalent binary step and the length along the local z-axis of each cylinder is the thickness of the equivalent binary step, as shown in Figure 4. For the optical design in nonsequential mode, we need to define x, y, z position of each object.

Figure 5 shows the difference in FFT PSF cross-sectional curves between the refractive, diffractive, and the designed four-step binary germanium lens.

3. Effect of Discrete Levels Width Error on the Diffractive Optical Element Performance

Imprecision in the metallic mask fabrication process can cause the width of the discrete levels to differ from the theoretical target. This can have an adverse effect on the optical performance. To understand how this effect degrades

TABLE 2: Specifications of the diffractive lens (mm).

Surface	Type	Radius	Thickness	Glass	Diameter	Coeff. on ρ^2	Coeff. on ρ^4
OBJ	Standard	Infinity	Infinity		0.000		
STO	Standard	225.371	5.000	Germanium	33.097		
2	Binary 2	Infinity	73.339		32.787	−0.65554	8.97588
IMA	Standard	Infinity			13.323		

TABLE 3: Diameters and thickness of each binary zone.

Binary zone's number	Equivalent phase value (radian)	Radius of each binary zone (mm)	Diameter of each binary zone (mm)	Equivalent sag's thickness (μm)
1	$\pi/2$	11.148	22.295	0.833
2	π	13.089	26.177	1.667
3	$3\pi/2$	14.404	28.807	2.498
4	2π	15.426	30.851	3.333

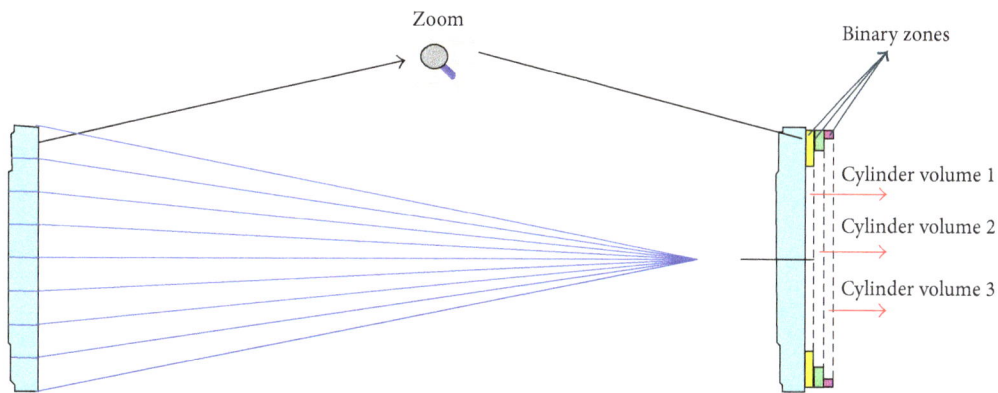

FIGURE 4: The binary diffractive lens with discrete phase levels.

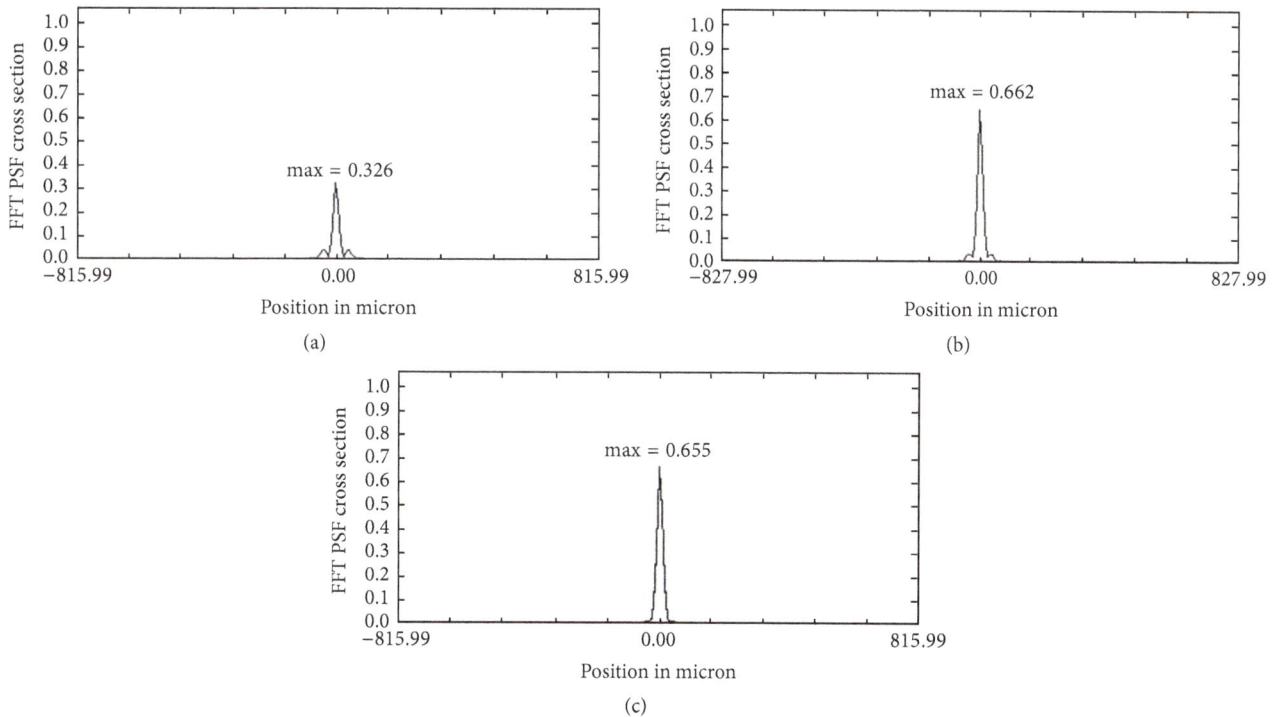

FIGURE 5: FFT PSF cross section of (a) refractive lens, (b) diffractive lens, and (c) designed binary lens.

TABLE 4: Optical design of the binary germanium lens in nonsequential mode.

Surface	Type	Radius	Thickness	Glass	Diameter	Exit lock Z
OBJ	Standard	Infinity	Infinity		Infinity	
STO	Standard	225.371	5.000	Germanium	33.097	
2	Standard	Infinity	0.000		32.787	
3	Nonsequential	Infinity			32.787	73.368
IMA	Standard	Infinity			13.342	

TABLE 5: Data in the nonsequential component editor.

Object number	Object 1	Object 2	Object 3	Object 4	Object 5	Object 6
Object type	Standard lens	Cylinder volume	Standard lens	Cylinder volume	Standard lens	Cylinder volume
Z position (mm)	0.000	0.000	0.000833	0.000833	0.001667	0.001667
Material	Germanium		Germanium		Germanium	
Front R (mm)	0.000	11.148	0.000	13.089	0.000	14.404
Z length (mm)	0.000	0.000833	0.000	0.000833	0.000	0.000833
Back R (mm)	16.500	11.148	16.500	13.089	16.500	14.404
Edge 1 (mm)	16.500	Not used	16.500	not used	16.500	Not used
Thickness (mm)	0.000833	Not used	0.000833	not used	0.000833	Not used
Clear 2 (mm)	16.500	Not used	16.500	not used	16.500	Not used
Edge 2 (mm)	16.500	Not used	16.500	not used	16.500	Not used

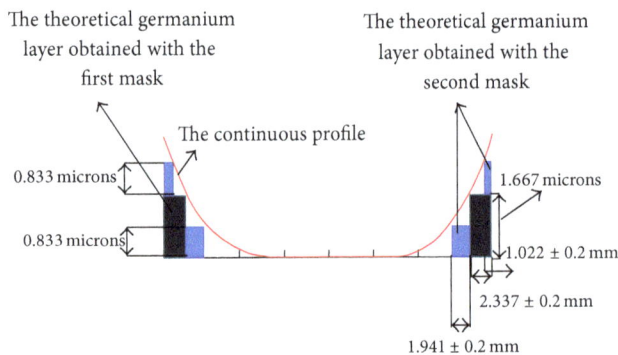

FIGURE 6: Germanium layers (binary zones) and the expected error in their width.

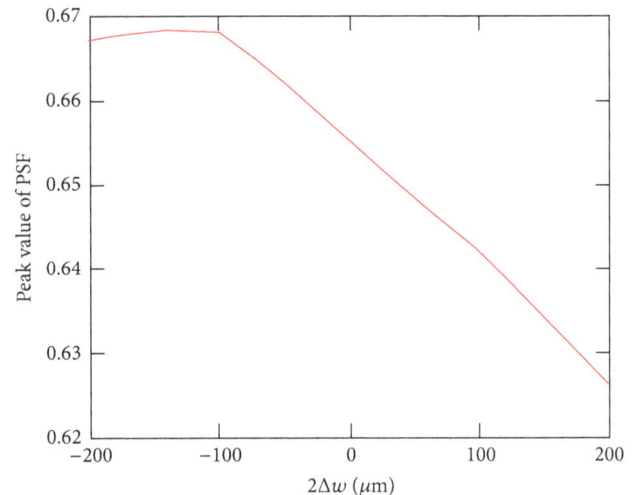

FIGURE 7: Peak value of PSF as a function of variation in zone width error.

performance and thus obtain a tolerance for fabrication errors, we studied how the peak value of the PSF of the designed lens changes as a function of discrete level or zone width variation. The variable Δw is introduced to specify the difference between the final and intended position of the boundaries of each binary zone (an expected error in the width of each binary zone will result due to the metallic masks fabrication accuracy 0.1 mm of the laser machine), as in Figure 6 [12]; the width of each binary zone is then changed by $2\Delta w$ ($2\Delta w = 0.2$ mm). The sign of Δw can be positive or negative corresponding to wider and narrower zones, respectively. In this study, it is assumed that Δw is equal for all zones, independent of their width.

4. Results and Discussion

Table 6 and Figure 7 show the variation in peak value of PSF as a function of $2\Delta w$. The change in $2\Delta w$ studied was limited to

200 μm ($\Delta w = 100 \, \mu$m, the metallic masks fabrication error caused by laser machine). The change 200 μm in $2\Delta w$ has the effect of lowering the PSF peak value (Table 6) by 5%, lowering the diffraction efficiency by 5% [15].

Figure 8 shows the FFT PSF cross section of the considered lens for the extreme error values and without error. It can be seen from Figure 8 that, for this particular diffractive binary lens, the axial resolution increases with increasing zone width, but this occurs at the expense of decreasing the PSF peak value. The metallic mask can be replaced with similar dimension masks which can be produced by three-dimensional printer (rapid prototype) with accuracy of 35 μm so the discrete levels width error will change from $-70 \, \mu$m to

TABLE 6: Peak value of PSF as a function of $2\Delta w$.

$2\Delta w$ (μm)	−200	−150	−100	−50	0.00	50	100	150	200
Peak value of PSF	0.667	0.6684	0.6681	0.662	0.655	0.648	0.642	0.634	0.626

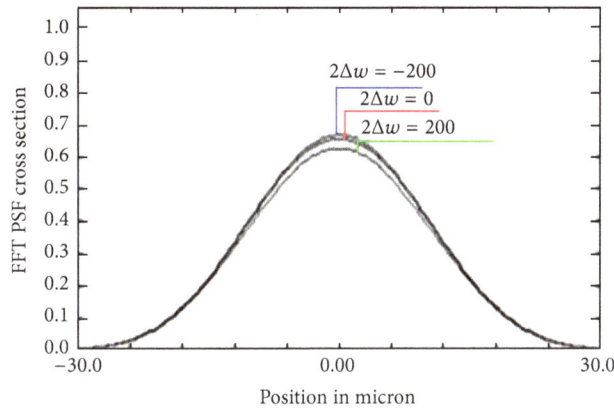

FIGURE 8: FFT PSF cross section.

70 μm which cause lowering in PSF peak value less than 2% then lowering in diffraction efficiency less than 2%; in this case within the 70 μm change in $2\Delta w$, the performance of the considered lens is still acceptable.

5. Conclusion

The effects of discrete levels width error developed by thin film deposition on the optical performance of diffractive binary germanium lens with four discrete levels have been analyzed using nonsequential mode in the optical design code ZEMAX. The thin film deposition technique errors considered are metallic mask fabrication errors. It was found by using the peak value of PSF as criterion that the metallic mask fabrication errors (100 μm laser machine accuracy) have a significant effect on the designed four-level binary germanium lens performance, and to reduce this effect it will be preferable to use another technique to fabricate the desired mask with more fabrication accuracy like, for example, by three-dimensional printer (35 μm is its accuracy).

Competing Interests

The authors declare that they have no competing interests.

Acknowledgments

This work was totally supported by HIAST (Higher Institute for Applied Sciences and Technology).

References

[1] H. Choi, W.-C. Kim, S.-H. Lee, N.-C. Park, and U. Y.-P. Park, "Effects of fabrication errors in the diffractive optical element on the modulation transfer function of a hybrid lens," *Journal of the Optical Society of America A: Optics and Image Science, and Vision*, vol. 25, no. 11, pp. 2764–2766, 2008.

[2] Zemax Product, http://www.zemax.com.

[3] G. J. Swanson, *Binary Optics Technology*, Massachusetts Institute of Technology, Cambridge, Mass, USA, 1989.

[4] A. D. Kathman and S. K. Pitalo, "Binary optics in lens design," in *International Lens Design Conference*, vol. 1354 of *Proceedings of SPIE*, 1990.

[5] Lambda Research Corporation, OSLO, version 6.2, 2001.

[6] N.-H. Kim and R. Zemax, "How Diffractive Surfaces are Modeled in Zemax," September 2005.

[7] Code V reference manual, CODE V version 7.10, Optical Research Associates, March 1987.

[8] T. Shirakawa, K. L. Ishikawa, S. Suzuki, Y. Yamada, and H. Takahashi, "Design of binary diffractive microlenses with subwavelength structures using the genetic algorithm," *Optics Express*, vol. 18, no. 8, pp. 8388–8391, 2010.

[9] V. Raulot, B. Serio, P. Gérard, P. Twardowski, and P. Meyrueis, "Modeling of a diffractive micro-lens by an effective medium method," in *Micro-Optics 2010, 77162J*, vol. 7716 of *Proceedings of SPIE*, May 2010.

[10] E. N. Glytsis, M. E. Harrigan, T. K. Gaylord, and K. Hirayama, "Effects of fabrication errors on the performance of cylindrical diffractive lenses: rigorous boundary-element method and scalar approximation," *Applied Optics*, vol. 37, no. 28, pp. 6591–6602, 1998.

[11] J. Jahns and S. J. Walker, "Two-dimensional array of diffractive microlenses fabricated by thin film deposition," *Applied Optics*, vol. 29, no. 7, pp. 931–936, 1990.

[12] M. Alshami, A. Wabby, and M. F. Mousselly, "Design and development of binary diffractive Germanium lens by thin film deposition," *Journal of the European Optical Society*, vol. 10, Article ID 15055, 2015.

[13] D. A. Pommet, E. B. Grann, and M. G. Moharam, "Effects of process errors on the diffraction characteristics of binary dielectric gratings," *Applied Optics*, vol. 34, no. 14, pp. 2430–2435, 1995.

[14] T. G. Jabbour, *Design, Analysis, and Optimization of Diffractive Optical Elements under High Numerical Aperture Focusing*, University of Central Florida, 2009.

[15] G. J. Swanson and W. B. Veldkamp, "Diffractive optical elements for use in infrared systems," *Optical Engineering*, vol. 28, no. 6, pp. 605–608, 1989.

21

Optical, Thermal, and Mechanical Properties of L-Serine Phosphate, a Semiorganic Enhanced NLO Single Crystal

K. Rajesh,[1] **A. Mani,**[2] **V. Thayanithi,**[2] **and P. Praveen Kumar**[2]

[1]*Department of Physics, Dhanalakshmi College of Engineering, Chennai 601301, India*
[2]*Department of Physics, Presidency College, Chennai 5, India*

Correspondence should be addressed to K. Rajesh; krishjayarajesh@gmail.com

Academic Editor: Mark Humphrey

Single crystals of L-serine phosphate (LSP) were grown by slow evaporation technique. The optical studies reveal the transparency of the crystal in the entire visible region. Grown crystal was subjected to single crystal XRD diffraction technique. Thermal studies of LSP confirm the thermal stability of the crystal and it is stable up to 210°C. The functional groups and optical behaviour of the crystal were identified from FT-IR and UV-Vis analysis. The crystals were also characterized by microhardness and photoconductivity to determine the mechanical strength and the optical conductivity. Laser damage threshold and nonlinear optical activity of the grown crystal were confirmed by Q-switched Nd : YAG laser beam.

1. Introduction

The impact of nonlinear optical crystals in science and technology has been recognized recently for several important applications [1, 2]. This includes sensors, waveguide, transmission, infrared detectors, polarizer, transducers, and image processing; apart from that, the growth of high quality NLO materials for optical switches, optical amplifiers, optical parametric oscillators, and frequency multipliers and mixtures opens a new direction in the field of material science for research. Due to the effectiveness in generating new frequencies from existing laser via harmonic generation, tremendous efforts have been made to identify new materials for such process.

During the past decades, organic and semiorganic materials remain the most widely used crystals for frequency conversion. Organic crystals have a large nonlinear coefficient compared to inorganic crystals. But organic crystals are very sensitive to the presence of intrinsic defects and phonon subsystem [3, 4]. Inorganic crystals have high mechanical and thermal stability than that of organic crystals [5, 6]. Semiorganic crystals are those which combine the positive aspects of organic and inorganic materials resulting in desired nonlinear optical properties [7]. Semiorganic crystals have shown potential applications in the field of nonlinear optics.

The importance of nonlinear optics in optical switching, optical memory storage devices, and telecommunication paved the path to identify a suitable and better NLO crystal [8].

Complexes of amino acids with organic and inorganic salts have been identified as promising materials for producing second harmonic generation (SHG) because of their bonding properties with the ions of organic and inorganic salts. Amino acids are playing vital role in the field of nonlinear optics [9]. L-Serine is one of the amino acid family crystals which are easily available in nature [10, 11]. The molecules of L-serine can combine with anionic, cationic, and overall neutral constituents. The study on growth of L-serine crystals from aqueous solution with hydrochloric acid [12], sodium fluoride [13], formic acid [14], sodium nitrate [15], and acetic acid [16] is reported in recent years.

Orthophosphoric acid is highly polar in nature. It is easily miscible with water. Crystal structure of L-serine phosphate ($C_3O_3NH_7 \cdot H_3PO_4$) was reported early [17]. In the present investigation, good optical quality L-serine phosphate (LSP) single crystal has been grown from aqueous solution by slow solvent evaporation method. Single crystal X-ray diffraction study has been carried out to confirm the crystalline nature of the grown crystal. FT-IR, UV-Vis-NIR, microhardness, thermal, and photoconductivity studies were carried out for the grown crystal. Second harmonic generation (SHG) and

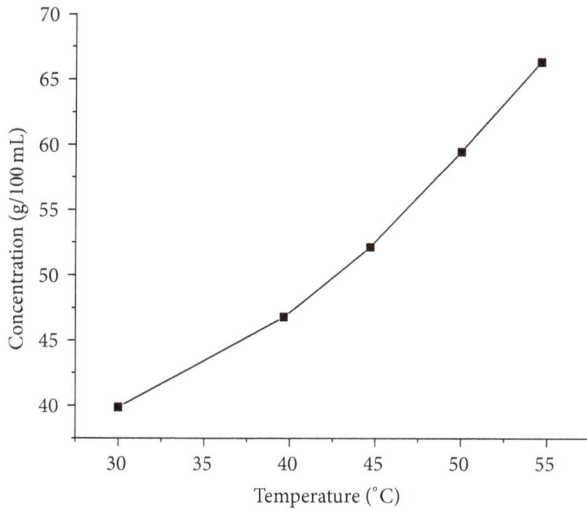

FIGURE 1: Solubility curve of LSP.

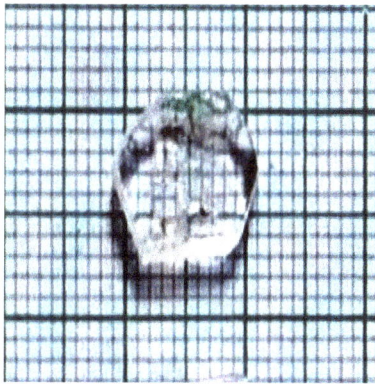

FIGURE 2: Photograph of as-grown LSP crystal.

laser damage threshold (LDT) studies were carried out for the LSP crystal.

2. Experimental

L-serine phosphate single crystal was formed in aqueous solution containing L-serine and phosphoric acid in equimolar ratio 1 : 1 at room temperature. The solution was stirred 9 hours continuously and filtered. The filtered solution is allowed for complete evaporation. The solubility of LSP in deionized water was assessed as a function of temperature in the range of 25°C–50°C. The solubility was gravimetrically analyzed and was found to increase with increase in temperature which is shown in Figure 1. Seed crystals were obtained in a period of 8 days. The quality of the crystal was improved by successive recrystallization process. Optically good quality single crystal was harvested after 35 days from the day of recrystallization. The grown crystal with dimension $12 \times 8 \times 2\,mm^3$ is shown in Figure 2.

3. Results and Discussion

3.1. X-Ray Diffraction Study. Single crystal X-ray diffraction studies were carried out using Enraf Nonius CAD-4/MACH

TABLE 1: XRD data of LSP crystal.

Cell parameters	Present work	Reported values
a	9.129 Å	9.134 (5) Å
b	9.407 Å	9.489 (5) Å
C	4.626 Å	4.615 (5) Å
Crystal system	Monoclinic	Monoclinic
$\alpha = \beta$	90°	90°
γ	98.46°	99.54 (5)°
Space group	$P2_1$	$P2_1$

3 diffractometer, with MoKα radiation to determine the lattice parameters and space group. The crystal belongs to monoclinic crystal system with the space group $P2_1$. Cell parameters and space group of the grown crystal are in good agreement with the reported value [17]. XRD data of the grown crystal with reported values are shown in Table 1.

3.2. FT-IR Study. In order to analyze the presence of functional groups in LSP crystal, Fourier transform infrared (FTIR) spectrum was recorded using the Perkin Elmer Infrared spectrophotometer. The fine powder sample of grown L-serine phosphate crystal was subjected to FT-IR analysis. The characteristic absorption peaks are observed in the range from 400 to 4000 cm^{-1} and are shown in Figure 3.

The strong absorption peak at 3426 cm^{-1} indicates the presence of amine in the grown crystal. NH_2 plane deformation of primary amine is observed at 1561 cm^{-1} and 1622 cm^{-1}. The P–O stretching frequency and deformation of orthophosphoric acid were identified at the peaks 1047 cm^{-1} and 496 cm^{-1}, respectively. This confirms the presence of phosphate ion in the grown crystal lattice [18]. The intense peaks at 2860 and 2950 cm^{-1} are due to C–H stretching [19]. The absorption peak at 1470 cm^{-1} corresponds to CH_3 antisymmetric deformation.

4. Linear and Nonlinear Optical Studies

4.1. UV-Visible Analysis. UV-Vis-NIR studies give important structural information of a given material because absorption of UV and visible light involves promotion of the electrons in π and n orbital from the ground state to higher energy states [20].

The optical absorption spectrum of LSP crystal is shown in Figure 4. The lower cutoff wavelength of the crystal is found as 214 nm. No absorption was found in the visible region of the UV-Vis spectrum. The absence of strong absorption in the entire visible range suggests that the grown LSP crystal is a useful material for the SHG applications [21].

Hence, it is concluded that the grown crystal can be used for optoelectronic applications. The optical absorption coefficient (α) of the crystal was calculated from the UV visible experimental data, using the formula

$$\alpha = \frac{1}{t} \log \frac{1}{T}, \qquad (1)$$

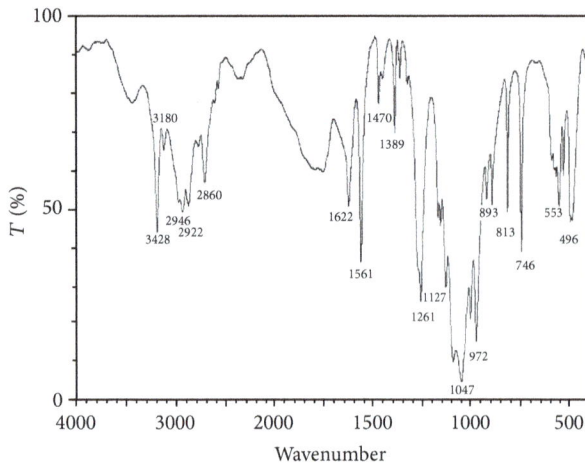

FIGURE 3: FT-IR spectrum of LSP crystal.

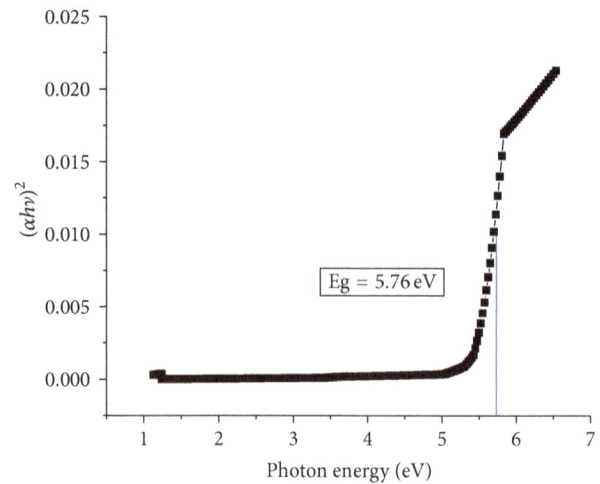

FIGURE 5: Optical band gap energy of LSP.

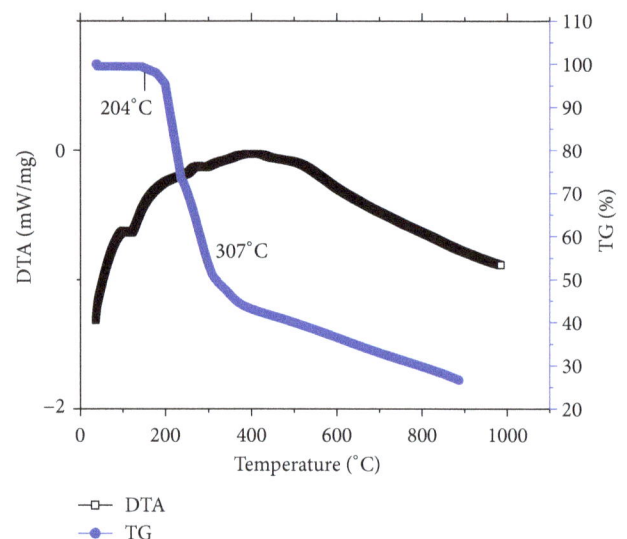

FIGURE 4: UV-Vis spectrum of LSP crystal.

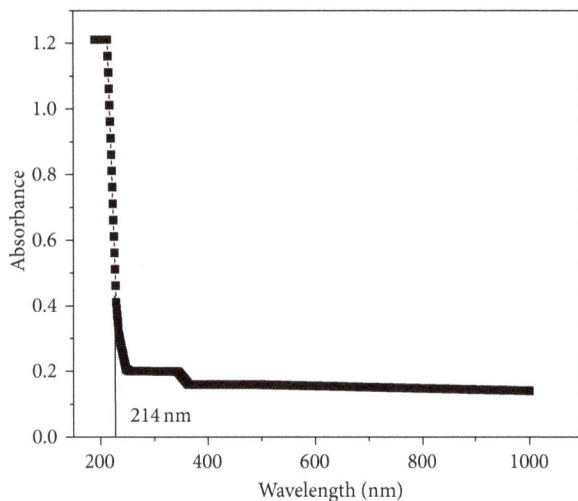

FIGURE 6: TGA and DTA traces of LSP crystal.

where t is the thickness of the crystal and T is the transmittance of the crystal. The optical band gap energy of the crystal can be calculated from Tauc's plot. The optical energy gap (E_g) of the grown LSP crystal was calculated as 5.76 eV from Tauc's plot of $h\nu$ versus $(\alpha h\nu)^2$ which is shown in Figure 5.

4.2. Nonlinear Optical (NLO) Studies. Kurtz powder technique was performed on LSP crystal to confirm the second harmonic generation efficiency [22]. Potassium dihydrogen phosphate (KDP) crystal was used as a reference material for second harmonic generation test. The output SHG signal of 79.7 mV for the LSP crystal was obtained for an input energy of 5 mJ/pulse, whereas the KDP crystal gave an output of 27.6 mV for the same input signal. Thus the SHG efficiency of the LSP crystal is 2.9 times greater than KDP crystal.

4.3. Laser Damage Threshold Test. LDT measurement of the LSP crystal has been carried out using a Q-switched Nd : YAG laser beam of wavelength 1064 nm with the pulse width of 10 ns. The repetition rate of the LDT measurement is 10 Hz.

The laser beam of focal length 1 mm is focused on the sample of 0.7 mm. The LDT value of the grown crystal is found to be 5.27 GW/cm^2.

Hence, the crystal has a high LDT value; it is observed that L-serine phosphate crystal can be used for high power frequency conversion application [23].

4.4. Thermal Analysis. Thermogravimetric analysis (TGA) and differential thermal analysis (DTA) of L-serine phosphate single crystal were carried out using a NETZSCH STA 409°C thermal analyzer. The sample was heated in the temperature ranges between 10°C and 1000°C at a heating rate of 10 K/min in nitrogen atmosphere.

The TGA and DTA traces of LSP crystal are shown in Figure 6. From the plot it is found that the first stage of decomposition of the grown crystal starts at 204°C and the crystal starts to decompose further in weight due to the liberation of volatile gases and compounds. It is concluded

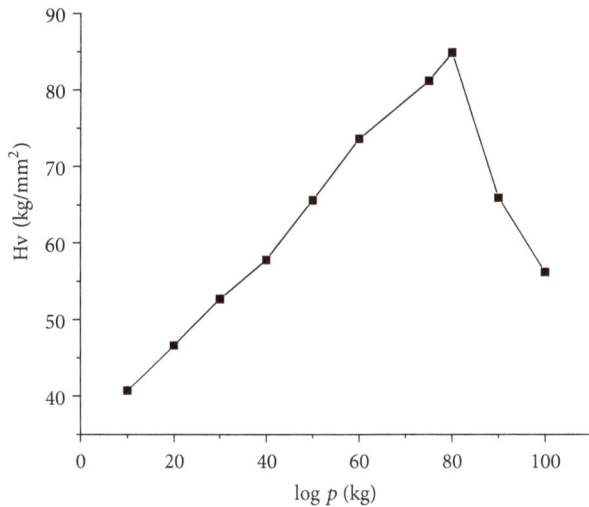

FIGURE 7: Hardness number plot of LSP.

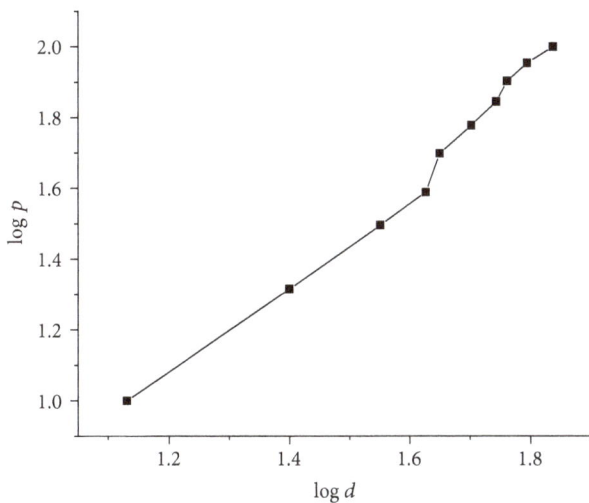

FIGURE 8: Plot of indentation (d) versus load (p).

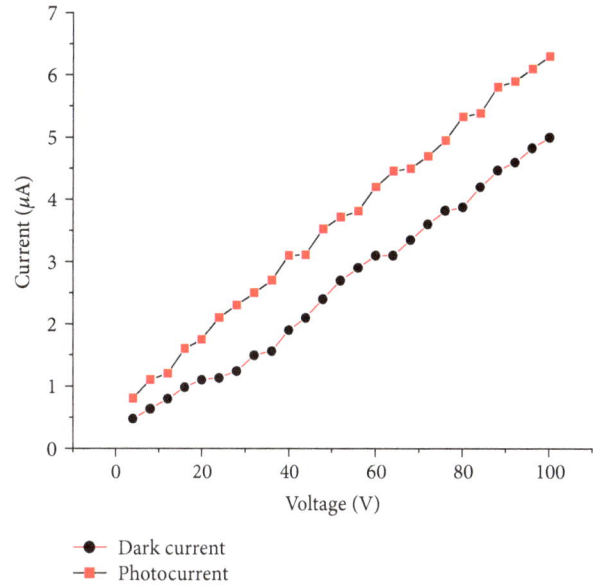

Dark current
Photocurrent

FIGURE 9: Photoconductivity plot of LSP.

that the LSP crystal is thermally stable up to 204°C. Almost 50% weight of the starting material is lost around 307°C. Further the crystal is fully decomposed around 900°C.

4.5. Vickers Microhardness Study. Mechanical stability of the crystal was found using Vickers microhardness tester. To get accurate results of hardness of the grown crystal, indentations were made on the LSP crystals with applied load ranging from 10 g to 100 g. The values of Vickers microhardness at different loads were calculated and a graph was drawn between hardness number and the applied load and is shown in Figure 7.

The hardness number was found to increase with the applied load up to 80 g and then decreases gradually due to the microcracks developed in the crystal.

The value of working hardness coefficient of the material was calculated from the slope of the plot between $\log p$ and $\log d$ (Figure 8) and it is found to be 1.17. According to

Onitsch, n lies between 1 and 1.6 for hard materials, and n is greater than 1.6 for soft materials [24]. Hence, it is concluded that the LSP crystal belongs to hard material category.

4.6. Photoconductivity Study. Photoconductivity studies were carried out for the LSP crystal using Keithley 485 Picoammeter at room temperature. The dark current (I_d) of the sample was measured using DC power supply and picoammeter. The power from the halogen lamp incident on the crystal is 100 W. Photocurrent of the sample was measured using halogen lamp containing iodine vapor. The light from halogen lamp is focused on the material using convex and the photocurrent is measured. DC supply is increased step by step from 10 V to 100 V and the photocurrent (I_p) was measured.

Figure 9 shows the variation of photocurrent and dark current as a function of applied field. It is observed from the plot that the dark current (I_d) and photocurrent (I_p) of the sample increase linearly with the applied field. The photocurrent is always higher than the dark current. This phenomenon is known as positive photoconductivity [25].

5. Conclusion

A potential semiorganic nonlinear optical single crystal of L-serine phosphate was grown by slow evaporation technique. Single crystal X-ray diffraction study confirms that the grown crystal belongs to monoclinic system with space group P2₁.

FT-IR studies confirm the various functional groups present in the crystal and they give the evident of anharmonic phonons in the vibrational spectrum of the crystal. These anharmonic phonons are responsible for nonlinearity in noncentrosymmetric crystals. The grown LSP has a wide transparency window in the entire visible region with a lower cutoff wavelength of 214 nm thus confirming the suitability of this material for various optoelectronic and photoelectric applications. The hardness of the material indicates that the material belongs to hard material category, which can be useful for the high frequency conversion devices. Thermal studies revealed that LSP crystal is thermally stable up to 204°C. Photoconductivity studies show that the material exhibits positive photoconductivity. Positive photoconductivity is achieved at anion interstitials. The powder SHG efficiency of LSP crystal is about 3 times that of KDP. It is concluded that LSP crystal is a suitable material for NLO applications with higher efficiency. Laser damage threshold value of the crystal is calculated from the experimental data and it is found to be 5.27 GW/cm^2. It is concluded that the grown LSP crystal can be used in potential applications in the field of optoelectronic sensors.

Conflict of Interests

The authors declare that there is no conflict of interests regarding the publication of this paper.

References

[1] P. N. Prasad and D. J. Williams, *Introduction of Nonlinear Optical Effects in Molecular and Polymers*, Wiley-Interscience, New York, NY, USA, 1991.

[2] Ch. Bosshard, M. Bösch, I. Liakatas, M. Jäger, and P. Günter, "Second-order nonlinear optical organic materials: recent developments," in *Nonlinear Optical Effects and Materials*, vol. 72 of *Springer Series in Optical Sciences*, pp. 163–299, Springer, Berlin, Germany, 2000.

[3] V. Krishnakumar, G. Eazhilarasi, R. Nagalakshmi, M. Piasecki, I. V. Kityk, and P. Bragiel, "Field-induced non-linear optical features of p-aminoazobenzene crystals," *The European Physical Journal: Applied Physics*, vol. 42, no. 3, pp. 263–267, 2008.

[4] V. Krishnakumar, S. Kalyanaraman, M. Piasecki, I. V. Kityk, and P. Bragiel, "Photoinduced second harmonic generation studies on Tris(thiourea)copper(I) perchlorate Cu(SC(NH$_2$)$_2$)$_3$(ClO$_4$)," *Journal of Raman Spectroscopy*, vol. 39, no. 10, pp. 1450–1454, 2008.

[5] V. G. Dimitrieve, G. G. Gurzodyan, and D. N. Nikogosyan, *Nonlinear Optical Crystals*, Springer, Berlin, Germany, 1991.

[6] V. J. Williams, *Nonlinear Optical Properties of Organic and Polymeric Materials*, American Chemical Society, Washington, DC, USA, 1993.

[7] S. Natarajan, G. Shanmugam, and S. A. M. B. Dhas, "Growth and characterization of a new semi organic NLO material: L-tyrosine hydrochloride," *Crystal Research and Technology*, vol. 43, no. 5, pp. 561–564, 2008.

[8] D. S. Chemla and J. Zyss, *Nonlinear Optical Properties of Organic Molecules and Crystals*, 1-2, Academic Press, New York, NY, USA, 1987.

[9] R. Ramesh Babu, N. Vijayan, R. Gopalakrishnan, and P. Ramasamy, "Growth and characterization of L-lysine mono-hydrochloride dihydrate (L-LMHCl) single crystal," *Crystal Research and Technology*, vol. 41, no. 4, pp. 405–410, 2006.

[10] S. Moitra and T. Kar, "Second harmonic generation of a new nonlinear optical material l-valine hydrobromide," *Journal of Crystal Growth*, vol. 310, no. 21, pp. 4539–4543, 2008.

[11] T. U. Devi, N. Lawrence, R. R. Babu, and K. Ramamurthi, "Growth and characterization of L-prolinium picrate single crystal: a promising NLO crystal," *Journal of Crystal Growth*, vol. 310, no. 1, pp. 116–123, 2008.

[12] M. Parthasarathy, M. Anantharaja, and R. Gopalakrishnan, "Growth and characterization of large single crystals of L-serine methyl ester hydrochloride," *Journal of Crystal Growth*, vol. 340, no. 1, pp. 118–122, 2012.

[13] G. R. Dillip, C. M. Reddy, and B. D. P. Raju, "Growth and characterization of non-linear optical material," *Journal of Minerals and Materials Characterization and Engineering*, vol. 10, no. 12, pp. 1103–1110, 2011.

[14] P. Krishnan, K. Gayathri, and G. Anbalagan, "Growth and characterization of L-serine formate nonlinear optical single crystal," *AIP Conference Proceedings*, vol. 1512, no. 1, pp. 906–907, 2013.

[15] S. D. Zulifiqar Ali Ahamed, G. R. Dillip, L. Manoj, P. Raghavaiah, B. Raghavaiah Deva, and P. Raju, "Growth and characterization of a new NLO material: L-serine sodium nitrate," *Photonics Letters of Poland*, vol. 2, no. 4, pp. 183–185, 2010.

[16] K. Rajesh, P. P. Kumar, A. Zamara, and A. Thirugnanam, "Growth, optical, mechanical and electrical properties of L-serine acetate: a promising semiorganic nonlinear optical crystal," in *Proceedings of the International Conference on Recent Trends in Applied Physics and Material Science (RAM '13)*, vol. 1536 of *AIP Conference Proceedings*, pp. 759–760, Bikaner, India, February 2013.

[17] Yu. I. Smolin, A. E. Lapshin, and G. A. Pankova, "Crystal structure of L-serine phosphate," *Crystallography Reports*, vol. 50, no. 1, pp. 58–60, 2005.

[18] J. R. Lehr, E. T. Brown, A. W. Frazier, J. P. Smith, and R. D. Thrasher, *Crystallographic Properties of Fertilizer Compounds*, Chemical Engineering Bulletin, National Fertilizer Development Center, Muscle Shoals, Ala, USA, 1967.

[19] D. W. Mayo, F. A. Miller, and R. W. Hannah, *Course Notes on the Interpretation of Infrared and Raman Spectra*, John Wiley & Sons, Hoboken, NJ, USA, 2003.

[20] M. Prakash, D. Geetha, and M. Lydia Caroline, "Crystal growth and characterization of L-phenylalaninium trichloroacetate—a new organic nonlinear optical material," *Physica B: Condensed Matter*, vol. 406, no. 13, pp. 2621–2625, 2011.

[21] V. Venkataramanan, S. Maheswaran, J. N. Sherwood, and H. L. Bhat, "Crystal growth and physical characterization of the semiorganic bis(thiourea) cadmium chloride," *Journal of Crystal Growth*, vol. 179, no. 3-4, pp. 605–610, 1997.

[22] S. K. Kurtz and T. T. Perry, "A powder technique for the evaluation of nonlinear optical materials," *Journal of Applied Physics*, vol. 39, article 3798, 1968.

[23] N. Vijayan, G. Bhagavannarayana, T. Kanagasekaran, R. R. Babu, R. Gopalakrishnan, and P. Ramasamy, "Crystallization of benzimidazole by solution growth method and its characterization," *Crystal Research and Technology*, vol. 41, no. 8, pp. 784–789, 2006.

[24] E. M. Onitsch, "The present status of testing the hardness of materials," *Microscope*, vol. 95, pp. 12–14, 1950.

[25] R. H. Bube, *Photoconductivity of Solids*, John Wiley & Sons, New York, NY, USA, 1981.

22

Sagnac Interferometer Based Generation of Controllable Cylindrical Vector Beams

Cristian Acevedo,[1] Angela Guzmán,[2] Yezid Torres Moreno,[1] and Aristide Dogariu[2]

[1]*Grupo de Óptica y Tratamiento de Señales (GOTS), Escuela de Física, Universidad Industrial de Santander, Carrera 27 Calle 9, A.A. 678, Bucaramanga, Colombia*

[2]*CREOL, The College of Optics and Photonics, University of Central Florida, P.O. Box 162700, 4000 Central Florida Boulevard, Orlando, FL 32816-2700, USA*

Correspondence should be addressed to Cristian Acevedo; cristian_rvd@yahoo.com

Academic Editor: Wojtek J. Bock

We report on a novel experimental geometry to generate cylindrical vector beams in a very robust manner. Continuous control of beams' properties is obtained using an optically addressable spatial light modulator incorporated into a Sagnac interferometer. Forked computer-generated holograms allow introducing different topological charges while orthogonally polarized beams within the interferometer permit encoding the spatial distribution of polarization. We also demonstrate the generation of complex waveforms obtained by combining two orthogonal beams having both radial modulations and azimuthal dislocations.

1. Introduction

It is well known that cylindrical vector beams are solutions of vector wave equation [1]. These solutions have cylindrical symmetry in both amplitude and polarization [2]. The unique optical properties of such cylindrical vector beams make them attractive for applications such as optical tweezers [3–6], single molecular imaging [7, 8], and plasmon excitation [9]. Motivated by these applications, various methods for generating cylindrical vector beams have been proposed [10–20]. These methods can be generally divided into active and passive approaches [2]. For active schemes, the resonator laser is modified such that the laser beam oscillates in a desired cylindrical vector mode [10–12]. In passive approaches the wavefront of a traditional laser beam (i.e., Gaussian beam) is transformed into the desired mode [13–15]. A typical passive scheme uses a spatial light modulator (SLM) because it allows a flexible way to generate and modify the selected cylindrical vector beam [16–18].

In this paper we present a novel approach for generating the desired cylindrical vector beams with controlled polarization configuration by using an optically addressable SLM incorporated into a Sagnac interferometer [19, 20].

This scheme enables interfering light beams that can be easily adjusted and are spatially variant in both phase and polarization.

2. Experimental Setup

The experimental arrangement for generating cylindrical vector beams is shown in Figure 1. A linearly polarized light from an Argon laser (JDS Uniphase, Power 100 mW) with wavelength 532 nm in the transverse electromagnetic ground state mode TEM00 is attenuated, filtered, and collimated. The attenuator is used to control the input intensity to the Sagnac interferometer. The collimated beam is directed by a first metallic mirror (M1) towards a half-wave plate (HWP), after which the beam is redirected by a second metallic mirror (M2) towards a 50/50 polarizing beam splitter (PBS) cube. There, the incoming beam is split into a horizontally polarized beam (*p*-component) and a vertically polarized beam (*s*-component). In the half-wave plate, the linear polarization of the laser beam is adjusted to an angle of 45° to ensure that the intensity ratios of the *s*- and *p*-components are 1 : 1. In Figure 1, the polarizing beam splitter (PBS), the two metallic mirrors (M3 and M4), and the optically addressable

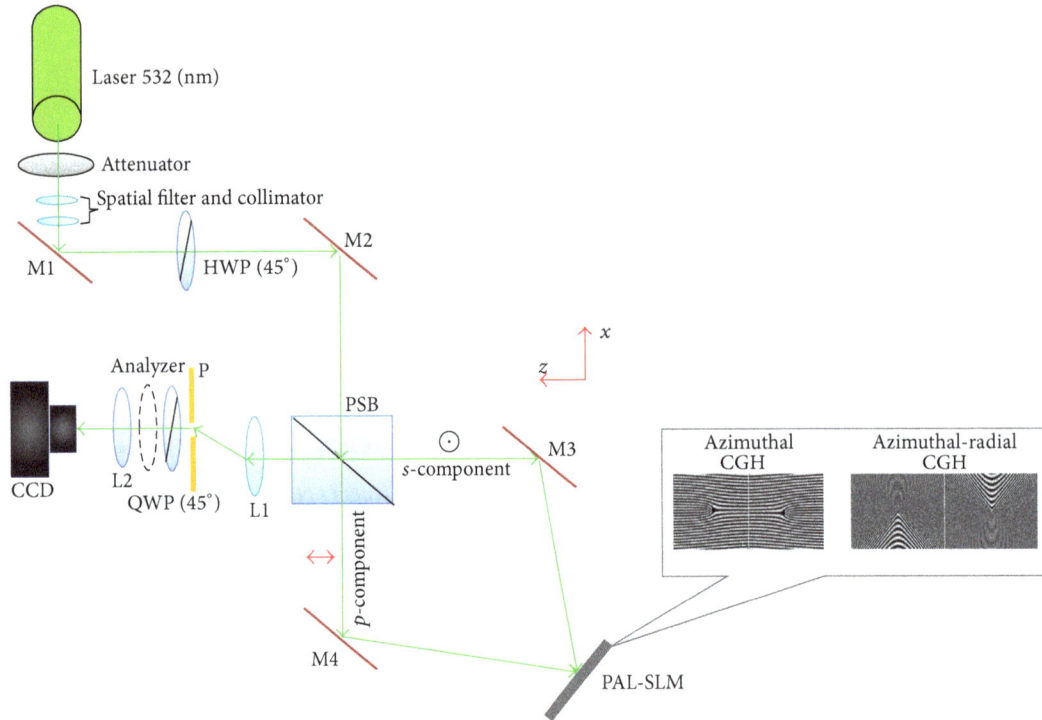

FIGURE 1: Schematic of experimental setup: attenuator; spatial filter; collimator; M, mirror; HWP, half-wave plate; PBS, polarizing beam splitter; PAL-SLM, optically addressable spatial light modulator; L, lens; P, iris diaphragm; QWP, quarter-wave plate; analyzer; and CCD. The inset shows the patterns of two compound holograms displayed on PAL-SLM for azimuthal (left) and azimuthal-radial (right) vector beams.

PAL-SLM (Hamamatsu, Model PPMX8267) compose a Sagnac interferometer. The optical paths for the s- and p-components in the interferometer are equal; that is, ab + bc + cd + da = ad + dc$'$ + c$'$b + ba. Two fork computer-generated holograms (CGHs), assembled by opposite sides, are displayed on SLM window at position c (and c$'$, resp.). The liquid crystal display LCD-coupled PAL-SLM module has a resolution of 1024 × 768 pixels and it is controlled in real time using Matlab software routines to compute the forked CGHs. In order to improve the diffraction efficiency of the orthogonally polarized beams, the LCD has been adjusted parallel to the polarization direction of the readout light in the phase modulation mode [21]. Each of the two components, that is, s- and p-, is diffracted by the hologram consisting of a 2 × 300 × 300-pixel window. Since the SLM is optimized only for diffraction of a horizontally polarized beam, the vertically polarized beam first passes a small half-wave plate (arranged at a 45° angle) in front of the computer hologram of s-component that rotates its horizontal polarization to vertical state. Finally the two back-diffracted beams (s- and p-components) are combined at the Sagnac interferometer output.

The corresponding holograms displayed on the SLM are calculated using an amplitude reflection function given by [21]

$$R(x,y) = \frac{1}{2}r_0\left[1 + \cos\left(\frac{2\pi y}{D} + \delta(x,y)\right)\right], \quad (1)$$

where r_0 is the modulation depth, $D = 1/\sqrt{f_y^2}$ is the spatial period, and $\delta(x,y)$ denotes the relative phase between the orthogonal components due to the two adjacent SLM holograms. The ±1 orders of the s- and p-components are recombined at the PBS and later filtered by an iris diaphragm (P) placed at the focal image plane of the Fourier lens L1 as shown in the figure. The selected orders pass through a quarter-wave plate (QWP) making an angle of 45° with the horizontal axis. It transforms the two orthogonal linear polarization components of the beam, into two orthogonal circularly polarized states. Finally, the output vector beam is sent through an analyzer (rotatable linear polarizer) to a CCD camera placed close to the image plane of the Fourier lens L2. The proposed setup is very stable and highly robust against noise like vibrations or air currents, due to the fact that the two orthogonal components travel along similar paths in the Sagnac interferometer. For this reason it is possible to use this proposed interferometric method with a very short coherence length of the laser beam, for example, micrometers.

3. Results and Discussions

For the two adjacent CGHs placed on the PAL-SLM, the p-mode that is transmitted directly by the PBS at the input of the Sagnac interferometer reads $R_p = R$, while the s mode reads $R_s = (1/2)r_0[1 + \cos(2\pi/D + \delta_-(x,y))]$, where $\delta_- = \delta(-x,y)$, because the circulation of the two beams in the Sagnac interferometer is in opposite directions. The ±1

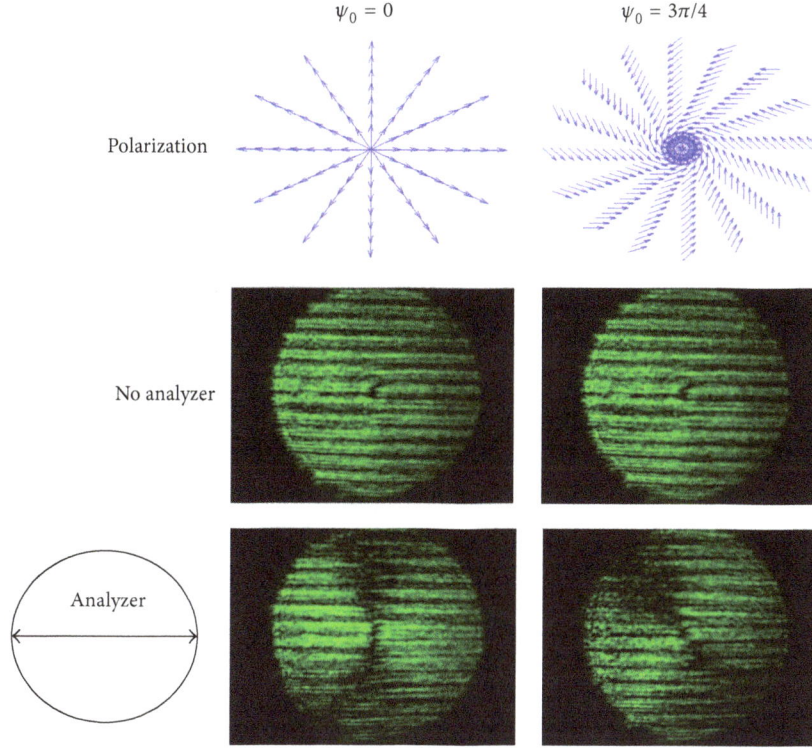

FIGURE 2: Cylindrical vector beams generated with integer topological charge $m = 1$ and constant phase differences of $\psi_0 = 0$ and $\psi_0 = 3\pi/4$.

diffracted orders of the p- and s-components then pass the same interferometric arrangement with phase distributions $\exp(\pm\delta)$ and $\exp(\pm\delta_- + \psi)$, respectively. Here ψ is a constant and controllable phase difference between components. The two beams are overlapped at the rear port of the PBS of the following form: the +1 order of p-component and +1 order of the s-component or −1 order of p- and s-components. The iris diaphragm selects and passes through the QWP only one of the superimposed orders, for instance, the positive orders mentioned above. The resulting optical field immediately behind the QWP can be described as

$$\overrightarrow{E_s} = \left|E_0\right| e^{i\delta} \begin{pmatrix} 1 \\ i \end{pmatrix},$$

$$\overrightarrow{E_p} = \left|E_0\right| e^{i\delta_- + \psi} \begin{pmatrix} 1 \\ -i \end{pmatrix}, \tag{2}$$

where E_0 is a constant factor. After recombination of two optical fields into a single one at the rear focal plane of Fourier lens L2, we write the optical field as follows:

$$\vec{E} = \overrightarrow{E_s} + \overrightarrow{E_p}$$

$$= 2E_0 e^{i(\delta - \delta_- + \psi)/2} \begin{pmatrix} \cos\left(\dfrac{(\delta - \delta_- + \psi)}{2}\right) \\ -\sin\left(\dfrac{(\delta - \delta_- + \psi)}{2}\right) \end{pmatrix}. \tag{3}$$

This beam is a combination of orthogonal base vectors, left- and right-hand circularly polarized components, and

describes a locally linear polarization distribution [1], where the orientation of the direction of polarization can depend on the spatial position. Equation (3) tells us that arbitrary polarization beams can be created with a well-defined phase distribution δ and by adjusting the phase difference ψ. The additional phase distribution $\exp((\delta - \delta_- + \psi)/2)$ is taken into account if we designed $\delta_- = -\delta$. We begin with the cylindrical vector beams with an integer helical phase distribution, which is represented by $\delta = m\theta$ (where θ and m are the azimuth angle in the polar coordinate system and the integer topological charge, resp.). The cylindrical vector beam of the resulting field is $[\cos(m\theta - m\pi - \psi/2), \sin(m\theta - m\pi - \psi/2)]^T$. This field consists of two components corresponding to left- and right-hand circular polarizations with different integer topological charge values.

First, we examine the case for $m = 1$. Figure 2 illustrates the result intensity distribution of the experimental beams generated with constant phase differences of $\psi = 0$ and $\psi = 3\pi/4$, respectively. When the analyzer is not in use, the intensity distributions are the same for all values of the constant phase difference and a dark spot exists at the center of the beam due to the singularity of the optical field. When the analyzer is used, the extinction pattern appears owing to the polarization distribution in the beam cross section. Figure 3 shows a collection of cylindrical vector beams generated with integer topological charges from $m = 2$ to $m = 5$ and constant phase difference $\psi = 0$. We note that the size of the central dark spot in the intensity pattern grows as m increases. When an analyzer is inserted, with transmission axis to zero or ninety degrees, then the

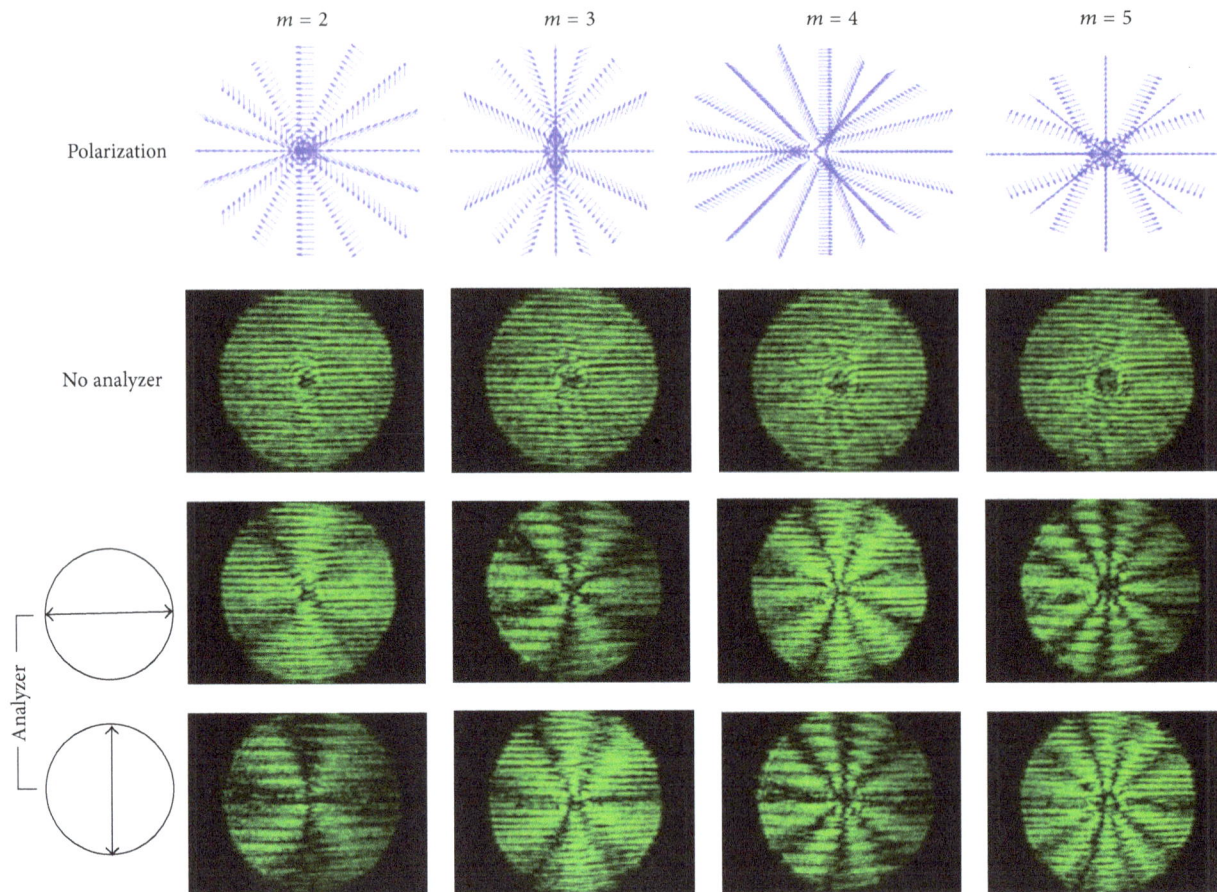

FIGURE 3: Cylindrical vector beams generated with integer topological charges from $m = 1$ to 5, all beams are generated with the same constant phase difference $\psi = 0$.

extinction pattern appears again in the intensity distribution. The dark regions correspond to regions where the beam polarization is perpendicular to the analyzer. Figures 2 and 3 show that the number of both bright and dark fringes is $2m$.

We have also investigated cylindrical vector beams with noninteger topological charge M (where M have integer part, m, and remainder, μ [22]). For that, we choose as phase distribution $\delta = M\theta$, in (3). Figure 4 shows the experimental results obtained for the values of $M = 1.5$ and $M = 2.5$, when $\psi = 0$. The intensity pattern when there is no analyzer consists basically in the breaking of the cylindrical symmetry in the intensity distribution due to the noninteger M. The dark stripe, which originates by the alternating addition of helicoidal beams with the same integer topological charge [22, 23] and different direction of polarization, starts from the center of the intensity pattern, along the $+y$ direction. Without the analyzer, the central dark spot for $M = 2.5$ is bigger than for $M = 1.5$, as expected [23]. With the analyzer in place, one observes three extinction fringes for $M = 1.5$ while for $M = 2.5$ five different fringes appear.

The next experiment demonstrates the generation of cylindrical vector beams with azimuthal dislocation and radial modulation in the phase term of (3). In order to create this phase structure, we consider the arbitrary spatial distribution δ, like the contribution of the two terms, a helical

phase distribution and a radial phase distribution, related to θ and ρ, respectively: $\delta = n\pi\rho/\rho_0 + m\theta$. Here ρ is the radial coordinate, ρ_0 is the size of the cross beam section, and n is a radial number in analogy with m, the azimuthal number. The CGHs for the p- and s-components with values for $m = 1$ and $n = 1$ are shown in the right side of the inset in Figure 1. The corresponding experimental cylindrical vector beams for three different positive radial numbers, with values of $n = 0.5$, 1.0 and $n = 1.5$, are presented in Figure 5. The azimuthal number has been fixed in $m = 1$. The intensity distribution of the three cylindrical vector beams without analyzer is the same. If the analyzer is inserted, the intensity distribution pattern exhibits the Archimedean Spirals where the number of the arms is independent of n.

Finally, if we use negative radial numbers, for example, $n = -0.5, -1.0$ and $n = -1.5$, and the analyzer is inserted, then the chirality in the intensity distribution pattern of the Archimedean Spirals changes to left-handed screw with the same number of arms as shown in Figure 6.

4. Conclusion

We have demonstrated an original method for generating cylindrical vector beams of arbitrary structure and polarization distribution. By means of a Sagnac interferometer,

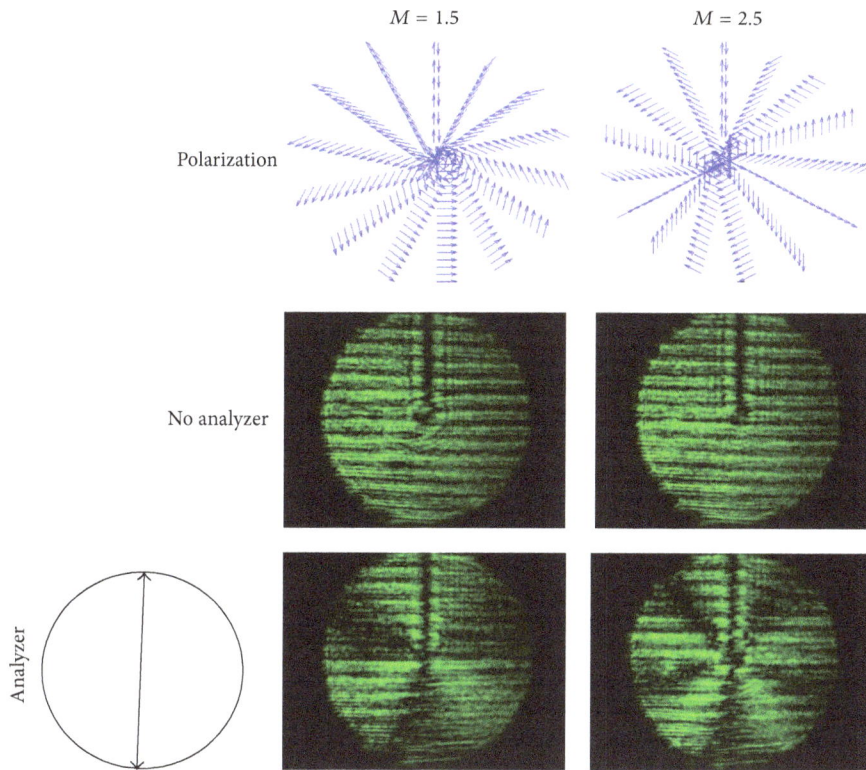

FIGURE 4: Cylindrical vector beams generated for a noninteger topological charge $M = 1.5$ and $M = 2.5$ and constant phase difference of $\psi = 0$ for both.

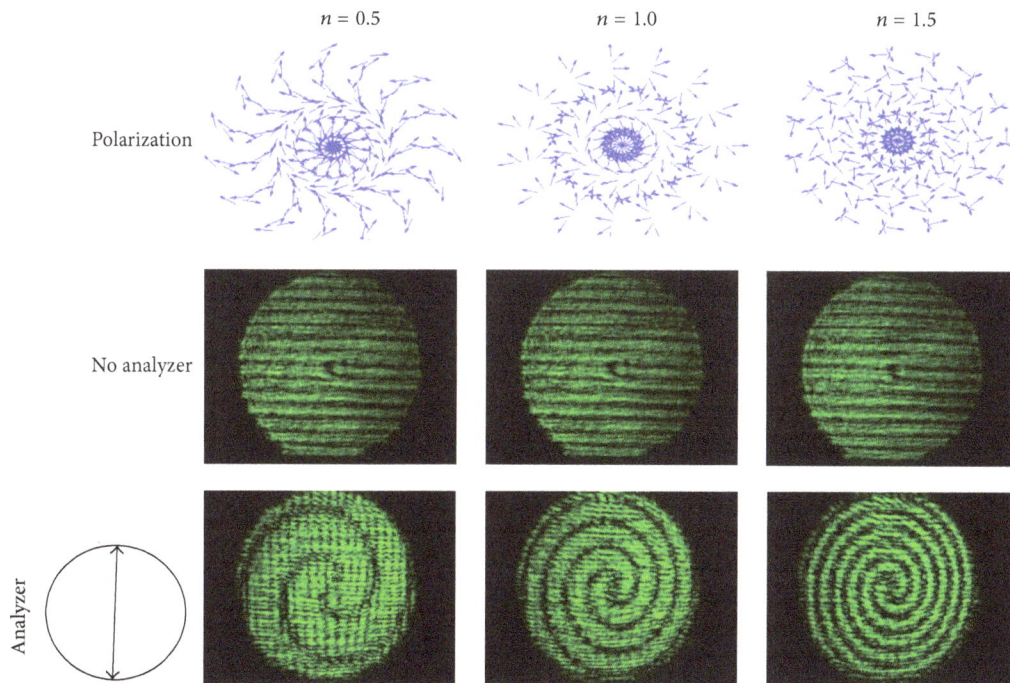

FIGURE 5: Cylindrical vector beams with topological charge $m = 1$ and radial numbers $n = 0.5, 1.0$ and $n = 1.5$.

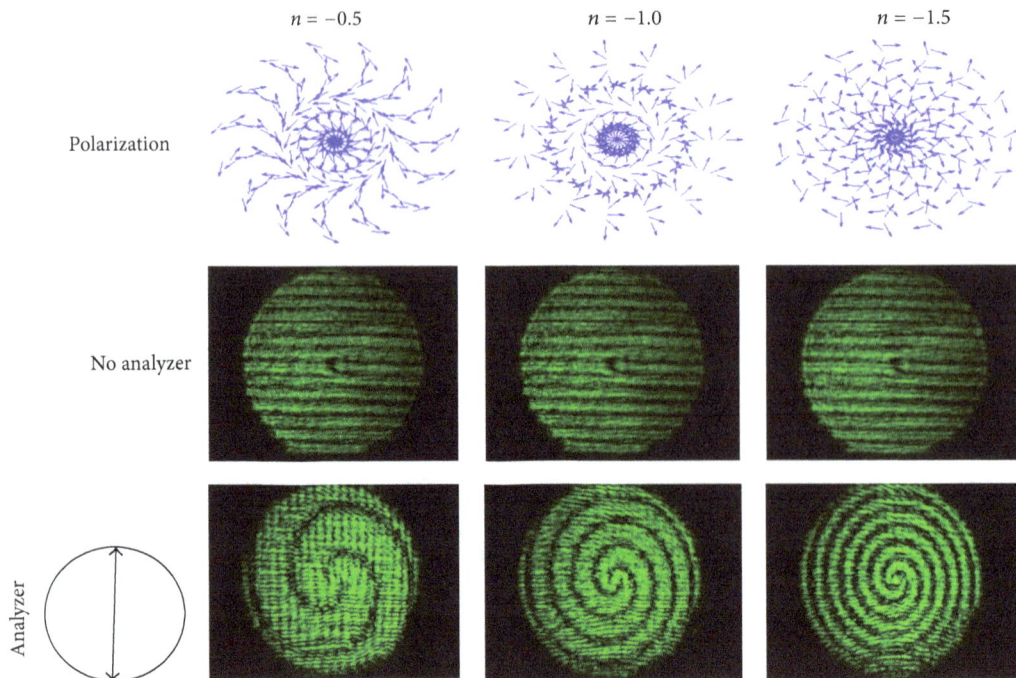

FIGURE 6: Cylindrical vector beams with topological charge $m = 1$ and radial numbers $n = -0.5, -1.0$ and $n = -1.5$.

we have demonstrated that any desired polarization can be realized by the appropriate designing CGHs which are projected adjacent to one another, onto a PAL-SLM. The superposition of two orthogonally and coaxially polarized beams is obtained at the output of the Sagnac interferometer. The first diffraction order of the Fourier transform of that output is then analyzed. This experimental setup is stable against environmental disturbances like mechanical vibrations; it is highly reliable and easily reconfigurable. We note however that the energy efficiency of using CGHs to tailor the cylindrical vector beams obtained is relatively low. Nevertheless, in the same experimental geometry, one could substitute the adjacent CGHs by holograms with high diffraction efficiency. This new method provides both a convenient way to investigate the physics of the propagation of cylindrical vector beams and practical means to explore new applications.

Competing Interests

The authors declare that they have no competing interests.

Acknowledgments

Support was provided by the Vicepresidencia of Research and Services to Universidad Industrial de Santander (funding Projects 5191/5803 and 5708, both of the institutional program for consolidation of research groups, years 2012 and 2013, resp.). Colciencias was supported through Project 110256934957, "Optics devices for quantum key distribution, high dimensionality systems based in orbital angular momentum of light." Funds were obtained from the National Call for the Bank of Projects in Science, Technology and Innovation 2012.

References

[1] D. G. Hall, "Vector-beam solutions of Maxwell's wave equation," *Optics Letters*, vol. 21, no. 1, pp. 9–11, 1996.

[2] Q. Zhan, "Cylindrical vector beams: from mathematical concepts to applications," *Advances in Optics and Photonics*, vol. 1, no. 1, pp. 1–57, 2009.

[3] Q. Zhan, "Trapping metallic Rayleigh particles with radial polarization," *Optics Express*, vol. 12, no. 15, pp. 3377–3382, 2004.

[4] T. A. Nieminen, N. R. Heckenberg, and H. Rubinsztein-Dunlop, "Forces in optical tweezers with radially and azimuthally polarized trapping beams," *Optics Letters*, vol. 33, no. 2, pp. 122–124, 2008.

[5] M. G. Donato, S. Vasi, R. Sayed et al., "Optical trapping of nanotubes with cylindrical vector beams," *Optics Letters*, vol. 37, no. 16, pp. 3381–3383, 2012.

[6] D. Naidoo, M. Fromager, K. Ait-Ameur, and A. Forbes, "Radially polarized cylindrical vector beams from a monolithic microchip laser," *Optical Engineering*, vol. 54, no. 11, article 111304, 2015.

[7] T. J. Gould, J. R. Myers, and J. Bewersdorf, "Total internal reflection STED microscopy," *Optics Express*, vol. 19, no. 14, pp. 13351–13357, 2011.

[8] Z. Man, C. Min, S. Zhu, and X.-C. Yuan, "Tight focusing of quasi-cylindrically polarized beams," *Journal of the Optical Society of America A*, vol. 31, no. 2, pp. 373–378, 2014.

[9] P. B. Phua, W. J. Lai, Y. L. Lim et al., "Mimicking optical activity for generating radially polarized light," *Optics Letters*, vol. 32, no. 4, pp. 376–378, 2007.

[10] G. Machavariani, Y. Lumer, I. Moshe, A. Meir, S. Jackel, and N. Davidson, "Birefringence-induced bifocusing for selection of

radially or azimuthally polarized laser modes," *Applied Optics*, vol. 46, no. 16, pp. 3304–3310, 2007.

[11] M. P. Thirugnanasambandam, Y. Senatsky, and K.-I. Ueda, "Generation of radially and azimuthally polarized beams in Yb:YAG laser with intra-cavity lens and birefringent crystal," *Optics Express*, vol. 19, no. 3, pp. 1905–1914, 2011.

[12] R. Dorn, S. Quabis, and G. Leuchs, "Sharper focus for a radially polarized light beam," *Physical Review Letters*, vol. 91, no. 23, Article ID 233901, 2003.

[13] Y. Kozawa and S. Sato, "Generation of a radially polarized laser beam by use of a conical Brewster prism," *Optics Letters*, vol. 30, no. 22, pp. 3063–3065, 2005.

[14] A. Flores-Pérez, J. Hernández-Hernández, R. Jáuregui, and K. Volke-Sepúlveda, "Experimental generation and analysis of first-order TE and TM Bessel modes in free space," *Optics Letters*, vol. 31, no. 11, pp. 1732–1734, 2006.

[15] K. C. Toussaint Jr., S. Park, J. E. Jureller, and N. F. Scherer, "Generation of optical vector beams with a diffractive optical element interferometer," *Optics Letters*, vol. 30, no. 21, pp. 2846–2848, 2005.

[16] C. Maurer, A. Jesacher, S. Fürhapter, S. Bernet, and M. Ritsch-Marte, "Tailoring of arbitrary optical vector beams," *New Journal of Physics*, vol. 9, article 78, 2007.

[17] X.-L. Wang, Y. Li, J. Chen, C.-S. Guo, J. Ding, and H.-T. Wang, "A new type of vector fields with hybrid states of polarization," *Optics Express*, vol. 18, no. 10, pp. 10786–10795, 2010.

[18] H. Chen, J. Hao, B.-F. Zhang, J. Xu, J. Ding, and H.-T. Wang, "Generation of vector beam with space-variant distribution of both polarization and phase," *Optics Letters*, vol. 36, no. 16, pp. 3179–3181, 2011.

[19] X.-L. Wang, J. Ding, W.-J. Ni, C.-S. Guo, and H.-T. Wang, "Generation of arbitrary vector beams with a spatial light modulator and a common path interferometric arrangement," *Optics Letters*, vol. 32, no. 24, pp. 3549–3551, 2007.

[20] V. G. Niziev, R. S. Chang, and A. V. Nesterov, "Generation of inhomogeneously polarized laser beams by use of a Sagnac interferometer," *Applied Optics*, vol. 45, no. 33, pp. 8393–8399, 2006.

[21] N. Fukuchi, Y. Biqing, Y. Igasaki, N. Yoshida, Y. Kobayashi, and T. Hara, "Oblique-incidence characteristics of a parallel-aligned nematic-liquid-crystal spatial light modulator," *Optical Review*, vol. 12, no. 5, pp. 372–377, 2005.

[22] J. B. Götte, K. O'holleran, D. Preece et al., "Light beams with fractional orbital angular momentum and their vortex structure," *Optics Express*, vol. 16, no. 2, pp. 993–1006, 2008.

[23] M. V. Berry, "Optical vortices evolving from helicoidal integer and fractional phase steps," *Journal of Optics A: Pure and Applied Optics*, vol. 6, no. 2, pp. 259–268, 2004.

23

Analytical Approach to Polarization Mode Dispersion in Linearly Spun Fiber with Birefringence

Vinod K. Mishra

US Army Research Laboratory, Aberdeen, MD 21005, USA

Correspondence should be addressed to Vinod K. Mishra; vkmishr@gmail.com

Academic Editor: Gang-Ding Peng

The behavior of Polarization Mode Dispersion (PMD) in spun optical fiber is a topic of great interest in optical networking. Earlier work in this area has focused more on approximate or numerical solutions. In this paper we present analytical results for PMD in spun fibers with triangular spin profile function. It is found that in some parameter ranges the analytical results differ from the approximations.

1. Introduction

The Polarization Mode Dispersion (PMD) is a well-known phenomenon in optical fibers and its role in the propagation of light pulse in various kinds of optical fibers has been a subject of intensive investigation [1–6] in the past. Its physical origin lies in the birefringence property of an optical fiber so that the orthogonal modes of the light electromagnetic wave acquire different propagation speeds resulting in a phase difference between them. The optical fiber at granular level is nonhomogeneous and also has other defects accumulated during the manufacturing process. Due to these issues, the birefringence in a physical fiber becomes random as pointed out by Foschini and Poole in [7]. In addition, Menyuk and Wai [8] have also considered the nonlinear effects arising from higher order susceptibility that also becomes important under certain physical conditions.

Sometime ago, Wang et al. [1] derived expressions for the Differential Group Delay (DGD) of a randomly birefringent fiber in the Fixed Modulus Model (FMM) in which the DGD has both modulus and the phase. The FMM assumes that the modulus of the birefringence vector is a random variable. They presented analytical results with the following assumptions: (i) the spin function is periodic (a sine function) and (ii) the periodicity length (p) of the fiber is much smaller than the fiber correlation length (L_F) or $p \ll L_F$. Later they also generalized the FMM and presented the Random Modulus Model (RMM), which includes the randomness in

the direction of the birefringence vector. But then the RMM equations could only be solved numerically.

The present work is a contribution to the analytical calculations within FMM and so is only valid for a short fiber distance. This limitation arises because beyond that distance the birefringence randomness [7] becomes dominant. In the present work the full implications of the FMM have been explored under the following conditions: (i) The $p \ll L_F$ approximation has been relaxed, (ii) a nonzero twist has been included, and (iii) the periodic spin rate has been replaced with a constant spin rate. We give the analytical solutions of the exact FMM equations under these conditions and also present some numerical results based on them showing the effect of different physical conditions. The analytical methods are those applicable to the coupled mode theory calculations adapted to the optical fibers [9].

2. Theoretical Analysis

2.1. The Model with Periodic Spin Function. The starting point is the well-known vector equation describing the change in the Jones local electric field vector $\vec{A}(\omega, z)$ with the angular frequency ω and distance z along a twisted fiber. Consider

$$\begin{bmatrix} \dfrac{dA_1(z)}{dz} \\ \dfrac{dA_2(z)}{dz} \end{bmatrix} = \frac{i}{2}(\Delta\beta) \begin{bmatrix} 0 & e^{2i\Theta(z)} \\ e^{-2i\Theta(z)} & 0 \end{bmatrix} \begin{bmatrix} A_1(z) \\ A_2(z) \end{bmatrix}. \quad (1)$$

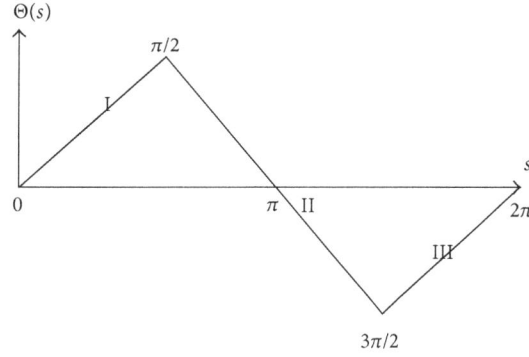

FIGURE 1: The 3-segment approximation to the periodic sine function.

Here $\Delta\beta(\omega)$ is the birefringence and

$$\Theta(z) = \frac{\alpha_0}{\eta} \sin(\eta z) \tag{2}$$

is the periodic spin profile function with spin magnitude α_0 and angular frequency of spatial modulation η.

The boundary conditions are

$$A_1(0) = 1,$$
$$\frac{dA_1(0)}{dz} = 0, \tag{3a}$$
$$A_2(0) = 0,$$
$$\frac{dA_2(0)}{dz} = i\left(\frac{\Delta\beta}{2}\right). \tag{3b}$$

Let $s = \eta z$ be a dimensionless variable. We use $(d/dz) = \eta(d/ds)$ to rewrite (1). Consider

$$\begin{bmatrix} A_{1s}(s) \\ A_{2s}(s) \end{bmatrix} = ia \begin{bmatrix} 0 & e^{2ic\sin s} \\ e^{-2ic\sin s} & 0 \end{bmatrix} \begin{bmatrix} A_1(s) \\ A_2(s) \end{bmatrix}. \tag{4}$$

The subscripts denote differentiation ($A_{1s} = dA_{1s}/ds$, $A_{2s} = dA_{2s}/ds$). Also, $a = (\Delta\beta/2\eta)$ and $c = (\alpha_0/\eta)$ are dimensionless constants.

We express all parameters in terms of the lengths given as beat length ($L_B = 2\pi/\Delta\beta$), spin period ($\Lambda = 2\pi/\eta$), and coupling length ($l_0 = 2\pi/\alpha_0$). Then we can write $a = \Lambda/2L_B$, $c = L_B/l_0$.

The new boundary conditions are

$$A_1(0) = 1,$$
$$A_{1s}(0) = 0, \tag{5a}$$
$$A_2(0) = 0,$$
$$A_{2s}(0) = ia. \tag{5b}$$

These equations ((1) or equivalently (4)) do not have analytical solutions.

In the perturbative approximation (see Appendix B), an analytical result has been derived earlier [1]. In the present work we derive analytic solutions by replacing the sine function by linear segments and compare them to the perturbative solutions for the same segments.

2.2. Linear Segment Approximation to the Periodic Spin Function: Analytical Solutions for the Jones Amplitude Equations

The Model. The periods of the straight line segments shown in Figure 1 approximate the periodic sine function. Here a single period with 3-segment approximation is shown in Figure 1.

The field amplitudes for a given segment satisfy the following equations:

$$\begin{bmatrix} A_{1s}{}^{(m)}(s) \\ A_{2s}{}^{(m)}(s) \end{bmatrix} = ia \begin{bmatrix} 0 & e^{2i\theta_m(s)} \\ e^{-2i\theta_m(s)} & 0 \end{bmatrix} \begin{bmatrix} A_1{}^{(m)}(s) \\ A_2{}^{(m)}(s) \end{bmatrix}. \tag{6}$$

The superscript and subscript m both indicate the segments for which the coupled equations hold. The limits of segments are given below.

We require that the endpoints of $\theta_m(s)$ should be the same as that of the sine-function spin profile $\Theta(s)|_{\text{spin}} = c\sin s$ for all segments. Define $\tilde{c} = (2c/\pi)$ so that the endpoint conditions for segments hold.

For $n = 1$, Segment I ($0 \le s \le \pi/2$),

$$\theta_1(s) = \tilde{c}s,$$
$$\Theta(s = 0)|_{\text{spin}} = 0 = \theta_1(s = 0), \tag{7}$$
$$\Theta\left(s = \frac{\pi}{2}\right)\Big|_{\text{spin}} = c = \theta_1\left(s = \frac{\pi}{2}\right).$$

For $n = 2$, *Segment II* ($\pi/2 \leq s \leq 3\pi/2$),

$$\theta_2(s) = -\tilde{c}s + 2c,$$

$$\Theta\left(s = \frac{\pi}{2}\right)\bigg|_{\text{spin}} = c = \theta_2\left(s = \frac{\pi}{2}\right), \tag{8}$$

$$\Theta\left(s = \frac{3\pi}{2}\right)\bigg|_{\text{spin}} = -c = \theta_2\left(s = \frac{3\pi}{2}\right).$$

For $n = 3$, *Segment III* ($3\pi/2 \leq s \leq 2\pi$),

$$\theta_3(s) = \tilde{c}s - 4c,$$

$$\Theta\left(s = \frac{3\pi}{2}\right)\bigg|_{\text{spin}} = -c = \theta_3\left(s = \frac{3\pi}{2}\right), \tag{9}$$

$$\Theta(s = 2\pi)|_{\text{spin}} = 0 = \theta_3(s = 2\pi).$$

The General m-Segment Solutions. The solutions for the mth segment have the following general form:

$$
\begin{bmatrix}
e^{-i\theta_m(s)} A_1^{(m)}(s) \\
iae^{i\theta_m(s)} A_2^{(m)}(s)
\end{bmatrix}
$$

$$
= \begin{bmatrix}
a_1^{(m)} + ib_1^{(m)} & a_2^{(m)} + ib_2^{(m)} \\
\left\{-\theta_{m/s}b_1^{(m)} + q_m a_2^{(m)} + i\left(\theta_{m/s}a_1^{(m)} + q_m b_2^{(m)}\right)\right\} & \left\{-\left(\theta_{m/s}b_2^{(m)} + q_m a_1^{(m)}\right) + i\left(\theta_{m/s}a_2^{(m)} - q_m b_1^{(m)}\right)\right\}
\end{bmatrix}
\begin{bmatrix}
\cos q_m s \\
\sin q_m s
\end{bmatrix} \tag{10}
$$

with

$$q_m^2 = a^2 + \theta_m^2(s),$$

$$\theta_{m/s} = \frac{d\theta_m(s)}{ds}. \tag{11}$$

The exact solutions for the coupled equations in one segment are related to those in the previous adjacent segment by the following chain-relations among the coefficients.

Define $u = (q_{m-1}/q_m)$, $v = (\theta_{m/s} - \theta_{m-1/s})/q_m$, and then the chain-relations are given by

$$
\begin{bmatrix}
a_1^{(m)} \\
a_2^{(m)} \\
b_1^{(m)} \\
b_2^{(m)}
\end{bmatrix}
= \left\{
\begin{bmatrix}
t_1 & t_3 & 0 & 0 \\
t_2 & t_4 & 0 & 0 \\
0 & 0 & t_1 & t_3 \\
0 & 0 & t_2 & t_4
\end{bmatrix}
+ u
\begin{bmatrix}
t_4 & -t_2 & 0 & 0 \\
-t_3 & t_1 & 0 & 0 \\
0 & 0 & t_4 & -t_2 \\
0 & 0 & -t_3 & t_1
\end{bmatrix}
+ v
\begin{bmatrix}
0 & 0 & -t_2 & -t_4 \\
0 & 0 & t_1 & t_3 \\
t_2 & t_4 & 0 & 0 \\
-t_1 & -t_3 & 0 & 0
\end{bmatrix}
\right\}
\begin{bmatrix}
a_1^{(m-1)} \\
a_2^{(m-1)} \\
b_1^{(m-1)} \\
b_2^{(m-1)}
\end{bmatrix}. \tag{12}
$$

Here the matrix elements are

$$t_1 = \cos q_{m-1}s_{m-1} \cos q_m s_{m-1},$$

$$t_2 = \cos q_{m-1}s_{m-1} \sin q_m s_{m-1},$$

$$t_3 = \sin q_{m-1}s_{m-1} \cos q_m s_{m-1}, \tag{13}$$

$$t_4 = \sin q_{m-1}s_{m-1} \sin q_m s_{m-1}.$$

The matrix chain-relations can be written compactly by expressing the 4×4 matrices as outer products (denoted by the symbol $\otimes$) of two 2×2 matrices as

$$
\begin{bmatrix}
a_1^{(m)} \\
a_2^{(m)} \\
b_1^{(m)} \\
b_2^{(m)}
\end{bmatrix}
= \left\{
\left(
\begin{bmatrix}
t_1 & t_3 \\
t_2 & t_4
\end{bmatrix}
+ u
\begin{bmatrix}
t_4 & -t_2 \\
-t_3 & t_1
\end{bmatrix}
\right) \otimes
\begin{bmatrix}
1 & 0 \\
0 & 1
\end{bmatrix}
+ v
\begin{bmatrix}
t_2 & t_4 \\
-t_1 & -t_3
\end{bmatrix}
\otimes
\begin{bmatrix}
0 & -1 \\
1 & 0
\end{bmatrix}
\right\}
\begin{bmatrix}
a_1^{(m-1)} \\
a_2^{(m-1)} \\
b_1^{(m-1)} \\
b_2^{(m-1)}
\end{bmatrix}. \tag{14}
$$

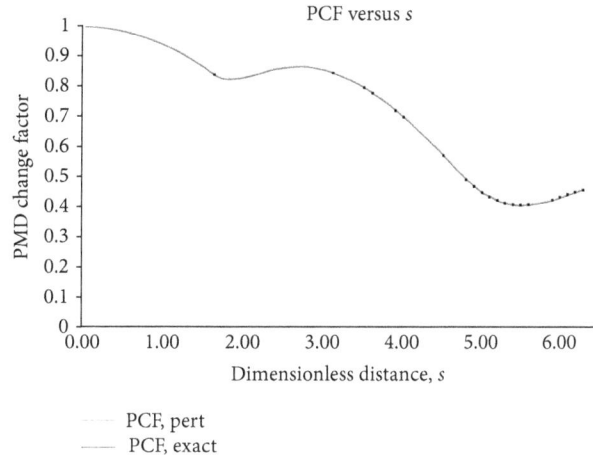

FIGURE 2: The PCF curves for a perturbative limit with $\Lambda = 1$ and $L_B = 12$.

2.3. Calculation of PMD Correction Factor (PCF). The sum of squares of the ω-differentiated amplitudes is similar to power and can be calculated by the following expression using expressions from Appendix A:

$$
\frac{\left|A_{1\omega}^{(m)}(s)\right|^2 + \left|A_{2\omega}^{(m)}(s)\right|^2}{(a_\omega/q)^2} = \left(\frac{1}{2}\right)\left[(1-n^2)\right.
$$

$$
\cdot \left\{\left(p_1^{(m)}\right)^2 + \left(p_2^{(m)}\right)^2 + \left(p_3^{(m)}\right)^2 + \left(p_4^{(m)}\right)^2\right\}
$$

$$
+ \left(p_5^{(m)}\right)^2 + \left(p_6^{(m)}\right)^2 + \left(p_7^{(m)}\right)^2 + \left(p_8^{(m)}\right)^2\right]
$$

$$
+ \left(\frac{1}{2}\right)\left[(1-n^2) \tag{15}\right.
$$

$$
\cdot \left\{\left(p_1^{(m)}\right)^2 + \left(p_2^{(m)}\right)^2 - \left(p_3^{(m)}\right)^2 - \left(p_4^{(m)}\right)^2\right\}
$$

$$
+ \left(p_5^{(m)}\right)^2 + \left(p_6^{(m)}\right)^2 - \left(p_7^{(m)}\right)^2 - \left(p_8^{(m)}\right)^2\right]
$$

$$
\cdot \cos 2qs + \left[(1-n^2)\left\{p_1^{(m)} p_3^{(m)} + p_2^{(m)} p_4^{(m)}\right\}\right.
$$

$$
+ p_5^{(m)} p_7^{(m)} + p_6^{(m)} p_8^{(m)}\right] \sin 2qs.
$$

Here m $(= 1, 2, 3)$ refers to segments in sequential manner.

For calculating the normalized PCF we need a similar expression for unspun-fiber given below:

$$
\frac{\left[\left|A_{1\omega}(s)\right|^2 + \left|A_{2\omega}(s)\right|^2\right]_{\text{unspun-fiber}}}{(a_\omega/q)^2} = (qs)^2. \tag{16}
$$

Then the expression for the PCF becomes

$$
\text{PCF}^{(m)}(s)
$$

$$
= \left[\frac{\left|A_{1\omega}^{(m)}(s)\right|^2 + \left|A_{2\omega}^{(m)}(s)\right|^2}{\left[\left|A_{1\omega}(s)\right|^2 + \left|A_{2\omega}(s)\right|^2\right]_{\text{unspun-fiber}}}\right]^{1/2}. \tag{17}
$$

The LHS is a function of parameters n and q and argument s. In general the expressions are quiet complicated, but for the first segment, the PCF is easily calculated and is given by

$$
\text{PCF}^{(1)}(s) = \sqrt{1 - n^2\left\{1 - \left(\frac{\sin qs}{qs}\right)^2\right\}}. \tag{18}
$$

3. Numerical Results

The physical constants $((\Delta\beta, \alpha_0, \eta)$ or equivalently $(L_B, l_0, \Lambda))$ and the parameters (n, q) appearing in the PCF expressions are related by

$$
q = \left(\frac{2\Lambda}{\pi l_0}\right)\left[1 + \left(\frac{\pi l_0}{4L_B}\right)\right]^{1/2},
$$
$$
n = \left[1 + \left(\frac{\pi l_0}{4L_B}\right)\right]^{-1/2}. \tag{19}
$$

We show results for sets of parameters in two extreme limits to emphasize the difference between the exact and perturbative calculations.

The Small-q Limit $(\Lambda < L_B)$. In this limit two sets of parameters were chosen to get small-q-values (less than 1). This corresponds to beat length being much larger than the spin period.

The resulting plots are given in Figures 2 and 3.

It is seen that the curves in Figure 2 for exact and perturbative calculations for small-q approximation are almost identical.

The curves in Figure 3 for exact and perturbative calculations are almost identical. Note that after $s = 5$ the two curves start diverging a little.

The Large-q Limit $(\Lambda > L_B)$. In this limit two sets of parameters were chosen to get large-q-values (much larger than 1). This corresponds to beat length being smaller than spin period.

The resulting plots are given in Figures 4 and 5.

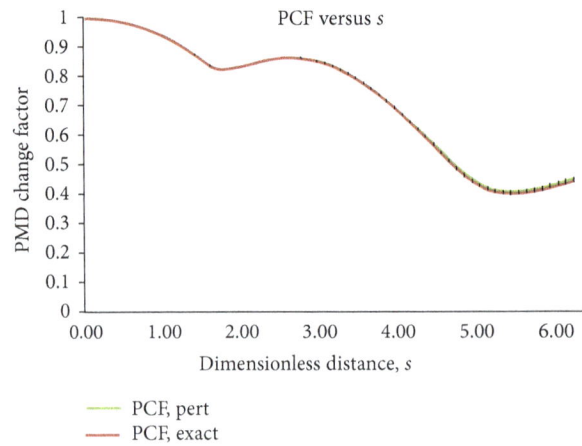

FIGURE 3: The PCF curve for a perturbative limit with $\Lambda = 1$ and $L_B = 5$.

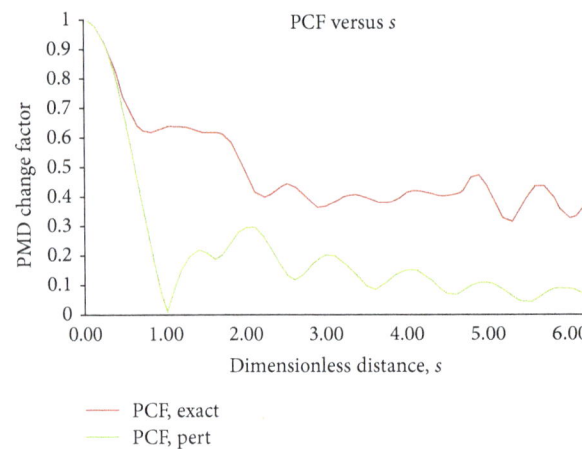

FIGURE 4: The PCF curve for a nonperturbative limit with $\Lambda = 5$ and $L_B = 1$.

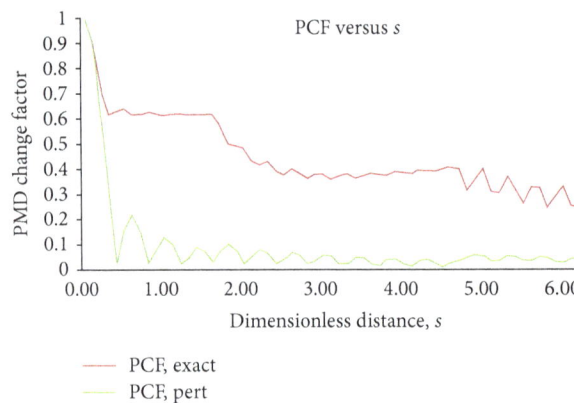

FIGURE 5: The PCF curves for a nonperturbative limit with $\Lambda = 12$ and $L_B = 1$.

The top and bottom curves in Figure 4 show exact and perturbative calculations, respectively. It is seen that perturbative approximation underestimates the PCF in this regime. The two start diverging significantly for value of s a little less than 1.

The top and bottom curves in Figure 5 show exact and perturbative calculations, respectively. It is seen that perturbative approximation underestimates the PCF in this regime. The two start diverging significantly for value of s a little beyond zero.

TABLE 1: PCF versus z plots.

Parameters: Λ, L_B, l_0 (in meters)	Values (n, q)	Comments
$(1, 12, 1)$	$(0.9978, 0.6379)$	$\Lambda \ll L_B$
$(1, 5, 1)$	$(0.9879, 0.6444)$	$\Lambda < L_B$

TABLE 2: PCF versus z plots.

Parameters: Λ, L_B, l_0 (in meters)	Values (n, q)	Comments
$(5, 1, 1)$	$(0.7864, 4.0475)$	$\Lambda > L_B$ (physical nonperturbative limit)
$(12, 1, 1)$	$(0.7864, 9.7139)$	$\Lambda \gg L_B$ (physical very nonperturbative limit)

4. Conclusions

The sine-function spin profile can be approximated in general by any number of segments. In this work a 3-segment approximation was chosen and analytical results for the PCF function were obtained. The PCF calculations were also repeated under the assumptions of the perturbative approximation made in [1]. As expected, it was shown that the perturbative approximation has limited validity compared to an exact calculation.

The 3-segment approximation given here can be extended to any number of segments for the spin function. The analytical results become very complicated very soon but they will approach the exact results as the number of segments increases. The method is also generalizable to an arbitrary spin function, which can be approximated by linear segments. This applies to almost all practically realizable spin functions. The exact analytic expressions for segment-approximated spin function and approximate numerical calculation of the exact spin function should complement one another to enhance our understanding of the underlying physics (Tables 1 and 2).

Appendix

A. Exact Calculation for Segments

A.1. The Specific 3-Segment Solutions. The details about solutions for 3 segments follow.

Segment I ($0 \leq s \leq \pi/2$). The equations are

$$\begin{bmatrix} A_{1s}^{(1)}(s) \\ A_{2s}^{(1)}(s) \end{bmatrix} = ia \begin{bmatrix} 0 & e^{2i\theta_1(s)} \\ e^{-2i\theta_1(s)} & 0 \end{bmatrix} \begin{bmatrix} A_1^{(1)}(s) \\ A_2^{(1)}(s) \end{bmatrix}. \quad \text{(A.1)}$$

The boundary conditions are

$$\left[A_1^{(1)}(s = 0) \right] = 1,$$
$$\left[A_{1s}^{(1)}(s = 0) \right] = 0, \quad \text{(A.2a)}$$
$$\left[A_2^{(1)}(s = 0) \right] = 0,$$
$$\left[A_{2s}^{(1)}(s = 0) \right] = ia. \quad \text{(A.2b)}$$

Let

$$n = \left(\frac{\tilde{c}}{q} \right) = \left[1 + \left(\frac{\pi l_0}{4 L_B} \right) \right]^{-1/2}, \quad \text{(A.3)}$$

and then the analytical solutions are similar to those given in Section 2.2. Consider

$$\begin{bmatrix} e^{-i\tilde{c}s} A_1^{(1)}(s) \\ \left(\frac{q}{a} \right) e^{i\tilde{c}s} A_2^{(1)}(s) \end{bmatrix} = \begin{bmatrix} 1 & -in \\ 0 & i \end{bmatrix} \begin{bmatrix} \cos qs \\ \sin qs \end{bmatrix}. \quad \text{(A.4)}$$

Comparison with general expression gives the following coefficients:

$$a_1^{(1)} = 1,$$
$$b_1^{(1)} = 0,$$
$$a_2^{(1)} = 0, \quad \text{(A.5)}$$
$$b_2^{(1)} = -n.$$

For calculating PCF, the amplitudes have to be differentiated with respect to ω, which will be denoted by subscript ω. Some useful relations needed for this are

$$\frac{d}{d\omega} \left(\frac{a}{q} \right) = n^2 \left(\frac{a_\omega}{q} \right),$$
$$a_\omega = \frac{da}{d\omega} = \frac{\gamma}{2\eta}, \quad \gamma = \frac{d(\Delta\beta)}{d\omega},$$
$$n_\omega = -n \left(\frac{a}{q} \right) \left(\frac{a_\omega}{q} \right), \quad \text{(A.6)}$$
$$q_\omega = a \left(\frac{a_\omega}{q} \right).$$

Then we can write

$$\begin{bmatrix} \left(\frac{q}{a} \right) e^{-i\tilde{c}s} A_{1\omega}^{(1)}(s) \\ e^{i\tilde{c}s} A_{2\omega}^{(1)}(s) \end{bmatrix}$$
$$= \left(\frac{a_\omega}{q} \right) \begin{bmatrix} p_1^{(1)} + ip_2^{(1)} & p_3^{(1)} + ip_4^{(1)} \\ p_5^{(1)} + ip_6^{(1)} & p_7^{(1)} + ip_8^{(1)} \end{bmatrix} \begin{bmatrix} \cos qs \\ \sin qs \end{bmatrix},$$
$$p_1^{(1)} = 0,$$
$$p_2^{(1)} = -nqs,$$
$$p_3^{(1)} = -qs,$$

$$p_4^{(1)} = n,$$

$$p_5^{(1)} = 0,$$

$$p_6^{(1)} = \left(1 - n^2\right) qs,$$

$$p_7^{(1)} = 0,$$

$$p_8^{(1)} = n^2.$$

$$(A.7)$$

Some interesting relations are found as

$$\Delta\beta = \left(\frac{4\pi q}{\Lambda}\right)\sqrt{1 - n^2},$$

$$z = \left(\frac{\Lambda}{2\pi}\right)s,$$

$$(A.8)$$

$$\alpha_0 = \left(\frac{2\pi^2 q^2}{\Lambda}\right)n\sqrt{1 - n^2}.$$

$$\begin{bmatrix} e^{-i(-\bar{c}s+2c)} A_1^{(2)}(s) \\ \left(\frac{q}{a}\right)e^{i(-\bar{c}s+2c)} A_2^{(2)}(s) \end{bmatrix} = \begin{bmatrix} 1 - n^2 + n^2\cos\pi q - in\sin\pi q & n\left(n\sin\pi q + i\cos\pi q\right) \\ -n\left(1 - \cos\pi q\right) & n\sin\pi q + i \end{bmatrix} \begin{bmatrix} \cos qs \\ \sin qs \end{bmatrix}. \qquad (A.11)$$

The ω-differentiated amplitudes are found as

$$\begin{bmatrix} \left(\frac{q}{a}\right)e^{-i(\bar{c}s-2c)} A_{1\omega}^{(2)}(s) \\ e^{i(\bar{c}s-2c)} A_{2\omega}^{(2)}(s) \end{bmatrix} = \left(\frac{a_\omega}{q}\right)$$

$$\cdot \begin{bmatrix} p_1^{(2)} + ip_2^{(2)} & p_3^{(2)} + ip_4^{(2)} \\ p_5^{(2)} + ip_6^{(2)} & p_7^{(2)} + ip_8^{(2)} \end{bmatrix} \begin{bmatrix} \cos qs \\ \sin qs \end{bmatrix},$$

$$p_1^{(2)} = n^2\left\{2\left(1 - \cos\pi q\right) - \pi q\sin\pi q + qs\sin\pi q\right\},$$

$$p_2^{(2)} = n\left\{\sin\pi q - \pi q\cos\pi q + qs\cos\pi q\right\},$$

$$p_3^{(2)} = n^2\left(-2\sin\pi q + \pi q\cos\pi q\right) - \left(1 - n^2\right.$$

$$\left. + n^2\cos\pi q\right)qs,$$

$$p_4^{(2)} = n\left\{-\left(\cos\pi q + \pi q\sin\pi q\right) + qs\sin\pi q\right\},$$

$$p_5^{(2)} = n\left\{\left(1 - 2n^2\right)\left(1 - \cos\pi q\right)\right.$$

$$\left. - \left(1 - n^2\right)\pi q\sin\pi q + \left(1 - n^2\right)qs\sin\pi q\right\},$$

$$p_6^{(2)} = \left(1 - n^2\right)qs,$$

$$p_7^{(2)} = n\left\{-\left(1 - 2n^2\right)\sin\pi q + \left(1 - n^2\right)\pi q\cos\pi q\right.$$

Segment II $(\pi/2 \leq s \leq 3\pi/2)$. The equations are

$$\begin{bmatrix} A_{1s}^{(2)}(s) \\ A_{2s}^{(2)}(s) \end{bmatrix} = ia \begin{bmatrix} 0 & e^{2i\theta_2(s)} \\ e^{-2i\theta_2(s)} & 0 \end{bmatrix} \begin{bmatrix} A_1^{(2)}(s) \\ A_2^{(2)}(s) \end{bmatrix}. \qquad (A.9)$$

The boundary conditions are

$$\left[A_1^{(1)}\left(s = \frac{\pi}{2}\right)\right] = \left[A_1^{(2)}\left(s = \frac{\pi}{2}\right)\right],$$

$$\left[A_{1s}^{(1)}\left(s = \frac{\pi}{2}\right)\right] = \left[A_{1s}^{(2)}\left(s = \frac{\pi}{2}\right)\right]. \qquad (A.10)$$

Similar expressions exist for $A_2^{(2)}(s)$. Using the chain-relations with $n = 2$, the analytical solutions are obtained:

$$+ \left(1 - n^2\right)\left(1 - \cos\pi q\right)qs\right\},$$

$$p_8^{(2)} = n^2.$$

$$(A.12)$$

Segment III $(3\pi/2 \leq s \leq 2\pi)$. The equations are

$$\begin{bmatrix} A_{1s}^{(3)}(s) \\ A_{2s}^{(3)}(s) \end{bmatrix} = ia \begin{bmatrix} 0 & e^{2i\theta_3(s)} \\ e^{-2i\theta_3(s)} & 0 \end{bmatrix} \begin{bmatrix} A_1^{(3)}(s) \\ A_2^{(3)}(s) \end{bmatrix}. \qquad (A.13)$$

The boundary conditions are

$$\left[A_1^{(2)}\left(s = \frac{3\pi}{2}\right)\right] = \left[A_1^{(3)}\left(s = \frac{3\pi}{2}\right)\right],$$

$$\left[A_{1s}^{(2)}\left(s = \frac{3\pi}{2}\right)\right] = \left[A_{1s}^{(3)}\left(s = \frac{3\pi}{2}\right)\right]. \qquad (A.14)$$

Similar expressions exist for $A_2^{(3)}(s)$. Using the chain-relations with $n = 3$, the analytical solutions are obtained:

$$\begin{bmatrix} e^{-i(\bar{c}s-4c)} A_1^{(3)}(s) \\ \left(\frac{q}{a}\right)e^{i(\bar{c}s-4c)} A_2^{(3)}(s) \end{bmatrix}$$

$$(A.15)$$

$$= \begin{bmatrix} 1 - n^2 + n^2\cos\pi q + in\left\{n^2\sin2\pi q + \left(1 - n^2\right)\left(\sin3\pi q - \sin\pi q\right)\right\} & n^2\sin2\pi q - in\left\{n^2\cos2\pi q + \left(1 - n^2\right)\left(1 + \cos3\pi q - \cos\pi q\right)\right\} \\ \left[n\left(\cos\pi q - \cos3\pi q\right) + in^2\left(\sin3\pi q - \sin2\pi q - \sin\pi q\right)\right] & \left[n\left(\sin\pi q - \sin3\pi q\right) + i\left\{1 - n^2 + n^2\left(\cos\pi q + \cos2\pi q - \cos3\pi q\right)\right\}\right] \end{bmatrix} \begin{bmatrix} \cos qs \\ \sin qs \end{bmatrix}.$$

The ω-differentiated amplitudes are found as

$$
\begin{bmatrix}
\left(\dfrac{q}{a}\right) e^{-i(\bar{c}s-4c)} A_{1\omega}{}^{(3)}(s) \\
e^{i(\bar{c}s-4c)} A_{2\omega}{}^{(3)}(s)
\end{bmatrix}
= \begin{pmatrix} a_\omega \\ q \end{pmatrix}
$$

$$
\cdot \begin{bmatrix}
p_1{}^{(3)} + i p_2{}^{(3)} & p_3{}^{(3)} + i p_4{}^{(3)} \\
p_5{}^{(3)} + i p_6{}^{(3)} & p_7{}^{(3)} + i p_8{}^{(3)}
\end{bmatrix}
\begin{bmatrix} \cos qs \\ \sin qs \end{bmatrix},
$$

$$
p_1{}^{(3)} = 2n^2 \left(1 - \cos 2\pi q - \pi q \sin 2\pi q\right) + n^2
$$

$$
\cdot\, qs \sin 2\pi q,
$$

$$
p_2{}^{(3)} = n\left[-3n^2 \sin 2\pi q - \left(1 - 3n^2\right)\left(\sin 3\pi q\right.\right.
$$

$$
\left. - \sin \pi q\right) + \pi q \left\{ 2n^2 \cos 2\pi q \right.
$$

$$
+ \left(1 - n^2\right)\left(3\cos 3\pi q - \cos \pi q\right)\right\} - \left\{ n^2 \cos 2\pi q \right.
$$

$$
\left.\left. + \left(1 - n^2\right)\left(1 - \cos \pi q + \cos 3\pi q\right)\right\} qs \right],
$$

$$
p_3{}^{(3)} = -2n^2 \left(\sin 2\pi q - \pi q \cos 2\pi q\right) - \left(1 - n^2 + n^2\right)
$$

$$
\cdot \cos 2\pi q\big) qs,
$$

$$
p_4{}^{(3)} = n\left[3n^2 \cos 2\pi q + \left(1 - 3n^2\right)\left(1 - \cos \pi q\right.\right.
$$

$$
+ \cos 3\pi q\big) + \pi q \left\{ 2n^2 \sin 2\pi q \right.
$$

$$
+ \left(1 - n^2\right)\left(3\sin 3\pi q - \sin \pi q\right)\right\} - \left\{ n^2 \sin 2\pi q \right.
$$

$$
\left.\left. + \left(1 - n^2\right)\left(\sin 3\pi q - \sin \pi q\right)\right\} qs \right],
$$

$$
p_5{}^{(3)} = n\left(1 - 2n^2\right)\left(\cos 3\pi q - \cos \pi q\right) + n\left(1 - n^2\right)
$$

$$
\cdot \left(\sin 3\pi q - \sin \pi q\right)\pi q + n\left(1 - n^2\right)\left(\sin \pi q\right.
$$

$$
- \sin 3\pi q\big) qs,
$$

$$
p_6{}^{(3)} = \left(1 - n^2\right) qs + n^2 \left[\left(2 - 3n^2\right)\left(\sin \pi q\right.\right.
$$

$$
+ \sin 2\pi q - \sin 3\pi q\big) + \left(1 - n^2\right)\left(3\cos 3\pi q\right.
$$

$$
- 2\cos 2\pi q - \cos \pi q\big)\pi q - \left(1 - n^2\right)\left(1 - \cos \pi q\right.
$$

$$
\left.\left. - \cos 2\pi q + \cos 3\pi q\right) qs \right],
$$

$$
p_7{}^{(3)} = n\left(1 - 2n^2\right)\left(\sin 3\pi q - \sin \pi q\right) + n\left(1 - n^2\right)
$$

$$
\cdot \left(\cos \pi q - 3\cos 3\pi q\right)\pi q + n\left(1 - n^2\right)\left(\cos 3\pi q\right.
$$

$$
- \cos \pi q\big) qs,
$$

$$
p_8{}^{(3)} = n^2 + n^2 \left[\left(2 - 3n^2\right)\left(1 - \cos \pi q - \cos 2\pi q\right.\right.
$$

$$
+ \cos 3\pi q\big) + \left(1 - n^2\right)\left(3\sin 3\pi q - 2\sin 2\pi q\right.
$$

$$
- \sin \pi q\big)\pi q
$$

$$
\left.+ \left(1 - n^2\right)\left(\sin \pi q + \sin 2\pi q - \sin 3\pi q\right) qs \right].
$$

$$
\text{(A.16)}
$$

B. Perturbative Calculation for Segments

The perturbative approach is based on the following assumptions:

    (i) The coupling between the polarization states is so small that the equations become decoupled.

    (ii) The top component is constant ($A_1{}^{(m)} = 1$, $m = 1, 2, 3$) and only the second component changes.

    (iii) The boundary conditions remain unchanged.

Under these assumptions the dimensionless constant q becomes $\bar{c}$, which is related to the physical lengths as

$$
\bar{c} = \frac{2}{\pi}\left(\frac{\Lambda}{l_0}\right). \tag{B.1}
$$

The new equations and their solutions take the following form.

Segment I ($0 \le s \le \pi/2$). Perturbative equations are as follows:

$$
\begin{bmatrix} A_{1s}{}^{(1)}(s) \\ A_{2s}{}^{(1)}(s) \end{bmatrix}
= ia \begin{bmatrix} 0 & e^{2i\bar{c}s} \\ e^{-2i\bar{c}s} & 0 \end{bmatrix}
\begin{bmatrix} 1 \\ 0 \end{bmatrix}. \tag{B.2}
$$

Solutions are as follows:

$$
A_2{}^{(1)}(s) = \left(\frac{a}{\bar{c}}\right) i e^{-i\bar{c}s} \sin \bar{c}s. \tag{B.3}
$$

The sum of squares of the ω-differentiated amplitudes is as follows:

$$
\left(\frac{\left|A_{1\omega}{}^{(1)}(s)\right|^2 + \left|A_{2\omega}{}^{(1)}(s)\right|^2}{(a_\omega/\bar{c})^2}\right)_{\text{pert}} = \frac{1}{2}\left(1 - \cos 2\bar{c}s\right) \tag{B.4}
$$

$$
= \sin^2 \bar{c}s.
$$

So

$$
\text{PCF}^{(1)}(s)_{\text{pert}}
$$

$$
= \left[\frac{\left(\left|A_{1\omega}{}^{(1)}(s)\right|^2 + \left|A_{2\omega}{}^{(1)}(s)\right|^2\right)_{\text{pert}}}{\left[\left|A_{1\omega}(s)\right|^2 + \left|A_{2\omega}(s)\right|^2\right]_{\text{unspun-fiber}}}\right]^{1/2} \tag{B.5}
$$

$$
= \frac{\sin \bar{c}s}{\bar{c}s}.
$$

Segment II ($\pi/2 \leq s \leq 3\pi/2$). Perturbative equations are as follows:

$$\begin{bmatrix} A_{1s}^{(2)}(s) \\ A_{2s}^{(2)}(s) \end{bmatrix} = ia \begin{bmatrix} 0 & e^{2i(-\tilde{c}s+2c)} \\ e^{-2i(-\tilde{c}s+2c)} & 0 \end{bmatrix} \begin{bmatrix} 1 \\ 0 \end{bmatrix}. \quad \text{(B.6)}$$

Solutions are as follows:

$$A_2^{(2)}(s) = e^{i(\tilde{c}s-2c)}\left(\frac{a}{\tilde{c}}\right)$$
$$\cdot\left[-(1-\cos 2c)\cos\tilde{c}s + (\sin 2c + i)\sin\tilde{c}s\right]. \quad \text{(B.7)}$$

The sum of squares of the ω-differentiated amplitudes is as follows:

$$\left(\left|A_{1\omega}^{(2)}(s)\right|^2 + \left|A_{2\omega}^{(2)}(s)\right|^2\right)_{\text{pert}} = \frac{1}{2}\left(\frac{a_\omega}{\tilde{c}}\right)^2\left\{(3 - 2\cos 2c) + (\cos 4c - 2\cos 2c)\cos 2\tilde{c}s + (\sin 4c - 2\sin 2c)\sin 2\tilde{c}s\right\}. \quad \text{(B.8)}$$

Expression for PCF is obtained as before.

Segment III ($3\pi/2 \leq s \leq 2\pi$). Perturbative equations are as follows:

$$\begin{bmatrix} A_{1s}^{(3)}(s) \\ A_{2s}^{(3)}(s) \end{bmatrix} = ia \begin{bmatrix} 0 & e^{2i(\tilde{c}s-4c)} \\ e^{-2i(\tilde{c}s-4c)} & 0 \end{bmatrix} \begin{bmatrix} 1 \\ 0 \end{bmatrix}. \quad \text{(B.9)}$$

$$+ i(-\sin 2c - \sin 4c + \sin 6c)\}\cos\tilde{c}s$$
$$+ \{(\sin 2c + \sin 4c - \sin 6c)$$
$$+ i(1 + \cos 2c + \cos 4c - \cos 6c)\}\sin\tilde{c}s]. \quad \text{(B.10)}$$

Solutions are as follows:

$$A_2^{(3)} = e^{i(-\tilde{c}s+4c)}\left(\frac{a}{\tilde{c}}\right)$$
$$\cdot\left[\{(-1 + \cos 2c + \cos 4c - \cos 6c)\right.$$

The sum of squares of the ω-differentiated amplitudes is as follows:

$$\left(\left|A_{1\omega}^{(3)}(s)\right|^2 + \left|A_{2\omega}^{(3)}(s)\right|^2\right)_{\text{pert}} \quad \text{(B.11)}$$
$$= \frac{1}{2}\left(\frac{a_\omega}{\tilde{c}}\right)^2\left\{(5 - 4\cos 4c) + (2\cos 10c - \cos 8c - 2\cos 6c)\cos 2\tilde{c}s + (2\sin 10c - \sin 8c - 2\sin 6c)\sin 2\tilde{c}s\right\}.$$

The PCF can be calculated as before.

Competing Interests

The author declares that he has no competing interests.

Acknowledgments

The author thanks Nick Frigo (formerly at AT&T Labs and now at United States Naval Academy) for getting him interested in this topic.

References

[1] M. Wang, T. Li, and S. Jian, "Analytical theory for polarization mode dispersion of spun and twisted fiber," *Optics Express*, vol. 11, no. 19, pp. 2403–2410, 2003.

[2] A. Pizzinat, B. S. Marks, L. Palmieri, C. R. Menyuk, and A. Gastarossa, "Influence of the model for random birefringence on the differential group delay of periodically spun fibers," *IEEE Photonics Technology Letters*, vol. 15, no. 6, pp. 819–821, 2003.

[3] A. Galtarossa, L. Palmieri, A. Pizzinat, B. S. Marks, and C. R. Menyuk, "An analytical formula for the mean differential group delay of randomly birefringent spun fibers," *Journal of Lightwave Technology*, vol. 21, no. 7, pp. 1635–1643, 2003.

[4] A. Galtarossa, L. Palmieri, and A. Pizzinat, "Optimized spinning design for low PMD fibers: an analytical approach," *Journal of Lightwave Technology*, vol. 19, no. 10, pp. 1502–1512, 2001.

[5] P. K. A. Wai and C. R. Menyuk, "Polarization mode dispersion, decorrelation, and diffusion in optical fibers with randomly varying birefringence," *Journal of Lightwave Technology*, vol. 14, no. 2, pp. 148–157, 1996.

[6] C. R. Menyuk and P. K. A. Wai, "Polarization evolution and dispersion in fibers with spatially varying birefringence," *Journal of the Optical Society of America B*, vol. 11, no. 7, p. 1288, 1994.

[7] G. J. Foschini and C. D. Poole, "Statistical theory of polarization dispersion in single mode fibers," *Journal of Lightwave Technology*, vol. 9, no. 11, pp. 1439–1456, 1991.

[8] C. R. Menyuk and P. K. A. Wai, "Elimination of nonlinear polarization rotation in twisted fibers," *Journal of the Optical Society of America B*, vol. 11, no. 7, pp. 1305–1309, 1994.

[9] N. J. Frigo, "A generalized geometrical representation coupled mode theory," *IEEE Journal of Quantum Electronics*, vol. QE-22, no. 11, pp. 2131–2140, 1986.

24

A Comprehensive Lighting Configuration for Efficient Indoor Visible Light Communication Networks

Thai-Chien Bui,[1] Suwit Kiravittaya,[1] Keattisak Sripimanwat,[2] and Nam-Hoang Nguyen[3]

[1]*Advanced Optical Technology (AOT) Laboratory, Department of Electrical and Computer Engineering, Faculty of Engineering, Naresuan University, Muang, Phitsanulok, Thailand*
[2]*ECTI Association, Klong Luang, Pathumthani, Thailand*
[3]*Faculty of Electronics and Telecommunications, VNU University of Engineering and Technology, Hanoi, Vietnam*

Correspondence should be addressed to Suwit Kiravittaya; suwitki@gmail.com

Academic Editor: Chen Chen

Design of an efficient indoor visible light communication (VLC) system requires careful considerations on both illumination and communication aspects. Besides fundamental factors such as received power and signal-to-noise ratio (SNR) level, studies on mobility scenarios and link switching process must be done in order to achieve good communication link quality in such systems. In this paper, a comprehensive lighting configuration for efficient indoor VLC systems for supporting mobility and link switching with constraint on illumination, received power, and SNR is proposed. Full connectivity in mobility scenarios is required to make the system more practical. However, different from other literatures, our work highlights the significance of recognizing the main influences of field of view angle on the connectivity performance in the practical indoor scenarios. A flexible link switching initiation algorithm based on the consideration of relative received power with adaptive hysteresis margin is demonstrated. In this regard, we investigate the effect of the overlap area between two light sources with respect to the point view of the receiver on the link switching performance. The simulation results show that an indoor VLC system with sufficient illumination level and high communication link quality as well as full mobility and support link switching can be achieved using our approach.

1. Introduction

Radio frequency is facing more difficulty to meet the rapidly growing demand for high data rate wireless transmission and ubiquitous network connectivity in indoor environment due to limited bandwidth and electromagnetic interference. Several next generation indoor wireless communications have been proposed recently [1]. In this regard, visible light communication (VLC) has been emerging to be a potential alternative and complementary technology to its radio frequency counterpart along with the recent advanced development of light emitting diode (LED) [2–4]. By using LEDs for dual-function of illumination and communication, VLC inherits the advantages of LED such as lower power consumption, longer lifetime, smaller size, and cooler operation. As a result, the physical layer of VLC has attracted much attention to researchers over the past few years that are focusing primarily on point-to-point communication for high data rate [4, 5]. However, how to configure those high speed point-to-point systems into real scenarios is remaining as a difficulty especially for indoor environment. It is due to the stringent requirement of illumination standard, communication performance, and mobility.

A typical indoor VLC system requires multiple LEDs with wide half power angle $\Phi_{1/2}$ which is mounted on the room ceiling in order to meet the illumination requirement and support mobility with continuous data transmission. However, with respect of communication performance, such VLC system suffers from intersymbol interference (ISI) which is caused by multipath propagation at high data rate [6]. In general, high level of ISI limits the achievable data rate and reduces the quality of received data by decreasing the level of signal-to-noise ratio (SNR). One can design an optimal VLC system using less number of LEDs but placing at reasonable positions as well as reducing the field-of-view (FOV) angle Ψ_c of receiver to optimize SNR level while still satisfying the illumination requirement and other constraints, that is, sufficient received power and power consumptions [7, 8]. Nevertheless, apart from the illumination level and link quality which is normally determined by SNR and received power level, connectivity and link switching performance are necessary in evaluating an indoor VLC system.

Connectivity is vitally important for mobility scenarios as it defines the communication coverage of the system viewing by FOV angle of the receiver in the receiver plane. The work in [9] demonstrated the influence of the luminance uniformity of the system on mobility. The other work in [10] proposed an angle diversity receiver to offer full mobility within a typical home with an area of $5 \times 5 \, \text{m}^2$. Obviously, full-connectivity is desired for any system since it helps continuous data connection in all the places in the room. However, improper position of LEDs as well as too small FOV angle at receivers often introduces blind spots and limits the connectivity performance in indoor VLC system. Moreover, changing the receiver plane leads to varying the connectivity as well since it changes the lighting area of an LED and view area of receiver. Hence, connectivity can be adjusted within a range of receiver plane and FOV angle by simply designing the lighting layout.

Link switching in VLC is also an important issue as it is understood as the "handover process" in cellular wireless communication. It is thus considered as an indispensable process to maintain and improve the communication link in the scenario of mobility or interference in multi-LED VLC system [11]. In the literature, techniques for vertical handover between VLC hot spot and RF base station have been proposed in many works such as [12, 13]. However, horizontal handover when user would like to switch from receiving information of one light source to the other has not attracted much attention. The work in [14] demonstrated the benefit of applying handover algorithm on the link in terms of band-width usage and data transmission rate. A novel prescanning and received signal strength (RSS) prediction technique is proposed to reduce link switching delay and unnecessary link switching ratio [15]. In designing the system, demonstrating the influence of different parameters of indoor VLC system on a link switching algorithm is required. To do this, the parameters related to the overlap area between LEDs are taken into account since switching process occurred in this area. As in cellular wireless systems, different algorithms used require different overlap area conditions. However, in general, this overlap area should be large enough for satisfying the link switching initiation conditions and seamless connectivity while switching process occurred, meaning that this overlap area depends on the initiation algorithm used and the link switching delay time.

In this paper, we design indoor VLC system by studying the effect of setting lighting positions and setting FOV angle on connectivity and link switching performance as well as illumination and link quality. The paper is structured as follows: In Section 2, the system model used in the present method and its basic performances are shown. In Section 3, the details of the connectivity and link switching conditions

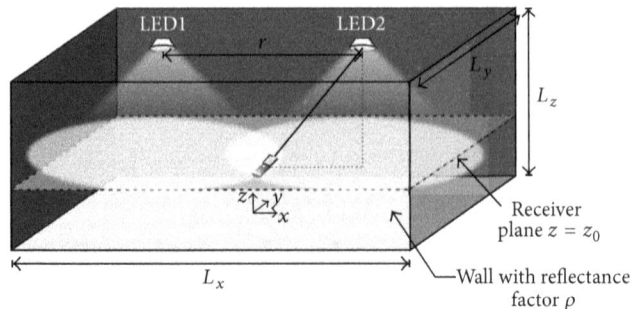

FIGURE 1: An indoor VLC system model with two LEDs tiled in the ceiling of a room dimension of $L_x \times L_y \times L_z$ (length, width, height).

TABLE 1: Simulation domain parameters.

Parameters and symbols	Values (unit)
Room size $L_x \times L_y \times L_z$	5 (m) $\times$ 2.5 (m) $\times$ 2.5 (m)
Distance between LEDs r	2.5 (m)
Receiver plane z_0	0.85 (m)
Half-power angle $\Phi_{1/2}$	70 ($^\circ$)
Transmitted optical power P_t	72 (W)
Center luminous intensity I_0	2628 (cd)
Reflectance factor of the walls ρ	0.8

are presented. In Section 4, simulation results to evaluate the performance of the configuration method are presented and discussed. Finally, Section 5 concludes the work.

2. System Model and Basic Performance

In a conventional indoor VLC, a regular and alignment distribution of LEDs is normally used since it gives uniform lighting, uniform overlap areas, and design simplicity. More-over, multiple-input multiple-output (MIMO) system can be realized with this LED configuration [16]. Consequently, demonstrating on the system with only two LEDs can be easily generalized to such MIMO systems with multiple LEDs. We therefore further simplify our VLC system model by considering only two LEDs installed on the ceiling of a room size of $L_x \times L_y \times L_z$ (length, width, height) as shown in Figure 1. The distance between the two LEDs is denoted as r. Each LED has a half-power angle $\Phi_{1/2}$, transmitted power P_t, and center luminous intensity I_0. It is assumed that VLC's receiver is a mobile terminal (MT) that can move on a plane $z = z_0$ above the floor. The reflectance factor of the wall is ρ. For later numerical simulation, all above parameters are set to typical values and summarized in Table 1.

2.1. Illumination Distribution with First Reflection. Illumination requirement should be considered as the first priority in indoor VLC, as it is the primary functionality of the system. In common indoor environment, illumination criterion is standardized by ISO (International Organization for Standardization). Referring to this set of standards, lightings for an indoor office workspace integrated VLC should be designed in the way so that illumination level between 300 and 1500 lx is achieved in all the places in the room [17]. To calculate illumination for a VLC system, it is assumed that each LED has a Lambertian radiation pattern, with the Lambert index m, depending on the half-power angle of LED $\Phi_{1/2}$ as $m = -1/\log_2(\cos \Phi_{1/2})$.

The line-of-sight (LOS) horizontal illumination E_{hor} at a point depends on Lambert index m, the angle of incidence ψ_d, the angle of irradiance ϕ_d, the center luminous intensity I_0, and the distance between LED and receiver's surface D_d [18]. In this paper, we make an assumption that the detector's surface is always in the vertical direction perpendicular to the plane of the ceiling. It means that for LOS link the angle of incidence ψ_d is always identical with the angle of irradiance ϕ_d. Hence, the formula of LOS horizontal illumination E_{hor} can be rewritten as

$$E_{\text{hor}} = I_0 \frac{\cos^{m+1}(\psi_d)}{D_d^2}. \tag{1}$$

If one considers light reflection at walls, the total illumination E_{total}, from the directed light and first reflection at a point, is given by $E_{\text{total}} = E_{\text{hor}} + E_{\text{ref}}$, where E_{ref} is the illumination level from reflections. For simplicity, only the first reflection from the wall is considered. In this case, the angle of irradiance ϕ_r is different from the angle of incidence ψ_r and E_{ref} can be calculated as follows [19]:

$$E_{\text{ref}} = \int_{\text{walls}} I_0 \frac{(m+1)\,\rho\cos^m\phi_r\,\cos\psi_r\,\cos\alpha\,\cos\beta}{2\pi^2} \frac{}{D_1^2 D_2^2} dA_{\text{wall}}, \tag{2}$$

where dA_{wall} is the reflective area of small region on the wall, α is the angle of irradiance to a reflective point, β is the angle of irradiance to the receiver, ψ_r is the angle of incidence, D_1 is the distance between the LED and the reflective point, and D_2 is the distance between the reflective point to the receiver. The schematic illustration of both LOS link and first reflection link is shown in Figure 2.

2.2. Received Power and SNR. Appropriated received power level is desired in any communication system, yet it normally comes with the penalty of power consumption or interference. In VLC, apart from increasing transmitted power and number of LEDs used which are known as normal ways to increase received power, there are still other ways such as adjusting the placement of LEDs in a proper way or decreasing the field of view of receiver. To calculate received power, the study of the channel is required. In an optical link, the channel DC gain on directed path is given as

$$H(0) = \begin{cases} \dfrac{(m+1)\,A}{2\pi D_d^2}\cos^{m+1}\psi_d T_s(\psi_d)\,g(\psi_d), & 0 \leq \psi_d \leq \Psi_c, \\ 0, & \psi_d > \Psi_c, \end{cases} \tag{3}$$

where A is the physical area of the detector in the receiver, which typically is a photodiode (PD), $T_s(\psi)$ is the gain of optical filter which is set as unity in this paper, $g(\psi_d)$ is the gain of an optical concentrator, and Ψ_c denotes the FOV angle of mobile terminal. The gain of optical concentrator $g(\psi_d)$ is given as [18]

$$g(\psi_d) = \begin{cases} \dfrac{n^2}{\sin^2(\Psi_c)}, & 0 \leq \psi_d \leq \Psi_c, \\ 0, & \psi_d > \Psi_c, \end{cases} \tag{4}$$

where n is the refractive index of optical concentrator. By using similar consideration, the channel DC gain on the first reflection is [20]

$$dH_{\text{ref}}(0) = \begin{cases} \dfrac{(m+1)\,A}{2\pi^2 D_1^2 D_2^2}\rho dA_{\text{wall}}\cos^m\phi_r\,\cos\psi_r\,\cos\alpha\,\cos\beta T_s(\psi_r)\,g(\psi_r), & 0 \leq \psi_r \leq \Psi_c, \\ 0, & \psi_r > \Psi_c. \end{cases} \tag{5}$$

The total received power P_r is the total power received from directed path $H(0)$ and first reflected path $dH_{\text{ref}}(0)$ which can be written as

$$P_r = \sum_i \left\{ P_t H(0) + \int_{\text{wall}} P_t dH_{\text{ref}}(0) \right\}, \tag{6}$$

where the summation is done over the index of LEDs ith. In this work, only two LEDs are considered. However, more LEDs can be treated within this scheme.

Generally, SNR determines the quality of the communication link in the sense that the higher SNR leads to better bit error rate (BER) performance. In VLC, ISI generated by optical path difference has main influenced to SNR. Basically, to treat ISI properly, a sophisticated process is needed [21].

For the sake of simplicity, we consider ISI as noise by adding its power to the noise power and assume the noise model as additive white Gaussian noise model. The on-off keying (OOK) modulation scheme is assumed with rectangular transmitted pulses of duration equal to the bit period T. The equation of SNR can be expressed as [19]

$$\text{SNR} = \frac{\gamma^2 P_{r\text{Signal}}^2}{\gamma^2 P_{r\text{ISI}}^2 + \sigma_{\text{noise}}^2}, \tag{7}$$

where γ is detector responsivity, σ_{noise}^2 is the noise variance, and $P_{r\text{Signal}}$ and $P_{r\text{ISI}}$ are the optical powers of signal and ISI, respectively. Considering the multipath case with ISI, the

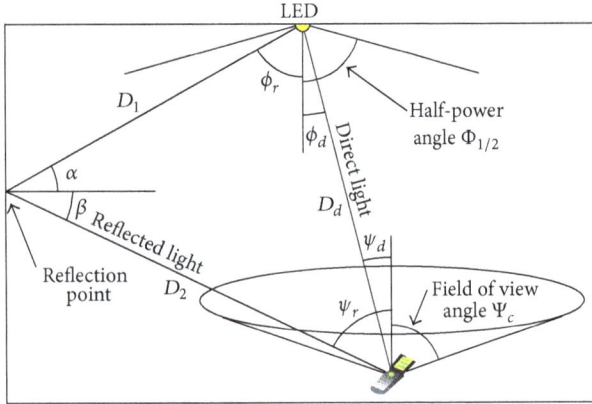

FIGURE 2: Propagation link with first reflection of a VLC system.

TABLE 2: System parameters.

Parameters and symbols	Values (unit)
Bit period T	10 (ns)
Gain of optical filter $T_s(\psi)$	1.0
Photodiode responsibility ρ	0.54 (A/W)
Refractive index of optical concentrator n	1.5
Detector area in PD A	1.0 (cm^2)
Constant related to noise C_1	1.696×10^{-11} (A^2/W)
Constant related to noise C_2	1.336×10^{-13} (A^2)

received power P_r from (6) can be calculated as the sum of the received power for the desired signal and ISI as [19]

$$P_{r\text{Signal}} = \sum_i \int_0^T h_i(t) \otimes X(t)\,dt,$$

$$P_{r\text{ISI}} = \sum_i \int_T^\infty h_i(t) \otimes X(t)\,dt, \qquad (8)$$

where $h_i(t)$ is the impulse response of the ith LED and $X(t)$ is transmitted optical pulse.

According to (7)-(8), more reflection from walls and opaque objects in VLC leads to increasing the ISI level. In case of ISI, where the pulse is broadening at the receiver, the transmitted symbol is more difficult to be demodulated correctly. Consequently, in the presence of reflection and ISI, the BER might be increased rapidly.

The noise variance originates from shot noise and thermal noises and it can be written in a simple form as

$$\sigma_{\text{noise}}^2 = C_1 \left(P_{r\text{Signal}} + P_{r\text{ISI}} \right) + C_2, \qquad (9)$$

where C_1 and C_2 are constants depending mainly on the properties of receiver specification. For our simulation, we adapt numerical values for calculating C_1 and C_2 from [19]. Above VLC system parameters are listed in Table 2.

In order to quantify our typical scenario, the calculations of illumination profile, received power, and SNR distribution are performed as shown in Figure 3. According to the results, it is obvious that this LED distribution gives the illumination

level satisfied the ISO standard (illumination between 300 and 1500 lx in all the places in the room). The illumination is highest at the area below each LED and it becomes lower moving towards the corners. Referring to Figure 3(b), the received power is −1 to 4 dBm in most of the places in the room. It is easy to see the abrupt change of the received power at the edge of the view overlap area where MT received lights from both LEDs. Obviously, this overlap area will be changed when the FOV of MT is changed (by simply tilting the receiver). According to the simulation result shown in Figure 3(c), SNR is highest at small areas nearby the wall including four corners where there is less or even no interference. In the view overlap area, SNR is decreased significantly despite the fact that the receivers receive more optical power from different light sources. This is because the total received power in the view overlap area is now the summation of the desired signal power and the ISI power. Moreover, the increasing of ISI power causes the decrease of the link quality in this area by reducing the SNR and increasing the BER. Remark that in the middle line of the room where the directed paths from two LEDs are equal the SNR level is at moderate level as shown in Figure 3(c).

3. Connectivity and Link Switching

3.1. VLC Connectivity. In terms of lighting function, the half-power angle $\Phi_{1/2}$ of the LED is an essential parameter in determining the coverage area and the uniformity illuminance. In general, this angle should stay wide for covering throughout the workspace and normally be a fixed parameter corresponding to the type of the LED. The coverage area of an LED can still be changed by varying the receiver plane. The field of view angle Ψ_c of the receiver must also be considered in terms of the communication function which is related to connectivity. This angle is defined as the angle of receivable signal rays from LED to receiver. There will be two possibilities that are $\Psi_c \leq \Phi_{1/2}$ or $\Psi_c > \Phi_{1/2}$. It has been figured out that reducing Ψ_c allows transmitting higher data transmission rate and gives better performance since optical gain is increased (see (4)) and ISI is decreased. Therefore, in this paper, we consider only the case that $\Psi_c \leq \Phi_{1/2}$. Obviously, in this case, the area on which the receiver can receive the signal is smaller than the coverage area made by the LED and that area is called the view area of the receiver which is given by $A_{\text{view\_area}} = \pi(h \tan(\Psi_c))^2$. In addition, the overlap area at the point view of the receiver which is made by two adjacent view areas is also smaller than the overlap area made by the LEDs and will be called as view overlap area, A_{vo}.

Connectivity is established when the received power at the PDs is greater than the receiver sensitivity (about −36 dBm) as reported in previous researches [9, 10]. However, as shown in Figure 3(b), the received power profile is ranging from about −1 to 4 dBm in most of the places in a typical indoor system. Consequently, connectivity in this paper is defined as the covering percentage of the light with respect to the view area defined by the FOV angle of the PDs. For all lighting configuration, it has been shown that full connectivity at which there is no blind spot throughout the room is desired. To receive full connectivity in conjunction

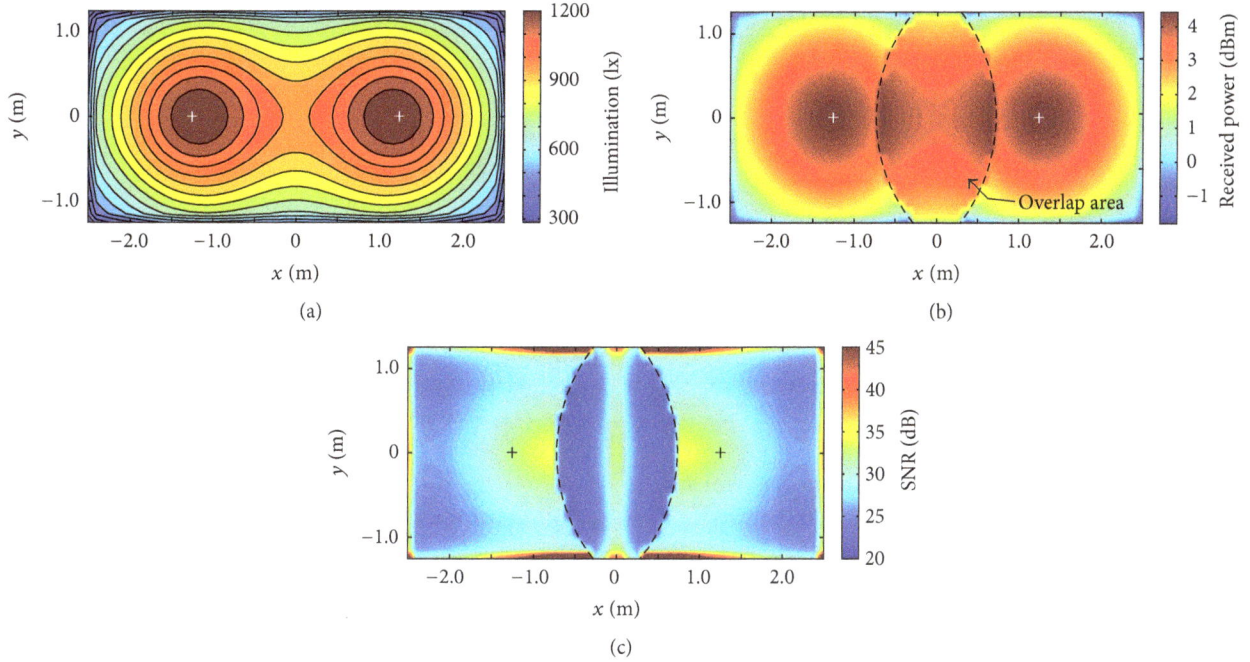

FIGURE 3: The profile of (a) illumination, (b) receiver power, and (c) SNR. The two LEDs are symmetrically arranged in a horizontal plane. LED1 and LED2 are at $(-1.25, 0, 2.5)$ and $(1.25, 0, 2.5)$. The receiver plane is at $z_0 = 0.85$ m.

with uniformity, the distance between LEDs must be taken into account; either the number of LEDs used or FOV has to be increased in order to fulfill the condition. From the above discussions, it is clear that too small FOV can lead to the rapid shrinking of MT view area whereby the blind spots occur in the workspace. However, there is a limitation by which increasing the FOV results in high ISI level. In this paper, the blind spot is defined in the plane of receiver not on the floor because receiver is rarely expected to lay on the floor. For the distance between LEDs, either too far or too near distance can lead to the occurrence of blind spots. The relation between this distance and FOV angle with respect to full connectivity under varying position of LEDs will be discussed along with the results from numerical simulations in the next section.

3.2. Link Switching Algorithm. Link switching in VLC is understood as the "handover process" in radio wireless communication and is considered as an indispensable process to maintain and improve the communication link in the scenario of mobility or interference in multiple LED VLC system. The work in [22] focused on optimizing the coverage area of the LED to achieve high average user net rate by taking into account the time spent for handover process. In other words, the research aimed at optimizing the half power angle of the LEDs. In this work, consideration is at both the transmitter and the receiver sides; however, it is assumed that the coverage area of the LED is fixed. Although the communication coverage of a cell in indoor VLC system is normally small and can be varied easily depending on the distance between LEDs, FOV angle, and the receiver plane, yet the speed of MT is normally slow as compared with cellular wireless communication. Consequently, VLC system

requires a fast and flexible link switching algorithm [15]. A strong candidate could be the algorithm of relative received power with hysteresis since this algorithm can decrease the amount of link switching and Ping-Pong rate [23]. Referring to this algorithm, when the MT moves from the coverage area of LED1 to the LED2, in the view overlap area d_{OA}, the link switching process is initiated if the received power of the LED2 exceeds the received power of LED1 at least by hysteresis margin H level (in dB) which is shown in

$$10 \log \left(\frac{P_2}{P_1} \right) \geq H, \tag{10}$$

where P_1 and P_2 represent the received power from the LED1 and LED2, respectively. For the sake of simplicity, P_1 and P_2 are assumed to be LOS received power and can be calculated from (3) as

$$P_1 = P_t \frac{(m+1)A}{2\pi D^2} \left(\cos^{m+1} \phi \right) \frac{n^2}{\sin^2 \Psi_c},$$
$$P_2 = P_t \frac{(m+1)A}{2\pi D'^2} \left(\cos^{m+1} \varphi \right) \frac{n^2}{\sin^2 \Psi_c}. \tag{11}$$

The value of H has been shown to be a critical selection factor for the link switching performances. It affects the amount of link switching and the view overlap area where link switching process occurred. If H is high, the amount of unnecessary link switching is decreased. Yet it requires large view overlap area in order to have successful link switching [24]. This paper will analyze the relation between hysteresis margin H (dB) and the view overlap area A_{vo} in respect of the success of link switching process under varying the distance between LED1 and LED2.

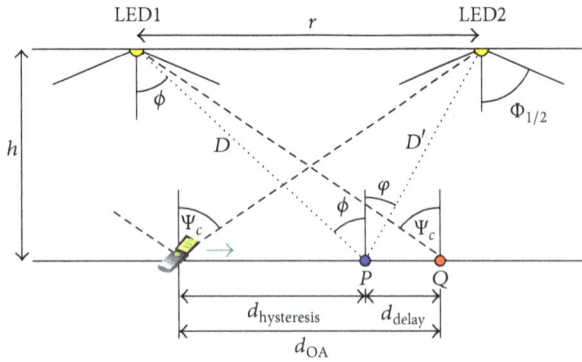

FIGURE 4: The view overlap distance of two LEDs.

TABLE 3: Additional system parameters.

Parameter and symbol	Value (unit)
Link switching delay time T_d	0.15 (s)
Speed of MT v	1 (m/s)
Hysteresis margin H	3–12 (dB)
Field of view angle of mobile terminal Ψ_c	48–60°

For configuration method, those relations are used to determine the minimum required view overlap distance and the maximum distance between LEDs in order to support link switching process. For system with given distance between LEDs, (14) can be used to determine a suitable hysteresis margin.

4. Numerical Results

To demonstrate different lighting layouts, the distance between LED1 and LED2 is varied. At first, LED1 and LED2 are placed in the center of the ceiling and then moved out evenly towards the walls with step of 0.2 m along length of the room. The full-connectivity requirement is checked first for each lighting layout; then all those lighting layouts satisfying connectivity condition will be calculated for average received power and received average SNR. At last, the relation between lighting layout and hysteresis margin is shown. All above simulations will be implemented for different FOV angle Ψ_c with condition $\Psi_c \leq \Phi_{1/2}$. All the parameters used in the simulations are given in Tables 1 and 2 in conjunction with Table 3.

4.1. Link Quality and Connectivity. Both high level and small fluctuation of received power and SNR are desirable when designing a VLC system. Yet when demonstrating a system with two LEDs, we expected to only consider the level of those two performances and leave the fluctuation issue for the VLC designers who apply our method to configure their systems. Consequently, we do the simulation on the average received power and SNR. Distance between LED1 and LED2 will be varied from 0.2 m to 4.8 m with step of 0.2 m since two LEDs should not be tilted at the same place in the center or right at the edge of the room. For each step, the connectivity and link quality will be calculated at set FOV angle. Figure 5 shows the average received power and average SNR level corresponding to the lighting layouts that give full-connectivity under set FOV angles. Referring to our simulation shown in Figure 5, when FOV angle Ψ_c is smaller than 48° there is no lighting layout providing full-connectivity at all; in other words, the blind area exists at any lighting layout when FOV angle is smaller than 48°. At $\Psi_c = 48°$, the distance between LEDs should only be between 2.4 and 2.6 m for full-connectivity. With increasing Ψ_c, the possibility of choosing the lighting layouts that satisfying full-connectivity is also increased; that is, at FOV 50° the distance can be between 2 and 3 m, and at FOV 55° distance of 1.5 to 4 m is acceptable. When FOV reaches 58° almost the distance between LED1 and LED2 gives full-connectivity.

We assume MT moves along the straight line which is connected by projection of two LEDs. By this way, MT can move the largest distance in the view overlap area, called view overlap distance, d_{OA}, as shown in Figure 4. It is assumed that MT can recognize and distinguish the signal powers from LED1 and LED2. When MT moves into the view overlap area, it continuously calculates the signal strengths from two LEDs. At point P, it is assumed that (10) is reached, meaning that the link switching is initiated at point P. From that point, the MT needs an additional distance in order to get successful link switching. This distance depends on the link switching delay time T_d and the speed of MT v as shown in

$$d_{delay} = T_d \times v. \tag{12}$$

Substituting (11) into (10) we get

$$\frac{D^2 \cos^{m+1}(\varphi)}{D'^2 \cos^{m+1}(\phi)} \geq 10^{H/10}. \tag{13}$$

Based on (13), after some geometry consideration (see Figure 4), the condition of the view overlap distance d_{OA} can be obtained as

$$d_{OA} \geq h \tan(\Psi_c)$$
$$- \sqrt{\frac{h^2 + (h \tan(\Psi_c) - T_d \times v)^2}{{}^{(m+3)/2}\sqrt{10^{H/10}}} - h^2} + T_d \times v. \tag{14}$$

The condition above is valid if and only if

$$\frac{h^2 + (h \tan(\Psi_c) - T_d \times v)^2}{{}^{(m+3)/2}\sqrt{10^{H/10}}} - h^2 \geq 0. \tag{15}$$

The relation among three parameters, hysteresis margin H (dB), view overlap distance d_{OA}, and the FOV of MT, are shown clearly in (14) and (15).

From (14), one can transform from view overlap distance, d_{OA}, to the distance between LEDs r by using the geometrical relation in Figure 4:

$$r \leq h \tan(\Psi_c) + \sqrt{\frac{h^2 + (h \tan(\Psi_c) - T_d \times v)^2}{{}^{(m+3)/2}\sqrt{10^{H/10}}} - h^2}$$
$$- T_d \times v. \tag{16}$$

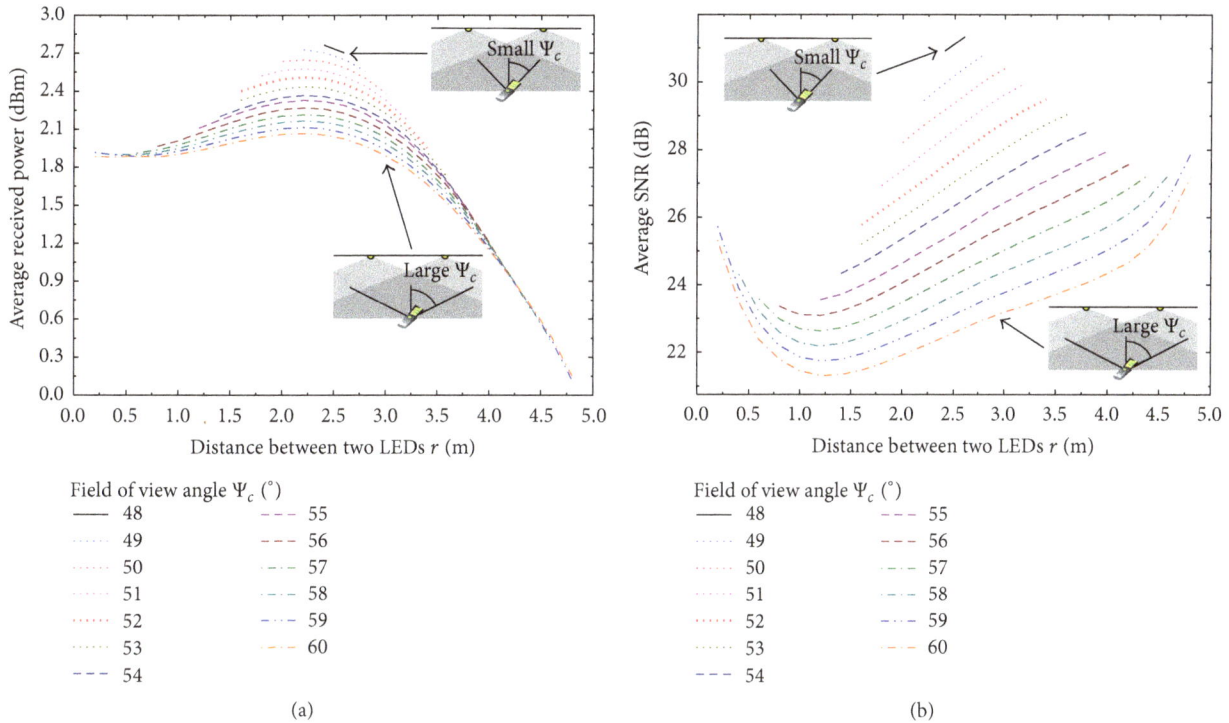

FIGURE 5: (a) Average received power and (b) average SNR level corresponding to the lighting layouts that give full-connectivity under different FOV angles.

Referring to Figure 5, for all the layouts that give full-connectivity, the average received power and average SNR at FOV 48° are the highest and descending as FOV increased. At FOV 48°, the average received power is about 2.8 dBm and average SNR is about 31 dB. This is due to the fact that when FOV increases the gain of optical concentrator at the detector decreases and it reduces received power afterwards. This reduction in received power together with increasing ISI leads to decreasing SNR level. When two LEDs are installed far from each other and nearby the edge of the room, the average receiver power is very small around 0.15 dBm. Furthermore, the trade-off between average received power and average SNR is also easily observed through Figure 5. When SNR is increased, the received power will be decreased and vice versa. This is true especially when the distance between LEDs is large and FOV is small. At large FOV, there will be some places of 2 LEDs that give high average received power and average SNR as well. FOV angle larger than 60° displays the same trend in our simulation (not shown). As for any communication system, both high level of received power and SNR are desirable. Consequently, in terms of link quality and connectivity, FOV of 48° and distance between LEDs around 2.5 m are recommended since this layout gives maximized received power and SNR as well.

4.2. Link Switching with Various Hysteresis Margins. In this section, we simulate the requirement of the view overlap distance between two LEDs for supporting link switching

as foregoing discussions to demonstrate the requirement in distance between two LEDs. Those relations are shown in (14)–(16), yet they are inequality formula. We simulate here the equality of those relations. Figure 6 displays the minimum view overlap distance, maximum distance between two LEDs, and FOV at different hysteresis margin for supporting link switching. Referring to Figure 6(a), there is a well satisfaction with the aforementioned theory; that is, when hysteresis margin H increases, the required view overlap distance is also increased. Moreover, the minimum FOV angle also increases at higher H. At hysteresis margin $H = 3$ dB FOV should be larger than 38°, because smaller FOV angle results in no layout that can satisfy the link switching conditions (see (15)). Similarly, when H equals 6, 9, or 12 dB, the minimum required FOV angles are 50°, 56°, and 63°, respectively. For any hysteresis margin, the minimum overlap distance stays quite stable as FOV increase. When H equals 3 dB, the minimum overlap distance is around 1.2 m for all FOV larger than 38° and around 1.6 m, 2.4 m, and 3.1 m for $H = 6$, 9, and 12 dB, respectively.

Figure 6(b) displays the maximum distance between two LEDs that can well satisfy the link switching condition at different hysteresis margin. From this figure, it is seen that, at all the hysteresis margin values, the increasing of FOV angle leads to an increase of maximum distance between LEDs. At the same FOV, high hysteresis margin often requires small maximum distance between two LEDs. At the FOV of 48° and the distance between two LEDs $r = 2.5$ m, link switching process can be applied for hysteresis margin of 3 dB.

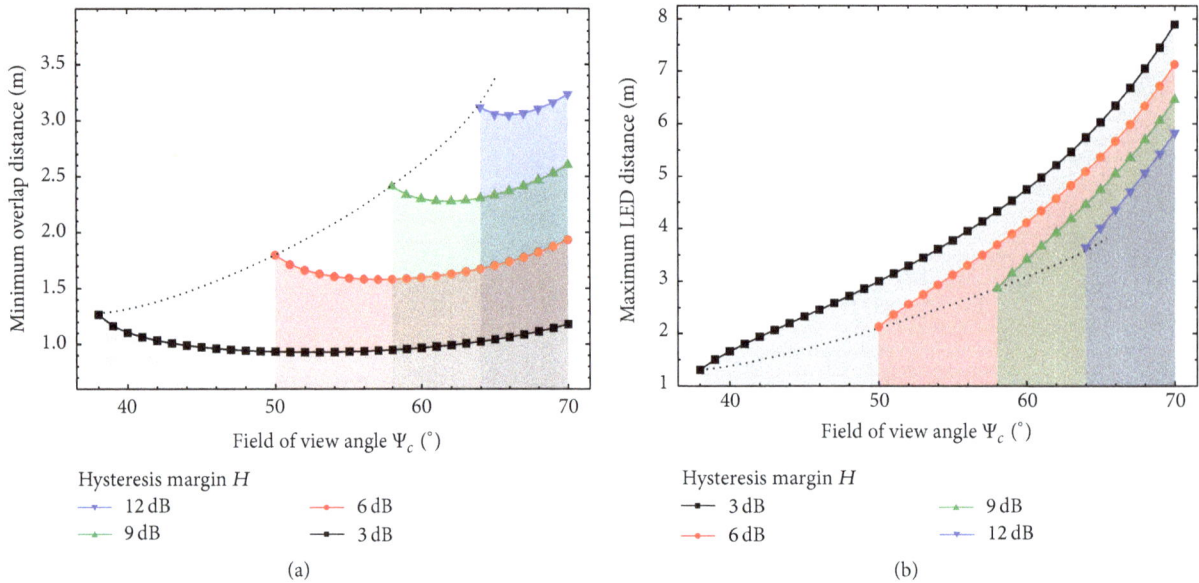

FIGURE 6: (a) Minimum overlap distance versus FOV and (b) maximum distance between 2 LEDs versus FOV at various hysteresis margin H for supporting link switching.

5. Conclusion

We discussed a comprehensive lighting configuration based on the illumination and communication aspects to fully integrate VLC system in indoor environment. It is important to consider the connectivity and link switching performance of an indoor VLC system in order to ensure high quality communication connection. Relationship of various parameters in VLC system, that is, the distance between LEDs, FOV angle of receiver, view area, view overlap area, and view overlap distance in respect of link quality, connectivity, and link switching process, is considered. Based on our simulation approaches, to achieve highest link quality, optimum connectivity, and fully supporting link switching process the suitable distance between two LEDs at specific FOV angle can be obtained. This work enhances the development of any practical VLC system and can be used to design effective MIMO VLC systems.

Competing Interests

The authors declare that there is no conflict of interests regarding the publication of this paper.

Acknowledgments

This work was supported by Naresuan University and VNU University of Engineering and Technology (UET).

References

[1] D. K. Borah, A. C. Boucouvalas, C. C. Davis, S. Hranilovic, and K. Yiannopoulos, "A review of communication-oriented optical wireless systems," *Eurasip Journal on Wireless Communications and Networking*, vol. 91, pp. 1–28, 2012.

[2] D. Karunatilaka, F. Zafar, V. Kalavally, and R. Parthiban, "LED based indoor visible light communications: state of the art," *IEEE Communications Surveys and Tutorials*, vol. 17, no. 3, pp. 1649–1678, 2015.

[3] P. H. Pathak, X. Feng, P. Hu, and P. Mohapatra, "Visible light communication, networking, and sensing: a survey, potential and challenges," *IEEE Communications Surveys & Tutorials*, vol. 17, no. 4, pp. 2047–2077, 2015.

[4] D. Wu, C. Chen, Z. Ghassemlooy, and W. Zhong, "Short-range visible light ranging and detecting system using illumination light emitting diodes," *IET Optoelectronics*, vol. 10, no. 3, pp. 94–99, 2016.

[5] J. Vucic, C. Kottke, S. Nerreter, K.-D. Langer, and J. W. Walewski, "513 Mbit/s visible light communications link based on DMT-modulation of a white LED," *Journal of Lightwave Technology*, vol. 28, no. 24, Article ID 5608481, pp. 3512–3518, 2010.

[6] J. H. Choi, S. W. Koo, and J. Y. Kim, "Influence of optical path difference on visible light communication systems," in *Proceedings of the 9th International Symposium on Communications and Information Technology (ISCIT '09)*, pp. 1247–1251, Incheon, Korea, September 2009.

[7] T.-H. Do and M. S. Yoo, "Received power and SNR optimization for visible light communication system," in *Proceedings of the 4th International Conference on Ubiquitous and Future Networks (ICUFN '12)*, pp. 6–7, Phuket, Thailand, July 2012.

[8] T.-H. Do and M. Yoo, "Optimization for link quality and power consumption of visible light communication system," *Photonic Network Communications*, vol. 27, no. 3, pp. 99–105, 2014.

[9] A. Burton, H. Le Minh, Z. Ghasemlooy, and S. Rajbhandari, "A study of LED lumination uniformity with mobility for visible light communications," in *Proceedings of the International Workshop on Optical Wireless Communications (IWOW '12)*, Pisa, Italy, October 2012.

[10] A. Burton, Z. Ghassemlooy, S. Rajbhandari, and S.-K. Liaw, "Design and analysis of an angular-segmented full-mobility visible light communications receiver," *Transactions on Emerging*

Telecommunications Technologies, vol. 25, no. 6, pp. 591–599, 2014.

[11] IEEE, "IEEE standard for local and metropolitan area networks. Part 15.7: short-range wireless optical communication using visible light," IEEE Standard 802.15.7, 2011.

[12] F. Wang, Z. Wang, C. Qian, L. Dai, and Z. Yang, "Efficient vertical handover scheme for heterogeneous VLC-RF systems," *IEEE/OSA Journal of Optical Communications and Networking*, vol. 7, no. 12, pp. 1172–1180, 2015.

[13] S. Liang, H. Tian, B. Fan, and R. Bai, "A novel vertical handover algorithm in a hybrid visible light communication and LTE system," in *Proceedings of the IEEE 85th Vehicular Technology Conference (VTC Fall '16)*, pp. 1–5, Montréal, Canada, January 2016.

[14] D. Wu, Z. Ghassemlooy, W.-D. Zhong, and C. Chen, "Cellular indoor OWC systems with an optimal lambertian order and a handover algorithm," in *Proceedings of the 7th International Symposium on Telecommunications (IST '14)*, pp. 777–782, IEEE, Tehran, Iran, September 2014.

[15] T. Nguyen, M. Z. Chowdhury, and Y. M. Jang, "A novel link switching scheme using pre-scanning and RSS prediction in visible light communication networks," *Eurasip Journal on Wireless Communications and Networking*, vol. 2013, article 293, 2013.

[16] P. F. Mmbaga, J. Thompson, and H. Haas, "Performance analysis of indoor diffuse VLC MIMO channels using angular diversity detectors," *Journal of Lightwave Technology*, vol. 34, no. 4, pp. 1254–1266, 2016.

[17] R. S. Berns, *Billmeyer and Saltzman's Principles of Color Technology*, John Wiley & Sons, New York, NY, USA, 2000.

[18] J. R. Barry, *Wireless Infrared Communications*, Kluwer Academic Press, Boston, Mass, USA, 1994.

[19] T. Komine and M. Nakagawa, "Fundamental analysis for visible-light communication system using LED lights," *IEEE Transactions on Consumer Electronics*, vol. 50, no. 1, pp. 100–107, 2004.

[20] F. R. Gfeller and U. Bapst, "Wireless in-house data communication via diffuse infrared radiation," *Proceedings of the IEEE*, vol. 67, no. 11, pp. 1474–1486, 1979.

[21] W. Hauk, F. Bross, and M. Ottka, "The calculation of error rates for optical fiber systems," *IEEE Transactions on Communications*, vol. 26, no. 7, pp. 1119–1126, 1978.

[22] S. Pergoloni, M. Biagi, S. Colonnese, R. Cusani, and G. Scarano, "Optimized LEDs footprinting for indoor visible light communication networks," *IEEE Photonics Technology Letters*, vol. 28, no. 4, pp. 532–535, 2016.

[23] S. Moghaddam, V. Tabataba, and A. Falahati, "New handoff initiation algorithm (optimum combination of hysteresis and threshold based methods)," in *Proceedings of the 52nd Vehicular Technology Conference (IEEE-VTS Fall '00)*, pp. 1567–1574, Boston, Mass, USA, September 2000.

[24] S. Lal and D. K. Panwar, "Coverage analysis of handoff algorithm with adaptive hysteresis margin," in *Proceedings of the International Conference on Information Technology (ICIT '07)*, December 2007.

25

Gas Bubbles Investigation in Contaminated Water Using Optical Tomography Based on Independent Component Analysis Method

**Mohd Taufiq Mohd Khairi, Sallehuddin Ibrahim,
Mohd Amri Md Yunus, and Mahdi Faramarzi**

*Department of Control and Mechatronics Engineering, Faculty of Electrical Engineering,
Universiti Teknologi Malaysia, 81310 Skudai, Johor, Malaysia*

Correspondence should be addressed to Sallehuddin Ibrahim; salleh@fke.utm.my

Academic Editor: Augusto Beléndez

This paper presents the results of concentration profiles for gas bubble flow in a vertical pipeline containing contaminated water using an optical tomography system. The concentration profiles for the bubble flow quantities are investigated under five different flows conditions, a single bubble, double bubbles, 25% of air opening, 50% of air opening, and 100% of air opening flow rates where a valve is used to control the gas flow in the vertical pipeline. The system is aided by the independent component analysis (ICA) algorithm to reconstruct the concentration profiles of the liquid-gas flow. The behaviour of the gas bubbles was investigated in contaminated water in which the water sample was prepared by adding 25 mL of colour ingredients to 3 liters of pure water. The result shows that the application of ICA has enabled the system to detect the presence of gas bubbles in contaminated water. This information provides vital information on the flow inside the pipe and hence could be very significant in increasing the efficiency of the process industries.

1. Introduction

Information about the flow regime in a process vessel or pipelines is important so as to design an accurate, safe, and low cost conveying system in various applications such as chemical engineering and nuclear engineering [1]. The flow regime identification is also vital in evaluating the performance of a process system since incorrect analysis from the system can lead to the reduction of production rates. Hence, an appropriate instrumentation system is essential in providing vital information on the flow patterns. The tomography technique has been selected extensively in the industrial field as a tool to provide information about the phase and spatial distribution without interrupting the process flow [2]. Before being applied in the industry, the technique has been successfully implemented in the medical field where it is responsible for capturing images of body tissue and detecting tumour [3]. A tomography system consists of several parts

such as sensors, signal conditioning circuit, a data acquisition system, and a computer. There are many types of sensors that have been used in process tomography, that is, electrical capacitance, electrical resistance, ultrasonic, and optical. In this paper, an optical sensing technique has been selected to be integrated into the tomography system due to its low cost, being straightforward, and having a better dynamic response [4]. Many researchers have published their result using the optical sensor in order to monitor the movement of gas, liquid, and solid. Idroas et al. [5] have successfully applied an optical tomography system for particle sizing identification using sample of beads and irregular shaped nut. Rzasa and Plaskowski [6] have presented gas bubbles measurement using an optical tomography system in a vertical aeration column. Ibrahim et al. [7] proposed a concentration measurement using halogen bulbs as light projectors to reconstruct images of small and large gas bubbles. This paper describes the analysis of gas bubble detection in a pipeline filled with contaminated water. Investigation on gas bubble behaviour in industry such as food and pharmaceutical is vital since the existence of unwanted bubbles can reduce the quality of product [8]. Previous research on gas bubbles measurement in the tomography field was conducted by Rzasa [9], Jin et al. [10], and Ayob et al. [11] which used pure water as a liquid. Generally, the unwanted gas bubbles are in the form of opaque liquid such as oil and paint. Therefore, this paper proposed a technique for imaging the gas bubbles in which the investigation was carried out using opaque liquid.

2. Mathematical Modelling

Mathematical modelling is an essential part in optical tomography. This section discusses the optical attenuation model and the application of independent component analysis. The first part will discuss the optical attenuation modelling for the presence of gas bubbles in contaminated water. The second part elaborated the application of ICA method to detect the distribution of gas bubbles.

2.1. Optical Attenuation Model. Light experienced attenuation when it traverses from one medium to another medium. This attenuating process is due to scattering and refraction of light when it travels through a different medium such as water and gas. This paper highlighted the absorption and attenuation model and neglected the light scattering effect since it is difficult to perform the modelling due to the random shape [12]. Beer-Lambert's Law is a popular law regarding the investigation of attenuation and absorption of the light energy. The law can be expressed as

$$V_R = V_T \exp\left[-\alpha_w l_w - \alpha_a l_a\right], \quad (1)$$

where V_R is the receiving sensor voltage, V_T is the voltage of the transmitter, α_w is the attenuation coefficient of water medium (mm^{-1}), α_a is the attenuation coefficient of air medium (mm^{-1}), l_w is the path length of water (mm), and l_a is the path length of air (mm). The reference attenuation coefficients for water and air are $\alpha_w = 0.0287$ mm^{-1} and $\alpha_a = 0.0142$ mm^{-1} [12]. The value of pipe diameter is $l_w = 100$ mm and the length of small bubble is assumed to be $l_a = 5$ mm. The value of V_T is being set to 5 volts (V) which was taken from the transmitter value programmed using the Peripheral Interface Controller (PIC) Microcontroller. Two cases are investigated: water with no gas bubble presence and water with gas bubble presence.

2.1.1. Water with No Gas Bubble Presence. The arrangement of the transmitter and receiver when there is no gas bubble is illustrated in Figure 1. Light from the transmitter traversed water without the interruption of bubble. The value of the attenuation coefficient for water is entered in (2) along with 100 mm of water path length. The value of the receiver voltage V_R is obtained by

$$V_R = 5 \exp\left[-(0.0287)(100) - (0)\right], \quad (2)$$

$$V_R = 0.2835 \, \text{V}. \quad (3)$$

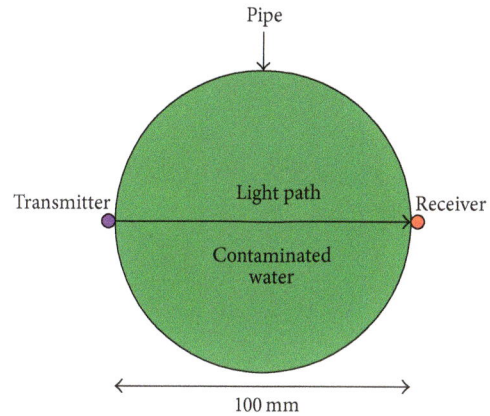

FIGURE 1: The light path in water when there is no gas bubble.

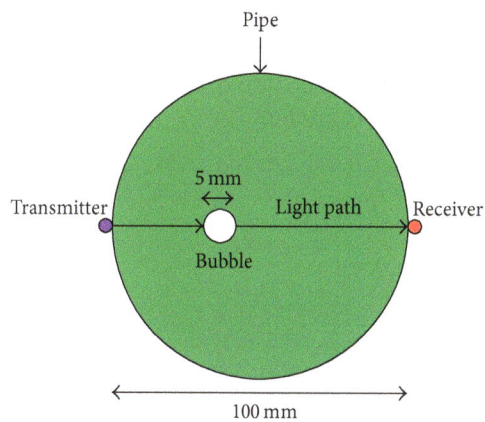

FIGURE 2: The condition of light path in water with gas bubble presence in contaminated water.

2.1.2. Water with Gas Bubble Presence. The aim of this modelling is to determine the change of parameters V_R and $\exp(-\alpha l)$ when a single gas bubble intercepts the transmitting light as shown in Figure 2. In this case, the illustration neglects the reflection and scattering effect when light beam passed through the bubble. The value of the receiving voltage V_R when light passed through bubble and water can be determined as follows:

$$V_R = 5 \exp\left[-(0.0287)(100 - 5) - (0.0142)(5)\right], \quad (4)$$

$$V_R = 0.3050 \, \text{V}. \quad (5)$$

The values of V_R when there is no gas bubble and when gas bubble is present are compared. The existence of bubble in water resulted in a higher value of V_R as indicated in (5) compared to when no bubble is present as indicated in (3). Light experienced less attenuation when gas bubble is present. When there is a bubble, light needs to pass two dissimilar mediums: water and gas. Hence the exponential value is higher compared to when no bubble exists. The values of V_R and $\exp(-\alpha l)$ are predicted to increase when more and

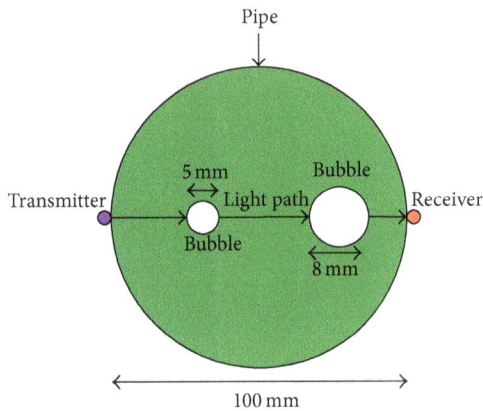

FIGURE 3: The light path when double model is present in contaminated water.

bigger size bubbles intercepted the light before it reaches the receiver's side. The illustration and an example of calculation of V_R for double bubble model can be seen in Figure 3. Related equations are given in

$$V_R = 5 \exp\left[-(0.0287)(100-13)-(0.0142)(5)\right.$$
$$\left. -(0.0142)(8)\right], \tag{6}$$

$$V_R = 0.3425\,\text{V}.$$

2.2. Independent Component Analysis. Blind source separation (BSS) is a technique which separates a set of individual source signals from their mixtures without any specific knowledge as with very limited information about the source signals [13, 14]. An adaptive system is required to solve this problem. Independent component analysis (ICA) is a method which is extensively used for extracting the mixture signals to individual signal [13]. The method is developed by Jutten and Herault at the end of 1980s for retrieving the original sound from the mixed sound signals in the microphone [15]. The method is different from the Principle Component Analysis (PCA) method where PCA is focused in finding the principle components which consist of linear combination of the observed variables, whereas ICA is concerned with discovering the independent components from the observed variables [16]. ICA has been applied in many areas such as telecommunications [17], image processing [18], gas-liquid phase separation in tomography [14], and biomedical applications [19]. A simple example on the application of ICA is shown in Figure 4. In Figure 4, the voice signals from two speakers are mixed together in each microphone. The crowded voices signals also add up and resulted in noise being produced. By using ICA, the original voice signal for both speakers can be acquired.

The general model of ICA is described as

$$X = AS, \tag{7}$$

where X is $n \times m$ matrix of mixture of source signals, A is $n \times d$ mixing matrix, and S is $d \times m$ matrix denoting a source signal. The aim of ICA is to get the value of mixing matrix (A)

and/or source signal matrix (S) given only realizations of the observation matrix of X. After the ICA process was executed, (10) converts to (11) where W is the unmixing matrix and equals A^{-1}:

$$\widehat{S} = WX. \tag{8}$$

The illustration in Figure 5 can assist in understanding the concept of ICA technique. After going through the ICA process, the estimation of source signal $\widehat{S}$ is produced and it looks similar to the original signal, $S(t)$. Moreover, the ICA has formed the separating matrix which was produced alongside the output signal at the output side.

Among the most widely used ICA algorithms are FastICA and Infomax [20]. FastICA provides recovery of independent sources by employing the higher order statistics and the estimation process is done one by one [13]. The algorithm separates the source signals based on their non-Gaussianity and it finds one independent component at a time instead of solving the mixing matrix [20]. The FastICA algorithm is as follows [21]:

(1) Choose an initial weight vector, w.

(2) Let

$$w^+ = E\left\{xg\left(w^T x\right)\right\} - E\left\{g'\left(w^T x\right)\right\}w, \tag{9}$$

where w^+ is a temporary variable used to calculate weight vector, $E\{\cdots\}$ is averaging over all column-vectors of matrix x, w^T is a transpose of weight vector, x is input data, and g is the derivative of nonquadratic used in the contrast function for solving ICA problem.

(3) Let

$$w = \frac{w^+}{\|w^+\|}. \tag{10}$$

(4) If not converged, go back to (7).

The expression "converged" means the value is either old or new for w in the same direction. The advantages of FastICA are that it is parallel and simple for computation process and requires small memory size [22]. For the Infomax algorithm, it is based on maximizing the information transferred in a network of nonlinear units and it is proposed by Bell and Sejnowski [23]. The algorithm is derived from a maximization principle between inputs and nonlinear outputs. The entropy can be described as

$$H\left(s_1, s_2\right) = H\left(s_1\right) + H\left(s_2\right) - I\left(s_1, s_2\right). \tag{11}$$

$H(s_1, s_2)$ is the conditional entropy where the entropy of s_1 is conditional on s_2. $H(s_1)$ and $H(s_2)$ are the entropy of s and $I(s_1, s_2)$ is mutual information. From this equation, maximizing the joint entropy of the outputs amount has led to minimising the mutual information [13]. The disadvantage of Infomax is that the function does not converge properly if step size is wrongly chosen [24].

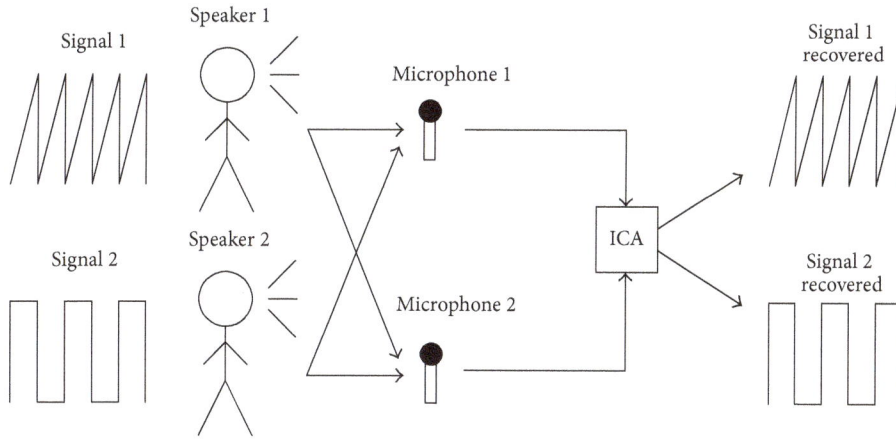

FIGURE 4: A simple example for understanding the ICA concept [13].

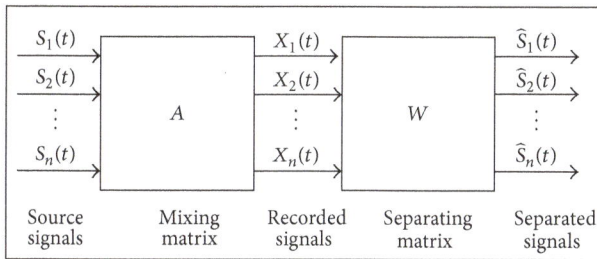

FIGURE 5: ICA block diagram.

This research emphasized the separation of the transmitters signals which were mixed in the receivers. In ICA, there is some ambiguity that should be handled properly. Firstly, the separated signals indicated that the amplitude of the signals is not identical as the source signals. Secondly, the separating matrix is formed in the arbitrary row [20]. A similar parameter from input and output that can be investigated in ICA is the frequency value. Therefore, each transmitter is set to have a different duty cycle in order to facilitate the rearrangement of the separating matrix row. An in-depth explanation about the rearrangement procedure is discussed in Section 4. For simplicity, the modelling for the algorithm is described using four transmitters and four receivers as shown in Figure 6 while the actual system consists of eighteen transmitters and eighteen receivers. This is because the explanation of the ICA concept using eighteen pairs of sensors is complex since the matrix equation is too long. Hence, in this section only four pairs of sensor and four absorption coefficients are chosen and are sufficient to show how ICA works. Transmitters 1, 2, 3, and 4 are represented as TX1, TX2, TX3, and TX4, respectively, while receivers 1, 2, 3, and 4 are represented as RX1, RX2, RX3, and RX4, respectively. The attenuation coefficients are represented as α_1, α_2, α_3, and α_4.

In this model, each receiver is supposed to detect light from all transmitters. Take, for example, receiver RX1. The four signals received at RX1 have been mixed where each light

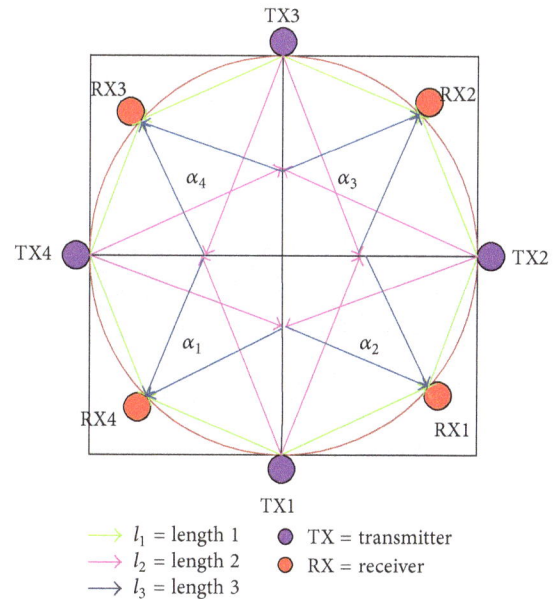

FIGURE 6: Model of four-sensor pair with four pixels.

has its own characteristics such as light length and absorption coefficient. The equation of receiver RX1 voltage V_{R1} can be written as

$$V_{R1} = V_{T1}\exp\left(-\alpha_2 l_1\right) + V_{T2}\exp\left(-\alpha_2 l_1\right)$$
$$+ V_{T3}\exp\left(-\alpha_3 l_2\right) + V_{T3}\exp\left(-\alpha_2 l_3\right) \quad (12)$$
$$+ V_{T4}\exp\left(-\alpha_1 l_2\right) + V_{T4}\exp\left(-\alpha_2 l_3\right),$$

where V_{T1} is transmitter TX1 voltage, V_{T2} is transmitter TX2 voltage, V_{T3} is transmitter TX3 voltage, and V_{T4} is transmitter TX4 voltage, while α_1 is attenuation coefficient for material 1, α_2 is attenuation coefficient for material 2, α_3 is attenuation coefficient for material 3, and α_4 is attenuation coefficient for material 4, whereas l_1 is length 1, l_2 is length 2, and l_3 is length 3. Voltage transmitter (V_T) is a source signal when we analyse the light received by receiver. For example, as shown in Figure 6, receiver 1 (RX1) detects four lights from transmitter 1 (TX1), transmitter 2 (TX2), transmitter 3 (TX3), and transmitter 4 (TX4). The distribution of bubbles in a pipeline is estimated based on the parameters of $\exp(-\alpha l)$ and receiver output voltage (V_R) as indicated in (16). Equation (12) can be arranged as

$$V_{R1} = V_{T1}\exp\left(-\alpha_2 l_1\right) + V_{T2}\exp\left(-\alpha_2 l_1\right)$$
$$+ V_{T3}\exp\left(-\alpha_3 l_2 - \alpha_2 l_3\right) \qquad (13)$$
$$+ V_{T4}\exp\left(-\alpha_1 l_2 - \alpha_2 l_3\right).$$

The equations for receiver RX2 voltage, receiver RX3 voltage, and receiver RX4 voltage are simplified as

$$V_{R2} = V_{T1}\exp\left(-\alpha_2 l_2 - \alpha_3 l_3\right) + V_{T2}\exp\left(-\alpha_3 l_1\right)$$
$$+ V_{T3}\exp\left(-\alpha_3 l_1\right) + V_{T4}\exp\left(-\alpha_4 l_2 - \alpha_3 l_3\right),$$
$$V_{R3} = V_{T1}\exp\left(-\alpha_1 l_2 - \alpha_4 l_3\right) + V_{T2}\exp\left(-\alpha_3 l_2 - \alpha_4 l_3\right)$$
$$+ V_{T3}\exp\left(-\alpha_4 l_1\right) + V_{T4}\exp\left(-\alpha_4 l_1\right), \qquad (14)$$
$$V_{R4} = V_{T1}\exp\left(-\alpha_1 l_1\right) + V_{T2}\exp\left(-\alpha_2 l_2 - \alpha_1 l_3\right)$$
$$+ V_{T3}\exp\left(-\alpha_4 l_2 - \alpha_1 l_3\right) + V_{T4}\exp\left(-\alpha_1 l_1\right).$$

Equations (13) to (14) are transformed into matrix as shown in (15), where X is a matrix of mixture of source signals, represented by the voltage receiver (V_R). $\exp(-\alpha l)$ is denoted as a mixing matrix (A) and voltage transmitter (V_T) is signified as a source signal (S):

$$
\begin{array}{ccc}
X & = & A \qquad\qquad\qquad\qquad\qquad S
\end{array}
$$

$$
\begin{bmatrix} V_{R1} \\ V_{R2} \\ V_{R3} \\ V_{R4} \end{bmatrix}
=
\begin{bmatrix}
e^{-\alpha_2 l_1} & e^{-\alpha_2 l_1} & e^{-\alpha_3 l_2 - \alpha_2 l_3} & e^{-\alpha_1 l_2 - \alpha_2 l_3} \\
e^{-\alpha_2 l_2 - \alpha_3 l_3} & e^{-\alpha_3 l_1} & e^{-\alpha_3 l_1} & e^{-\alpha_4 l_2 - \alpha_3 l_3} \\
e^{-\alpha_1 l_2 - \alpha_4 l_3} & e^{-\alpha_3 l_2 - \alpha_4 l_3} & e^{-\alpha_4 l_1} & e^{-\alpha_4 l_1} \\
e^{-\alpha_1 l_1} & e^{-\alpha_2 l_2 - \alpha_1 l_3} & e^{-\alpha_4 l_2 - \alpha_1 l_3} & e^{-\alpha_1 l_1}
\end{bmatrix}
\begin{bmatrix} V_{T1} \\ V_{T2} \\ V_{T3} \\ V_{T4} \end{bmatrix}. \qquad (15)
$$

ICA is utilized to separate the mixture of source signals where in (16) the matrix positioned of source signals (S) represented as V_T is interchanged with V_R denoted by matrix of mixture of source signals (X). The separating matrix (W) is formed to replace the mixing matrix (A) where theoretically W is obtained by the inversing process of A:

$$
\begin{array}{ccc}
\widehat{S} & = & W \qquad\qquad\qquad X
\end{array}
$$

$$
\begin{bmatrix} V_{T1} \\ V_{T2} \\ V_{T3} \\ V_{T4} \end{bmatrix}
=
\begin{bmatrix}
W_{1,1} & W_{1,2} & W_{1,3} & W_{1,4} \\
W_{2,1} & W_{2,2} & W_{2,3} & W_{2,4} \\
W_{3,1} & W_{3,2} & W_{3,3} & W_{3,4} \\
W_{4,1} & W_{4,2} & W_{4,3} & W_{4,4}
\end{bmatrix}
\begin{bmatrix} V_{R1} \\ V_{R2} \\ V_{R3} \\ V_{R4} \end{bmatrix}. \qquad (16)
$$

In this case, the parameter considered for computing the concentration profile of bubble flow is the matrix A. However, the matrix cannot be acquired directly from the ICA method. Hence, the related parameter regarding A is supported where the matrix W is obtained as an inverse process in order to find the matrix A.

3. Hardware Design

In the optical tomography system, eighteen transmitters and eighteen receivers are mounted around a 100-millimeter (mm) diameter pipe by using a sensor jig. The transparent pipe is made from an acrylic material which allows visual observation of the flow process. The transmitters are positioned next to the receivers since the system is modelled based on fan beam light projection. The infrared Light-Emitting Diode (LED) model TSAL6200 is used as the transmitters where it emits an invisible light. The invisible light has a smaller value of wavelength compared to the visible light in which the range is 750 nanometers (nm) and above. The photodiode model BPV10NF is selected due to its ability to detect the infrared light from the transmitter. A frequency of 50 Hertz (Hz) is chosen for the transmitting circuit frequency, and each transmitter is set using different pulse durations for switching time by using a Peripheral Interface Controller (PIC) Microcontroller. Transmitter 1 (TX1) is set to 1 millisecond (ms) and TX2 for 2 ms and followed until 18 ms for TX18. The purpose for doing this is to facilitate the determination of the transmitter's signal detected by the photodiode where it will be explained in Section 4. The outputs from the receivers are connected to a signal conditioning circuit in order to convert the receiving light into voltage and for voltage amplification. The amplified voltage signals are converted into the digital form using the data acquisition system (DAQ) model U2331 manufactured by Agilent Technologies Inc. It has 3 megasamples per second (MS/s) of sampling rate, 64 single ended/32 differential inputs for analog input channels, and 24-bit programmable TTL input/output channels. The experiment is performed offline. The voltage output from receiver is converted into a digital form using the DAQ and further analysis is performed using the LabVIEW software, which provided the graphical based system for the ICA algorithm. The concentration profiles of the bubble's condition are displayed on a computer. The total time to obtain the final result of concentration profile is around 1 minute. An illustration of the overall design of the optical tomography system is shown in Figure 7.

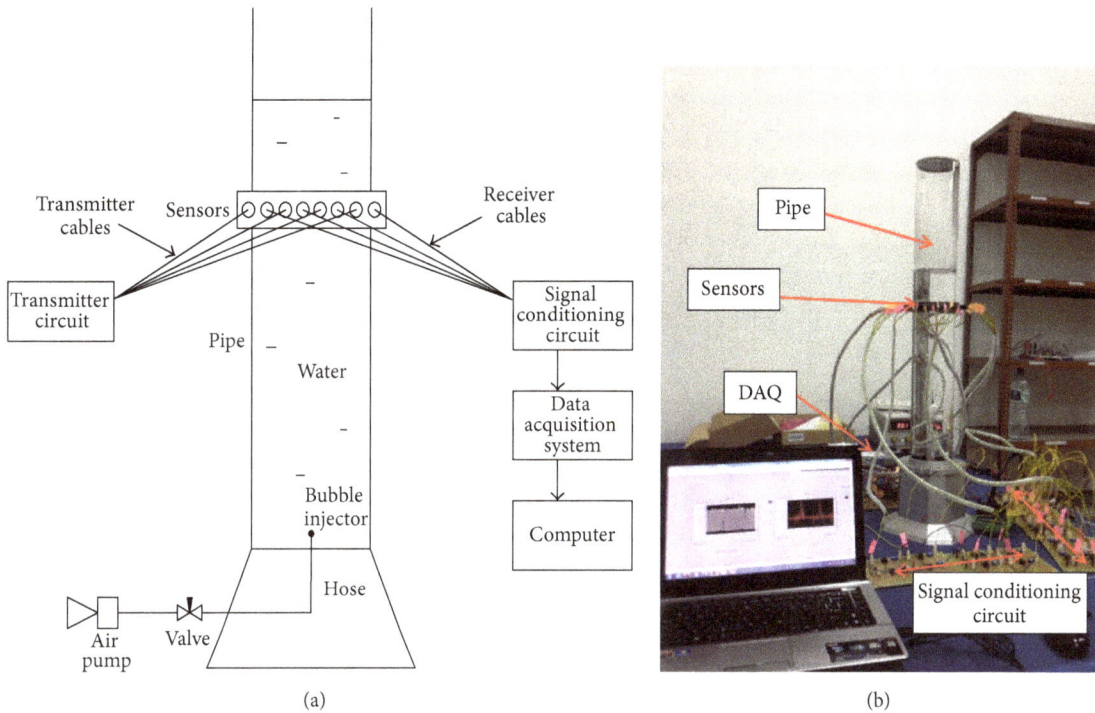

FIGURE 7: (a) The overall design of the optical tomography system. (b) The preparation of the apparatus.

4. Experimental Procedure

This section will elaborate the approach to calculate and analyse the result of bubble flow using ICA method. The steps of the experiment are according to these stages:

(1) Each transmitter has a unique value where it is programmed with different pulse duration value. Hence, the switching times are different from each other. It is assumed that there are eighteen signals from transmitters to be detected by each receiving sensor. Then, the ICA method is applied in order to separate the eighteen signals in order to get a single transmitter signal in each output indicator. The pulse duration value for the transmitter can be from 1 ms to 18 ms, which is set in the transmitter circuit.

(2) A transmitter's signal at each output indicator also brings along a row of unmixed matrix (W). The whole 18 rows for W are produced in arbitrary rows and rearrangement of the matrix's row is necessary according to the order of transmitters pulse duration. For example, a signal with 10 ms is shown in output indicator; hence the value at row 1 will be shifted to row 10. In other words, the rearrangement process for the row depends on the value of pulse duration displayed at each output indicator.

(3) The new matrix W is formed using the ICA method. The matrix of W is emphasized in the steps since it contains an important parameter that has a relationship with the mixing matrix (A). In the ICA theory, $W = A^{-1}$. Hence, the W matrix will be inversed to

get the value of A which consists of matrix 18×18 and expansion of $\exp(-\alpha l)$. The matrix of A is named as the turbidity factor (M).

(4) The 18 columns and 18 rows contained in matrix A represented 18 receivers and 18 transmitters, respectively. Each value is represented as $M_{e,f}$ where e is the transmitter number and f is the receiver number. For example, by referring to Table 1, $M_{2,1}$ has the value 2.782. It indicates that when light is transmitted from transmitter 2 (TX2) to receiver 1 (RX1), it has a turbidity factor (M) of 2.782.

(5) The value at each pixel (P) contained in the cross section of the pipe has to be determined. The P value is the affiliation between M and light's path length, where light crossed a pixel. An example of the light path along the pixels can be seen in Figure 8. For light transmitted from transmitter TX2 to receiver RX1, two pixels {1, 11} and {1, 12} are traversed by the light. The light's lengths in both pixels are measured as 0.4 cm and 0.5 cm, respectively, with the total light's length as 0.9 cm. The P value is computed in order to know the quantity of light that has crossed pixels {1, 11} and {1, 12}. The calculation started from pixel {1, 6}, followed by other pixels contained in the pipe's cross section by using

$$P_{r,c} = \frac{u}{t} \times M_{e,f}, \tag{17}$$

where $P_{r,c}$ is the pixel at row (r) and column (c), u is the path length of light that has crossed the pixel in centimeter (cm), t is the total length from

TABLE 1: Inverse matrix of 25% of air opening flow in 25 mL of water.

Row	Column																	
	1	2	3	4	5	6	7	8	9	10	11	12	13	14	15	16	17	18
1	0.369	1.706	0.692	0.077	0.05	0.335	0.139	0.918	0.439	0.041	0.828	0.08	0.248	1.114	1.123	2.595	1.706	0.978
2	2.782	5.009	2.06	0.295	0.039	0.208	0.003	1.384	0.275	0.481	2.333	1.678	1.607	2.722	1.29	7.143	5.219	3.903
3	1.647	3.761	2.121	0.125	0.077	0.097	0.791	1.407	0.381	0.89	0.918	1.247	0.714	2.083	1.345	4.635	4.479	2.775
4	2.812	1.763	0.797	0.114	0.109	0.12	0.183	0.06	0.336	0.007	0.321	0.063	0.641	0.465	0.898	3.436	1.275	1.264
5	0.321	2.998	1.212	0.008	0.002	0.19	0.032	1.099	0.573	0.088	1.181	0.684	0.93	1.829	2.363	4.242	2.499	1.654
6	0.002	1.648	0.746	0.081	0.153	0.443	0.224	1.074	0.654	0.002	1.25	0.852	0.573	1.227	1.394	2.245	2.667	1.212
7	0.07	0.985	0.393	0.284	0.372	0.597	0.123	0.505	0.2	0.653	0.605	0.639	0.355	1.782	1.353	1.436	2.448	0.119
8	2.568	5.099	2.003	0.489	0.395	0.606	0.078	1.302	0.607	1.392	1.466	1.332	0.569	2.464	2.074	7.335	6.613	4.337
9	4.519	7.131	3.013	0.326	0.326	0.671	0.324	1.897	0.004	1.225	2.591	0.988	0.916	3.111	0.617	10.786	8.416	5.695
10	2.934	4.501	1.44	0.354	0.095	0.247	0.06	0.612	0.085	0.736	1.272	0.833	0.869	1.574	0.137	6.034	3.741	2.812
11	2.637	4.568	2.215	0.534	0.266	0.588	0.119	0.526	0.103	1.031	1.181	0.859	0.663	1.693	0.033	6.054	4.261	2.341
12	1.282	3.864	1.875	0.413	0.537	0.764	0.101	0.453	0.186	0.717	0.911	0.657	0.534	1.25	0.518	4.831	3.379	2.337
13	1.018	2.277	1.045	0.259	0.206	0.586	0.087	0.013	0.362	0.622	0.334	0.524	0.693	0.846	0.385	2.59	1.54	1.07
14	2.523	3.722	1.544	0.228	0.079	0.347	0.141	0.611	0.217	1.141	1.227	0.951	0.76	2.282	0.543	5.418	3.854	2.447
15	0.728	2.226	0.936	0.335	0.011	0.093	0.009	1.954	0.252	0.266	1.098	0.733	0.439	1.708	1.698	2.726	2.803	1.828
16	1.02	1.797	0.933	0.088	0.711	0.184	0.143	0.411	0.121	0.548	0.036	0.226	1.185	0.876	0.347	2.357	3.662	1.622
17	1.164	0.334	0.131	0.12	0.159	0.976	0.235	0.464	0.373	1.086	0.175	0.105	0.725	0.198	1.029	0.802	1.385	0.677
18	0.243	4.09	1.562	0.768	0.805	0.812	1.465	2.311	0.716	1.571	0.867	1.576	0.25	2.802	2.89	4.063	6.35	3.268

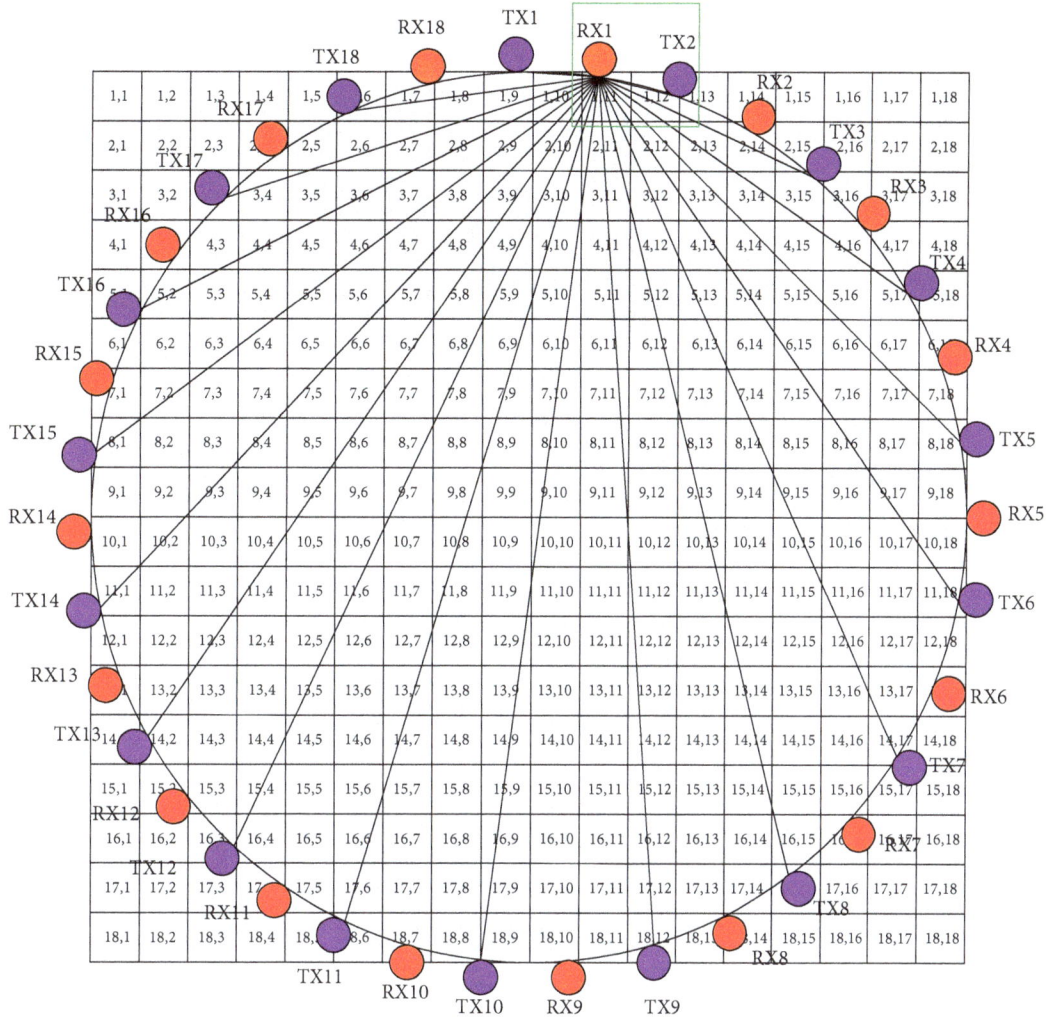

FIGURE 8: The analysis of transmitting light to receiver RX1.

the transmitter to the receiver, and $M_{e,f}$ is the turbidity factor. An example for calculating the P value for pixel $\{1, 12\}$ can be shown as

$$P_{1,12} = \frac{0.5}{0.9} \times 2.782, \tag{18}$$

$$P_{1,12} = 1.546. \tag{19}$$

(6) From Figure 8, pixel $\{1, 12\}$ is intercepted by four lights, which are from transmitters TX2, TX3, TX4, and TX5. By referring to Table 1, the M value for light from transmitter TX3 to receiver RX1 is $M_{3,1} = 1.647$. Then, the P value is recalculated using (18) based on its light's path length. The calculation process for getting the P value continued to other light sources, namely, transmitter TX4 and transmitter TX5.

(7) The transmitter's light also could cross pixel $\{1, 12\}$ when it is transmitted to receiver RX2 until receiver RX18. Hence, it is necessary to recognise whether light

from the transmitters has crossed the pixel or not. If it has, the P values from all transmitters are totalled up and divide into the number of lights that has passed through pixel $\{1, 12\}$. The result from this calculation is named as the average value (K). The K value is the final value to be computed in order to estimate the location of bubbles at certain pixels. Detail of the calculation is shown in Table 2.

5. Result and Discussion

This section presents the concentration profiles of gas bubbles flow in the vertical pipe filled with contaminated water. Five types of bubble flowing conditions are selected which are a single bubble flow, double bubble flow, and 25%, 50%, and 100% of air opening flow. Table 3 shows the gas bubble condition and concentration profile for a single bubble and double bubbles flow in contaminated water. The sample of contaminated water is prepared by adding 25 mL of green

TABLE 2: The calculation of average value for pixels {1, 11} and {1, 12} in 25% of air opening flow in 25 mL of colour ingredient.

Pixels	Turbidity factor (M_{e_f}) of pixel	Pixel's value (P)	Total	Number of pixels	Average value (K)
{1, 11}	$M_{1,1} + M_{2,1} + M_{4,1} + M_{5,1} + M_{6,1} + M_{7,1} + M_{8,1} + M_{9,1} + M_{10,1} + M_{11,1} + M_{12,1} + M_{13,1} + M_{14,1} + M_{15,1} + M_{16,1} + M_{17,1} + M_{18,1} + M_{1,2} + M_{1,3} + M_{1,4} + M_{2,16} + M_{2,15} + M_{2,17} + M_{2,18} + M_{3,1}$	$0.1107 + 1.236 + 0.244 + 0.2295 + 0.027 + 0.00016 + 0.0049 + 0.138 + 0.231 + 0.14467 + 0.133 + 0.0534 + 0.034 + 0.0615 + 0.0205 + 0.035 + 0.052 + 0.025 + 0.3656 + 0.0706 + 0.0013 + 0.018 + 0.4843 + 0.58 + 0.697$	5.0000	25	0.2000
{1, 12}	$M_{2,1} + M_{3,1} + M_{4,1} + M_{5,1} + M_{1,2} + M_{2,2} + M_{2,3} + M_{2,4} + M_{2,5} + M_{2,6} + M_{2,7} + M_{2,8} + M_{2,9} + M_{2,10} + M_{2,11} + M_{2,12} + M_{2,13} + M_{2,14} + M_{2,15} + M_{2,16} + M_{2,17} + M_{2,18}$	$1.546 + 0.366 + 0.2296 + 0.011 + 0.5009 + 0.5961 + 0.076 + 0.0067 + 0.00066 + 0.0055 + 0.00011 + 0.044 + 0.0083 + 0.0143 + 0.07 + 0.0519 + 0.0706 + 0.164 + 0.0896 + 0.6053 + 0.5798 + 0.7000$	5.5022	22	0.2501

TABLE 3: The gas bubble condition and concentration profile for a single bubble and double bubble in contaminated water.

Gas bubble flow	Gas bubble condition	Concentration profile
A single bubble		
Double bubbles		

food colouring to the pipe which contains 3 liters of pure water filled in the pipe. The limitation of this research is that the experiment cannot be performed in volume higher than 25 mL such as 30 mL or 35 mL of green contaminated water. This is due to the fact that the ICA cannot separate the source signals properly since the water sample is too dark.

5.1. 25% of Air Opening. The experiments were conducted by opening up a quarter of the valve to allow the 25% volume of gas entering upwards in the vertical pipe as shown in Figure 9. The transmitter circuit is switched on, before the LabVIEW software is run to apply the ICA algorithm. The K value is computed using the Matlab software by entering the inverse matrix (A) acquired by the ICA process.

Table 4 shows the result of K value for 25% of air opening flow. The location of bubbles is estimated by analysing the value of K. By referring to the mathematical model that has been discussed in considering the presence of gas bubbles, the value of receiver voltage V_R is getting high compared to V_R in homogeneous phase flow. The high value of V_R is related to $\exp(-\alpha l)$ which varied when gas bubbles exist. The properties of $\exp(-\alpha l)$ are the outcome for A matrix and have influenced

FIGURE 9: The condition of 25% of air opening.

the calculation of K parameter. The range of values of K is considered from 0.1 and above. Therefore, the bubbles are supposed to be located in the pixels which contain at least 0.1.

TABLE 4: *K* value for 25% of air opening flow.

0.000	0.000	0.000	0.000	0.166	0.190	0.181	0.125	0.159	0.200	0.250	0.589	0.000	0.000	0.000	0.000	0.000
0.000	0.000	0.194	0.480	0.129	0.147	0.156	0.124	0.131	0.131	0.107	0.132	0.270	0.446	0.303	0.000	0.000
0.000	0.093	0.211	0.155	0.111	0.098	0.079	0.067	0.088	0.092	0.067	0.116	0.130	0.162	0.110	0.110	0.000
0.000	0.211	0.140	0.106	0.106	0.100	0.098	0.068	0.082	0.071	0.067	0.073	0.075	0.052	0.068	0.069	0.000
0.000	0.159	0.098	0.096	0.105	0.099	0.069	0.076	0.073	0.072	0.089	0.059	0.088	0.059	0.045	0.049	0.000
0.154	0.128	0.095	0.077	0.102	0.082	0.089	0.089	0.095	0.076	0.059	0.075	0.052	0.041	0.044	0.043	0.015
0.124	0.095	0.110	0.111	0.077	0.091	0.090	0.089	0.074	0.071	0.063	0.083	0.061	0.054	0.039	0.048	0.062
0.140	0.097	0.099	0.099	0.084	0.079	0.062	0.130	0.059	0.093	0.049	0.070	0.058	0.047	0.044	0.029	0.057
0.121	0.068	0.091	0.074	0.092	0.094	0.069	0.077	0.061	0.074	0.056	0.064	0.052	0.048	0.041	0.035	0.024
0.194	0.069	0.071	0.077	0.098	0.033	0.127	0.017	0.093	0.056	0.061	0.070	0.054	0.034	0.032	0.032	0.035
0.157	0.051	0.073	0.089	0.040	0.083	0.058	0.097	0.066	0.074	0.045	0.072	0.045	0.038	0.028	0.026	0.052
0.099	0.062	0.074	0.049	0.078	0.075	0.066	0.076	0.081	0.060	0.055	0.068	0.048	0.047	0.034	0.031	0.055
0.064	0.054	0.053	0.062	0.069	0.072	0.083	0.072	0.066	0.087	0.065	0.071	0.047	0.050	0.041	0.050	0.030
0.076	0.068	0.045	0.068	0.064	0.077	0.053	0.105	0.079	0.058	0.081	0.079	0.043	0.067	0.039	0.027	0.000
0.069	0.094	0.072	0.061	0.041	0.065	0.062	0.045	0.104	0.049	0.074	0.088	0.061	0.052	0.013	0.000	0.000
0.000	0.074	0.081	0.068	0.067	0.054	0.048	0.054	0.056	0.085	0.064	0.055	0.087	0.073	0.000	0.000	0.000
0.000	0.000	0.083	0.147	0.098	0.052	0.070	0.055	0.036	0.088	0.092	0.058	0.132	0.056	0.000	0.000	0.000
0.000	0.000	0.000	0.000	0.069	0.075	0.140	0.082	0.049	0.075	0.148	0.088	0.000	0.000	0.000	0.000	0.000

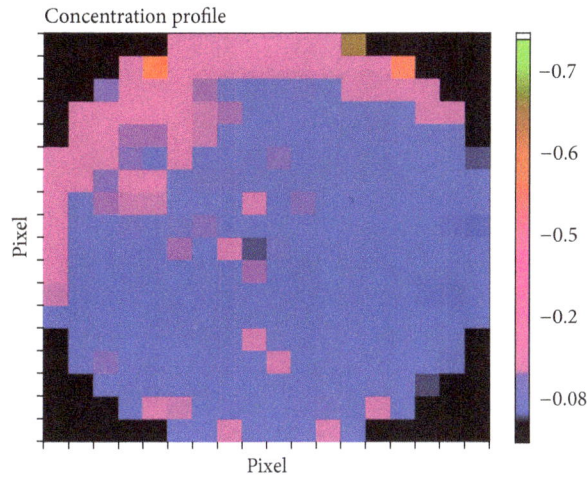

FIGURE 10: The concentration profile for 25% of air opening.

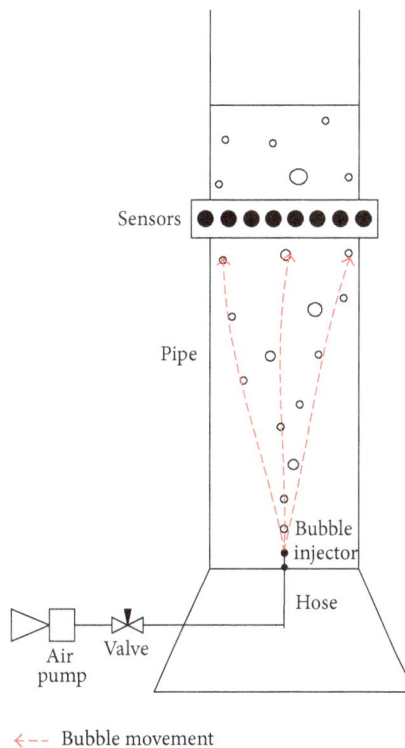

←-- Bubble movement

FIGURE 11: Bubbles movement upwards from the bottom.

The result in Table 4 is transformed to concentration profile as shown in Figure 10 in order to visualize bubble location. The pink colour in the concentration profile indicates that some bubbles are inclined to rise towards to the border of the pipe although the bubble injector is installed at the centre of the pipe base. The illustration for the bubble's movements can be seen in Figure 11. Other than that, the high value of pixels located near the pipe's periphery is probably due to the existence of receivers around the pipe's periphery. Hence, most of the light is likely to cross the pixels around the pipe's periphery. Although the average value is calculated,

the K value at the pipe's periphery is still higher compared to other locations.

5.2. 50% of Air Opening. Half of the valve is opened in this experiment as in Figure 12(a) to ensure that more bubbles are inserted into the pipe than the first experiment. The concentration profile for the medium bubble flow is shown in Figure 12(b). The result of K value is higher compared to the 25% of air opening where the range for bubble is from 0.12 to 0.5. It is due to the light's energy experiencing more attenuation when the bubbles interrupted the light path as modelled in (4) to (6). Moreover, the bubbles also became

(a)   (b)

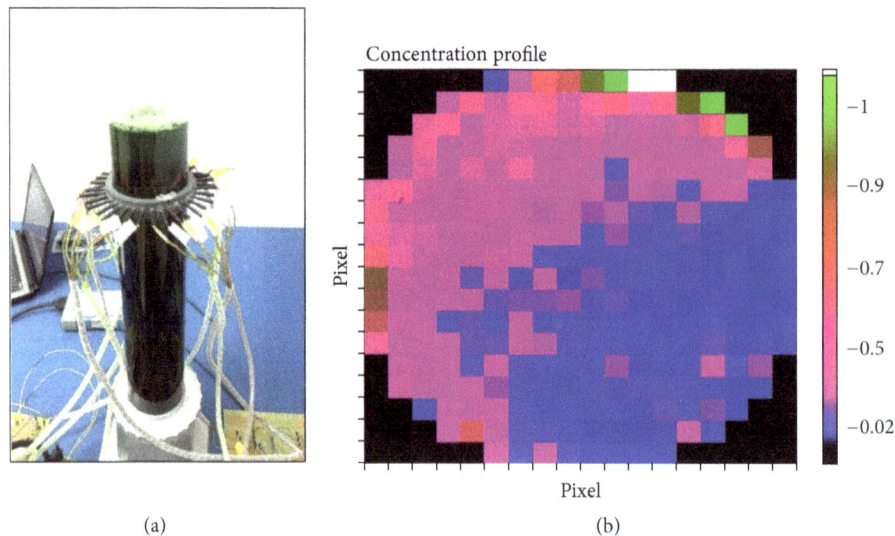

FIGURE 12: (a) The condition of 50% of air opening. (b) The concentration profile for 50% of air opening.

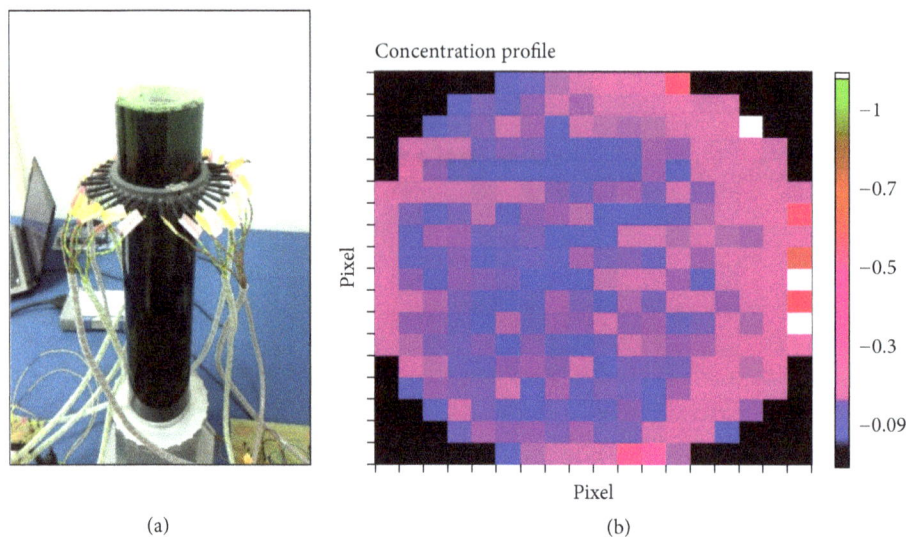

(a)   (b)

FIGURE 13: (a) The condition of 100% of air opening. (b) The concentration profile for 100% of air opening.

close to each other when more bubbles went upward since the space around the pipe is limited and compact. Thus, the collision between bubbles occurred and generated the additional energy which resulted in the value of exp $(-\alpha l)$ increasing.

5.3. 100% of Air Opening. In this experiment, the valve is fully open and the bubble flow condition is shown in Figure 13(a). The range of bubble is from 0.2 to 0.5, and this represents an increase compared to medium bubble flow. The location of bubble can be seen in Figure 13(b) which shows that most of the area in the cross section of the pipe is occupied by gas bubbles. Thus, all sensors in the system can detect the presence of bubble in high bubble flow condition. This condition also creates more collision between the bubbles than medium bubble flow since the free space in the pipe has been reduced by the addition of gas bubbles. More collisions of bubbles have influenced the reading in terms of the increasing value of K.

6. Conclusion

In this paper, an optical tomography system using the ICA method has been presented to estimate the distribution of bubbles in a vertical pipeline. The system proved its ability to obtain information on the gas bubbles concentration profile in the pipe without disturbing the process flow. The concentration profiles for the bubble flow quantities are investigated under five different flows conditions: a single bubble, double bubbles, 25% of air opening, 50% of air opening, and 100% of air opening flow rates. The ICA method has a great potential in improving the process efficiency by visualizing the concentration profiles inside the pipelines.

Appendix

See Tables 5–9.

TABLE 5: K value of 25 mL.

0.000	0.000	0.000	0.000	0.000	0.115	0.156	0.134	0.200	0.192	0.137	0.232	0.099	0.000	0.000	0.000	0.000	0.000
0.000	0.000	0.000	0.059	0.108	0.094	0.099	0.070	0.099	0.068	0.063	0.075	0.077	0.099	0.089	0.000	0.000	0.000
0.000	0.000	0.107	0.084	0.049	0.092	0.054	0.082	0.077	0.047	0.054	0.079	0.065	0.049	0.084	0.039	0.000	0.000
0.000	0.070	0.072	0.054	0.070	0.058	0.075	0.072	0.051	0.084	0.057	0.057	0.059	0.064	0.057	0.054	0.013	0.000
0.000	0.088	0.068	0.061	0.079	0.034	0.085	0.062	0.061	0.053	0.060	0.045	0.062	0.054	0.057	0.040	0.061	0.000
0.204	0.130	0.073	0.058	0.042	0.076	0.054	0.053	0.039	0.057	0.048	0.049	0.053	0.053	0.037	0.056	0.061	0.024
0.173	0.114	0.048	0.047	0.062	0.035	0.062	0.050	0.071	0.073	0.037	0.051	0.028	0.060	0.065	0.052	0.053	0.049
0.211	0.089	0.059	0.052	0.055	0.067	0.047	0.056	0.060	0.050	0.058	0.040	0.052	0.058	0.038	0.049	0.048	0.077
0.220	0.091	0.068	0.087	0.053	0.090	0.061	0.072	0.050	0.066	0.049	0.041	0.047	0.041	0.037	0.035	0.044	0.073
0.147	0.094	0.065	0.058	0.042	0.062	0.068	0.054	0.049	0.020	0.060	0.064	0.056	0.057	0.044	0.046	0.062	0.129
0.203	0.106	0.079	0.065	0.072	0.070	0.067	0.038	0.046	0.053	0.053	0.056	0.046	0.040	0.045	0.046	0.049	0.132
0.316	0.154	0.092	0.067	0.060	0.050	0.040	0.042	0.038	0.025	0.048	0.042	0.042	0.026	0.029	0.054	0.042	0.070
0.264	0.161	0.088	0.077	0.055	0.059	0.044	0.036	0.034	0.041	0.034	0.037	0.036	0.051	0.060	0.051	0.064	0.025
0.000	0.291	0.091	0.083	0.079	0.061	0.056	0.048	0.056	0.060	0.048	0.042	0.067	0.041	0.050	0.049	0.090	0.000
0.000	0.083	0.201	0.105	0.049	0.048	0.051	0.068	0.055	0.053	0.057	0.053	0.051	0.062	0.065	0.089	0.082	0.000
0.000	0.000	0.121	0.125	0.078	0.054	0.042	0.046	0.051	0.058	0.070	0.074	0.053	0.059	0.053	0.047	0.000	0.000
0.000	0.000	0.000	0.055	0.137	0.093	0.062	0.065	0.066	0.055	0.061	0.050	0.045	0.046	0.048	0.000	0.000	0.000
0.000	0.000	0.000	0.000	0.358	0.091	0.067	0.060	0.057	0.066	0.094	0.046	0.000	0.000	0.000	0.000	0.000	0.000

0.000	0.000	0.000	0.000	0.000	0.172	0.092	0.136	0.217	0.249	0.229	0.560	0.411	0.000	0.000	0.000	0.000	0.000
0.000	0.000	0.080	0.178	0.114	0.104	0.108	0.094	0.079	0.099	0.108	0.164	0.122	0.223	0.220	0.139	0.000	0.000
0.000	0.090	0.116	0.081	0.111	0.082	0.078	0.142	0.133	0.061	0.114	0.139	0.104	0.092	0.252	0.176	0.140	0.000
0.000	2.730	0.149	0.129	0.149	0.114	0.090	0.052	0.077	0.141	0.120	0.102	0.139	0.106	0.111	0.175	0.198	0.000
0.095	0.152	0.135	0.111	0.115	0.109	0.102	0.086	0.121	0.077	0.075	0.138	0.146	0.120	0.102	0.134	0.179	0.115
0.193	0.116	0.095	0.099	0.111	0.095	0.080	0.078	0.066	0.038	0.126	0.096	0.161	0.127	0.114	0.101	0.157	0.207
0.123	0.096	0.067	0.088	0.077	0.100	0.060	0.102	0.073	0.077	0.121	0.078	0.112	0.049	0.101	0.125	0.173	0.430
0.115	0.069	0.077	0.085	0.048	0.087	0.103	0.054	0.088	0.043	0.075	0.126	0.070	0.113	0.065	0.085	0.158	0.721
0.092	0.089	0.049	0.083	0.058	0.330	0.077	0.095	0.064	0.114	0.055	0.122	0.090	0.145	0.110	0.085	0.107	0.526
0.087	0.082	0.069	0.034	0.071	0.116	0.047	0.067	0.056	0.138	0.116	0.110	0.053	0.080	0.064	0.096	0.161	0.304
0.081	0.056	0.073	0.057	0.130	0.074	0.115	0.099	0.109	0.132	0.088	0.062	0.056	0.079	0.094	0.123	0.150	0.170
0.206	0.064	0.098	0.105	0.077	0.134	0.104	0.067	0.125	0.128	0.106	0.068	0.057	0.099	0.118	0.121	0.154	0.058
0.000	0.250	0.141	0.077	0.059	0.083	0.089	0.076	0.120	0.065	0.085	0.087	0.084	0.146	0.134	0.120	0.183	0.000
0.000	0.630	0.211	0.109	0.064	0.090	0.114	0.056	0.131	0.092	0.045	0.110	0.103	0.117	0.124	0.159	0.050	0.000
0.000	0.000	0.140	0.123	0.118	0.089	0.124	0.078	0.144	0.104	0.129	0.082	0.100	0.153	0.168	0.049	0.000	0.000
0.000	0.000	0.090	0.099	0.131	0.101	0.167	0.122	0.129	0.109	0.113	0.115	0.274	0.152	0.000	0.000	0.000	0.000
0.000	0.000	0.000	0.000	0.000	0.048	0.221	0.253	0.181	0.223	0.219	0.548	0.268	0.000	0.000	0.000	0.000	0.000

TABLE 6: K value of a single bubble flow.

TABLE 7: K value of double bubble flow.

0.000	0.000	0.000	0.000	0.000	0.044	0.091	0.076	0.085	0.105	0.072	0.118	0.091	0.000	0.000	0.000	0.000	0.000
0.000	0.000	0.000	0.017	0.058	0.065	0.068	0.046	0.052	0.057	0.041	0.058	0.104	0.115	0.050	0.000	0.000	0.000
0.000	0.000	0.030	0.056	0.048	0.059	0.039	0.045	0.040	0.043	0.044	0.068	0.122	0.078	0.098	0.049	0.000	0.000
0.000	0.086	0.055	0.050	0.038	0.049	0.052	0.050	0.046	0.036	0.071	0.080	0.073	0.084	0.070	0.121	0.039	0.000
0.000	0.090	0.048	0.063	0.068	0.035	0.046	0.050	0.058	0.078	0.081	0.055	0.084	0.078	0.057	0.076	0.104	0.000
0.158	0.075	0.055	0.046	0.025	0.052	0.039	0.051	0.069	0.067	0.070	0.050	0.079	0.055	0.054	0.054	0.108	0.028
0.555	0.059	0.067	0.056	0.053	0.044	0.065	0.039	0.040	0.073	0.046	0.064	0.063	0.057	0.062	0.068	0.076	0.129
0.089	0.069	0.063	0.044	0.044	0.078	0.041	0.040	0.043	0.052	0.061	0.040	0.093	0.075	0.057	0.062	0.071	0.149
0.119	0.067	0.045	0.079	0.048	0.041	0.028	0.037	0.039	0.055	0.074	0.094	0.057	0.055	0.040	0.068	0.062	0.096
0.070	0.061	0.072	0.050	0.042	0.065	0.065	0.155	0.053	0.021	0.058	0.065	0.041	0.057	0.047	0.071	0.044	0.113
0.109	0.087	0.062	0.058	0.058	0.067	0.038	0.050	0.043	0.073	0.050	0.080	0.056	0.064	0.058	0.063	0.056	0.094
0.508	0.126	0.078	0.061	0.060	0.044	0.044	0.062	0.150	0.037	0.042	0.062	0.063	0.056	0.050	0.061	0.071	0.131
0.196	0.122	0.106	0.081	0.089	0.089	0.077	0.059	0.041	0.049	0.045	0.066	0.046	0.078	0.100	0.082	0.123	0.082
0.000	0.195	0.096	0.075	0.068	0.071	0.095	0.068	0.075	0.084	0.085	0.063	0.094	0.070	0.089	0.104	0.094	0.000
0.000	0.091	0.221	0.122	0.085	0.071	0.065	0.095	0.082	0.060	0.066	0.108	0.083	0.098	0.145	0.131	0.091	0.000
0.000	0.000	0.134	0.191	0.135	0.124	0.081	0.090	0.102	0.089	0.073	0.114	0.075	0.131	0.118	0.084	0.000	0.000
0.000	0.000	0.000	0.139	0.218	0.161	0.153	0.108	0.111	0.115	0.135	0.117	0.149	0.206	0.138	0.000	0.000	0.000
0.000	0.000	0.000	0.000	0.000	0.185	0.121	0.154	0.140	0.204	0.213	0.284	0.222	0.000	0.000	0.000	0.000	0.000

TABLE 8: K value for 50% of air opening flow.

0.000	0.000	0.000	0.000	0.062	0.224	0.497	0.573	0.770	0.875	1.261	3.137	0.000	0.000	0.000	0.000	0.000
0.000	0.000	0.233	0.403	0.266	0.365	0.387	0.249	0.345	0.436	0.296	0.410	0.752	0.892	0.000	0.000	0.000
0.000	0.340	0.366	0.210	0.274	0.284	0.297	0.241	0.266	0.201	0.176	0.200	0.255	0.419	0.931	0.000	0.000
0.000	0.360	0.200	0.153	0.276	0.257	0.220	0.190	0.241	0.184	0.169	0.188	0.187	0.145	0.340	0.648	0.000
0.000	0.295	0.193	0.286	0.153	0.312	0.166	0.145	0.187	0.091	0.134	0.146	0.143	0.152	0.153	0.301	0.000
0.330	0.245	0.169	0.227	0.231	0.181	0.218	0.185	0.164	0.112	0.165	0.118	0.090	0.134	0.131	0.064	0.032
0.406	0.168	0.192	0.228	0.157	0.186	0.118	0.166	0.116	0.118	0.106	0.079	0.118	0.098	0.058	0.088	0.059
0.349	0.258	0.187	0.159	0.185	0.155	0.160	0.153	0.088	0.115	0.069	0.090	0.106	0.043	0.065	0.066	0.052
0.455	0.154	0.212	0.174	0.127	0.139	0.100	0.087	0.081	0.072	0.094	0.085	0.072	0.045	0.091	0.045	0.076
0.704	0.200	0.130	0.108	0.153	0.072	0.143	0.035	0.110	0.073	0.092	0.089	0.075	0.048	0.079	0.051	0.059
0.670	0.206	0.160	0.151	0.107	0.111	0.111	0.114	0.066	0.107	0.044	0.069	0.040	0.079	0.069	0.047	0.028
0.595	0.165	0.100	0.107	0.089	0.138	0.087	0.101	0.075	0.076	0.071	0.052	0.061	0.092	0.048	0.049	0.048
0.341	0.157	0.163	0.105	0.108	0.134	0.118	0.094	0.082	0.077	0.071	0.082	0.070	0.105	0.051	0.067	0.033
0.000	0.219	0.158	0.102	0.122	0.104	0.099	0.096	0.096	0.116	0.052	0.073	0.085	0.121	0.098	0.115	0.000
0.000	0.223	0.196	0.136	0.116	0.088	0.094	0.076	0.100	0.064	0.104	0.101	0.091	0.070	0.043	0.027	0.000
0.000	0.077	0.232	0.217	0.134	0.103	0.075	0.090	0.075	0.075	0.059	0.108	0.099	0.110	0.031	0.000	0.000
0.000	0.000	0.151	0.564	0.188	0.105	0.092	0.075	0.047	0.039	0.035	0.068	0.129	0.069	0.000	0.000	0.000
0.000	0.000	0.000	0.000	0.130	0.097	0.226	0.104	0.053	0.045	0.043	0.048	0.000	0.000	0.000	0.000	0.000

TABLE 9: K value for 100% of air opening flow.

0.000	0.000	0.000	0.000	0.000	0.362	0.685	0.478	0.756	1.076	0.704	1.158	0.513	0.000	0.000	0.000	0.000	0.000
0.000	0.000	0.000	0.246	0.287	0.383	0.432	0.323	0.342	0.345	0.336	0.385	0.339	0.443	0.207	0.000	0.000	0.000
0.000	0.000	1.011	0.413	0.299	0.354	0.300	0.183	0.274	0.323	0.226	0.186	0.385	0.193	0.331	0.174	0.000	0.000
0.000	0.315	0.406	0.238	0.217	0.237	0.230	0.188	0.251	0.212	0.176	0.247	0.238	0.337	0.222	0.343	0.088	0.000
0.000	0.253	0.290	0.267	0.274	0.166	0.185	0.252	0.176	0.130	0.238	0.155	0.181	0.280	0.301	0.246	0.287	0.000
0.682	0.453	0.304	0.175	0.078	0.211	0.139	0.190	0.201	0.177	0.241	0.144	0.172	0.152	0.178	0.270	0.392	0.190
0.490	0.253	0.193	0.180	0.180	0.158	0.126	0.227	0.142	0.215	0.153	0.176	0.191	0.155	0.124	0.116	0.234	0.570
0.450	0.204	0.186	0.133	0.120	0.183	0.104	0.212	0.162	0.244	0.187	0.172	0.202	0.154	0.151	0.165	0.155	0.636
0.506	0.207	0.195	0.094	0.066	0.175	0.115	0.121	0.156	0.197	0.126	0.217	0.144	0.125	0.156	0.139	0.151	0.342
0.258	0.180	0.201	0.127	0.099	0.146	0.193	0.065	0.182	0.120	0.159	0.170	0.099	0.202	0.138	0.127	0.183	0.256
0.188	0.238	0.127	0.130	0.107	0.165	0.179	0.121	0.134	0.132	0.182	0.175	0.139	0.165	0.184	0.148	0.158	0.201
0.141	0.170	0.177	0.160	0.105	0.198	0.170	0.158	0.148	0.102	0.138	0.143	0.168	0.177	0.118	0.126	0.174	0.190
0.055	0.136	0.212	0.161	0.097	0.238	0.135	0.149	0.144	0.154	0.101	0.190	0.119	0.170	0.201	0.134	0.168	0.112
0.000	0.156	0.161	0.151	0.112	0.198	0.193	0.146	0.126	0.162	0.133	0.114	0.162	0.140	0.158	0.154	0.184	0.000
0.000	0.080	0.144	0.204	0.140	0.213	0.154	0.179	0.168	0.172	0.154	0.146	0.152	0.146	0.160	0.266	0.146	0.000
0.000	0.000	0.059	0.207	0.171	0.245	0.133	0.184	0.141	0.129	0.186	0.170	0.157	0.194	0.173	0.131	0.000	0.000
0.000	0.000	0.000	0.344	0.508	0.318	0.158	0.136	0.133	0.165	0.200	0.169	0.190	0.327	0.060	0.000	0.000	0.000
0.000	0.000	0.000	0.000	0.000	0.501	0.207	0.246	0.240	0.282	0.336	0.377	0.309	0.000	0.000	0.000	0.000	0.000

Competing Interests

The authors declare that there are no competing interests regarding the publication of this paper.

Acknowledgments

The authors wish to acknowledge the assistance of the Ministry of Higher Education, Malaysia, and Universiti Teknologi Malaysia under the GUP Research Vote 05H67 for providing the funds and resources in carrying out this research.

References

[1] C. Yan, J. Zhong, Y. Liao, S. Lai, M. Zhang, and D. Gao, "Design of an applied optical fiber process tomography system," *Sensors and Actuators B: Chemical*, vol. 104, no. 2, pp. 324–331, 2005.

[2] D. L. George, J. R. Torczynski, K. A. Shollenberger, and T. J. O'Hern, "Validation of electrical-impedance tomography for measurements of material distribution in two-phase flows," *International Journal of Multiphase Flow*, vol. 26, no. 4, pp. 549–581, 2000.

[3] T. Dyakowski, L. F. C. Jeanmeure, and A. J. Jaworski, "Applications of electrical tomography for gas-solids and liquid-solids flows—a review," *Powder Technology*, vol. 112, no. 3, pp. 174–192, 2000.

[4] R. Abdul Rahim, C. K. Thiam, and M. H. F. Rahiman, "An optical tomography system using a digital signal processor," *Sensors*, vol. 8, no. 4, pp. 2082–2103, 2008.

[5] M. Idroas, R. Abdul Rahim, M. H. Fazalul Rahiman, M. N. Ibrahim, and R. G. Green, "Design and development of a CCD based optical tomography measuring system for particle sizing identification," *Measurement*, vol. 44, no. 6, pp. 1096–1107, 2011.

[6] M. R. Rzasa and A. Plaskowski, "Application of optical tomography for measurements of aeration parameters in large water tanks," *Measurement Science and Technology*, vol. 14, no. 2, pp. 199–204, 2003.

[7] S. Ibrahim, M. A. M. Yunus, R. G. Green, and K. Dutton, "Concentration measurements of bubbles in a water column using an optical tomography system," *ISA Transactions*, vol. 51, no. 6, pp. 821–826, 2012.

[8] R. M. Detsch and R. N. Sharma, "The critical angle for gas bubble entrainment by plunging liquid jets," *The Chemical Engineering Journal*, vol. 44, no. 3, pp. 157–166, 1990.

[9] M. R. Rzasa, "The measuring method for tests of horizontal two-phase gas-liquid flows, using optical and capacitance tomography," *Nuclear Engineering and Design*, vol. 239, no. 4, pp. 699–707, 2009.

[10] H. Jin, Y. Lian, S. Yang, G. He, and Z. Guo, "The parameters measurement of air–water two phase flow using the electrical resistance tomography (ERT) technique in a bubble column," *Flow Measurement and Instrumentation*, vol. 31, pp. 55–60, 2013.

[11] N. M. N. Ayob, M. H. F. Rahiman, Z. Zakaria, S. Yaacob, and R. A. Rahim, "Detection of small gas bubble using ultrasonic transmission-mode tomography system," in *Proceedings of the IEEE Symposium on Industrial Electronics & Applications (ISIEA '11)*, pp. 165–170, IEEE, Penang, Malaysia, October 2010.

[12] A. R. Daniels, *Dual modality tomography for the monitoring of constituent volumes in multi-component flows [Ph.D. thesis]*, Sheffield Hallam University, 1996.

[13] G. R. Naik and D. K. Kumar, "An overview of independent component analysis and its applications," *Informatica*, vol. 35, pp. 63–81, 2011.

[14] Y. Xu, H. Wang, Z. Cui, F. Dong, and Y. Yan, "Separation of gas-liquid two-phase flow through independent component analysis," *IEEE Transactions on Instrumentation and Measurement*, vol. 59, no. 5, pp. 1294–1302, 2010.

[15] C. Jutten and J. Herault, "Blind separation of sources, part I: an adaptive algorithm based on neuromimetic architecture," *Signal Processing*, vol. 24, no. 1, pp. 1–10, 1991.

[16] J. Chen and X. Z. Wang, "A new approach to near-infrared spectral data analysis using independent component analysis," *Journal of Chemical Information and Computer Sciences*, vol. 41, no. 4, pp. 992–1001, 2001.

[17] J. P. Huang and J. Mar, "Combined ICA and FCA schemes for a hierarchical network," *Wireless Personal Communications*, vol. 28, no. 1, pp. 35–58, 2004.

[18] S. Fiori, "Overview of independent component analysis technique with an application to synthetic aperture radar (SAR) imagery processing," *Neural Networks*, vol. 16, no. 3-4, pp. 453–467, 2003.

[19] C. J. James and C. W. Hesse, "Independent component analysis for biomedical signals (topical review)," *Physiological Measurement*, vol. 26, no. 1, pp. R15–R39, 2005.

[20] G. Wang, Q. Ding, and Z. Hou, "Independent component analysis and its applications in signal processing for analytical chemistry," *TrAC—Trends in Analytical Chemistry*, vol. 27, no. 4, pp. 368–376, 2008.

[21] A. Hyvärinen, "Fast and robust fixed-point algorithms for independent component analysis," *IEEE Transactions on Neural Networks*, vol. 10, no. 3, pp. 626–634, 1999.

[22] A. Hyvärinen and E. Oja, "Independent component analysis: algorithms and applications," *Neural Networks*, vol. 13, no. 4-5, pp. 411–430, 2000.

[23] A. J. Bell and T. J. Sejnowski, "An information-maximization approach to blind separation and blind deconvolution," *Neural Computation*, vol. 7, no. 6, pp. 1129–1159, 1995.

[24] J. H. Garvey, *Independent component analysis by entropy maximization (Infomax) [M.S. thesis]*, Naval Postgraduate School, Monterey, Calif, USA, 2007.

26

Irradiance Scintillation Index for a Gaussian Beam Based on the Generalized Modified Atmospheric Spectrum with Aperture Averaged

Chao Gao,[1] Yang Li,[2] Yiming Li,[1] and Xiaofeng Li[1]

[1]*School of Astronautics and Aeronautic, University of Electronic Science and Technology of China, 2006 Xiyuan Avenue, Chengdu 611731, China*
[2]*School of Accounting, Southwestern University of Finance and Economics, 555 Liutai Avenue, Chengdu 611130, China*

Correspondence should be addressed to Xiaofeng Li; lxf3203433@uestc.edu.cn

Academic Editor: Fortunato Tito Arecchi

This paper investigates the aperture-averaged irradiance scintillation index of a Gaussian beam propagating through a horizontal path in weak non-Kolmogorov turbulence. Mathematical expressions are obtained based on the generalized modified atmospheric spectrum, which includes the spectral power law value of non-Kolmogorov turbulence, the finite inner and outer scales of turbulence, and other optical parameters of the Gaussian beam. The numerical results are conducted to analyze the influences of optical parameters on the aperture-averaged irradiance scintillation index for different Gaussian beams. This paper also examines the effects of the irradiance scintillation on the performance of the point-to-point optical wireless communication system with intensity modulation/direct detection scheme.

1. Introduction

Optical wireless communication technology has drawn much attention for its significant technological challenges and prospective applications. It uses beams of laser propagating through the atmosphere to transmit data wirelessly at high speed [1, 2]. However, the atmosphere is full of numerous turbulence eddies, which has great degrading impacts on the performance of the communication system. Irradiance scintillation, one typical effect of atmospheric turbulence, results from stochastic redistribution of the optical energy within a cross section of the laser beam [3, 4]. The degree of irradiance scintillation for optical propagation on the receiving antenna can be characterized statistically by the irradiance scintillation index.

In the past few decades, various power spectrum models of refractive index have been proposed to analyze the irradiance scintillation index for different situations [5–15]. Generally speaking, these models can be classified into two categories: Kolmogorov models and non-Kolmogorov models. The former have a fixed power law value of 11/3

whereas the latter allow the power law value to vary in the range from 3 to 4. Practically, most non-Kolmogorov models can be generalized from corresponding Kolmogorov models [16]. Among the non-Kolmogorov models, the generalized modified atmospheric spectrum not only considers the finite inner and outer scales of turbulence, but also features the small rise at a high wave number, which is clearly seen in temperature data recorded by sensors [17, 18]. These properties make the generalized modified atmospheric spectrum useful in the investigation of the irradiance scintillation index along the radial and longitudinal propagation [10, 11, 13].

In this study, the generalized modified atmospheric spectrum is used to investigate the aperture-averaged irradiance scintillation index of a Gaussian beam in weak non-Kolmogorov turbulence along a horizontal path. The Gaussian beam, whose transverse electric field and intensity are normally distributed, is a typical kind of electromagnetic wave. The rest of the paper is organized as follows. Section 2 introduces the generalized modified atmospheric spectrum and the irradiance scintillation index of a Gaussian beam in weak turbulence. Section 3 presents a detailed expression reduction. The influences of the inner and outer scales of turbulence on the irradiance scintillation index of a Gaussian beam are analyzed in Section 4, followed by conclusions in Section 5.

2. Theoretical Models

2.1. Generalized Modified Atmospheric Spectrum. The generalized modified atmospheric spectrum takes the form [17]

$$\Phi_n(\kappa, \alpha) = A(\alpha) C_n^2 \kappa^{-\alpha} f(\kappa, \alpha, l_0, L_0), \quad (1)$$

where $\kappa \in [0, +\infty)$ is the angular wavenumber of the turbulence scale, $\alpha \in (3, 4)$ is the spectral power law value, C_n^2 is the generalized atmospheric structure parameter, and l_0 and L_0 are the inner and outer scales of turbulence. $A(\alpha)$ is a function related to α:

$$A(\alpha) = \frac{\Gamma(\alpha - 1)}{4\pi^2} \sin \frac{\pi(\alpha - 3)}{2}, \quad (2)$$

where $\Gamma(\cdot)$ is the gamma function.

Let

$$\kappa_0 = \frac{4\pi}{L_0},$$

$$\kappa_l = \frac{(\pi A(\alpha) C_\alpha)^{1/(\alpha-5)}}{l_0}, \quad (3)$$

where

$$C_\alpha = \frac{3 - \alpha}{3} \Gamma\left(\frac{3 - \alpha}{2}\right) + a \frac{4 - \alpha}{3} \Gamma\left(\frac{4 - \alpha}{2}\right)$$
$$- b \frac{3 + \beta - \alpha}{3} \Gamma\left(\frac{3 + \beta - \alpha}{2}\right), \quad (4)$$

and the constant coefficients a, b, and β in (4) are usually set as $a = 1.802$, $b = 0.254$, and $\beta = 7/6$. The term $f(\kappa, \alpha, l_0, L_0)$ in (1) is given by

$$f(\kappa, \alpha, l_0, L_0) = \left(1 - \exp\left(-\frac{\kappa^2}{\kappa_0^2}\right)\right) \exp\left(-\frac{\kappa^2}{\kappa_l^2}\right)$$
$$\cdot \left(1 + a\left(\frac{\kappa}{\kappa_l}\right) - b\left(\frac{\kappa}{\kappa_l}\right)^\beta\right). \quad (5)$$

For the convenience of mathematical analysis, (5) is rewritten as

$$f(\kappa, \alpha, l_0, L_0) = \sum_{i=1}^{3} \sum_{j=1}^{2} (-1)^{j-1} c_i \kappa^{p_i} \exp\left(-d_j^2 \kappa^2\right), \quad (6)$$

where the coefficients are $c_1 = 1, c_2 = a/\kappa_l, c_3 = -b/\kappa_l^\beta, p_1 = 0$, $p_2 = 1, p_3 = \beta, d_1 = \sqrt{1/\kappa_l^2}$, and $d_2 = \sqrt{1/\kappa_l^2 + 1/\kappa_0^2}$.

It must be pointed that the values of these coefficients a, b, and β are based on the experimental data for the classic Kolmogorov turbulence. Besides, the generalized modified atmospheric spectrum will turn into the generalized exponential spectrum if $a = b = 0$ or the generalized Kolmogorov turbulence if $a = b = l_0 = 0$ and $L_0 \to +\infty$.

2.2. Aperture-Averaged Irradiance Scintillation Index of a Gaussian Beam. The mathematical model of a Gaussian beam depends on the radius W_0 and the phase front radius R_0 at the transmitter. Let L be the propagation path length, and let k be the angular wavenumber of the Gaussian wave

$$k = \frac{2\pi}{\lambda}, \quad (7)$$

where λ is the wavelength of the Gaussian wave. Thus, the curvature parameter of the Gaussian wave at the transmitter Θ_0 and the Fresnel ratio of the Gaussian wave at the transmitter Λ_0 are defined as [19]

$$\Theta_0 = 1 - \frac{L}{R_0},$$
$$\Lambda_0 = \frac{2L}{kW_0^2}. \quad (8)$$

Θ, $\overline{\Theta}$, and Λ are nondimensional parameters of the Gaussian wave at the receiver:

$$\Theta = \frac{\Theta_0}{\Theta_0^2 + \Lambda_0^2},$$
$$\overline{\Theta} = 1 - \Theta, \quad (9)$$
$$\Lambda = \frac{\Lambda_0}{\Theta_0^2 + \Lambda_0^2}.$$

The aperture-averaged irradiance scintillation index of a Gaussian beam propagating through weak atmospheric turbulence in the absence of beam wander takes the form [8]

$$\sigma_I^2(D_G) = 8\pi^2 k^2 L \int_0^1 d\xi \int_0^{+\infty} d\kappa \times \kappa \Phi_n(\kappa, \alpha)$$
$$\cdot \exp\left(-\frac{L\kappa^2}{k(\Omega_G + \Lambda)} \left((1 - \overline{\Theta}\xi)^2 + \Lambda \Omega_G \xi^2\right)\right) \quad (10)$$
$$\cdot \left(1 - \cos\left(\frac{L\kappa^2}{k}\left(\frac{\Omega_G - \Lambda}{\Omega_G + \Lambda}\right)\xi(1 - \overline{\Theta}\xi)\right)\right),$$

where ξ is the normalized path coordinate

$$\xi = 1 - \frac{z}{L}, \quad (11)$$

z is the coordinate along the propagation direction, Ω_G is a nondimensional parameter characterizing the relative radius of the collecting lens

$$\Omega_G = \frac{16L}{kD_G^2} > \Lambda, \quad (12)$$

and D_G is the collecting lens diameter.

2.3. Optical Wireless Communication Link Performance. The aperture-averaged irradiance scintillation index is relevant to the point-to-point optical wireless communication system with intensity modulation/direct detection scheme.

The probability of fade, the mean signal-to-noise ratio, and the mean bit-error rate are often used to judge the performance of the link [15].

The probability of fade describes the percentage of time; the irradiance of the beam at the receiver is below certain threshold value I_T. For weak turbulence, the probability of fade is defined by

$$P_I\left(I < I_T\right)$$
$$= \frac{1}{2}\left(1 + \mathrm{erf}\left(\frac{0.5\sigma_I^2\left(D_G\right) - 0.23 F_T}{\sqrt{2\sigma_I^2}}\right)\right), \qquad (13)$$

where $\mathrm{erf}(\cdot)$ is the error function and F_T is the fade threshold parameter:

$$F_T = 10\log_{10}\left(\frac{\langle I \rangle}{I_T}\right). \qquad (14)$$

The fade parameter stands for the mean irradiance in decibels below the threshold value I_T.

The mean signal-to-noise ratio is a measure which compares the level of the received beam to the level of background radiance. For weak turbulence, the mean signal-to-noise ratio is defined by

$$\langle SNR \rangle = \frac{SNR_0}{\sqrt{1 + \sigma_I^2\left(D_G\right) \cdot SNR_0^2}}, \qquad (15)$$

where SNR_0 is the signal-to-noise ratio in the absence of turbulence.

The mean bit-error rate is the number of bit errors in unit time. For weak turbulence, the mean bit-error rate is defined by

$$\langle BER \rangle = \frac{1}{2}\int_0^{+\infty} p_I\left(u\right)\mathrm{erfc}\left(\frac{\langle SNR \rangle u}{2\sqrt{2}}\right)du, \qquad (16)$$

where $p_I(u)$ is the probability density function of lognormal distribution

$$p_I\left(u\right)$$
$$= \frac{1}{u\sqrt{2\pi\sigma_I^2\left(D_G\right)}}\exp\left(-\frac{\left(\ln u + (1/2)\sigma_I^2\left(D_G\right)\right)^2}{2\sigma_I^2\left(D_G\right)}\right) \qquad (17)$$

and $\mathrm{erfc}(\cdot)$ is the complementary error function.

3. Expression Reduction

This section mainly discusses the reduction of (10).

Without loss of generality, we define

$$P\left(\xi\right) = \frac{L}{k\left(\Omega_G + \Lambda\right)}\left(\left(1 - \overline{\Theta}\xi\right)^2 + \Lambda\Omega_G\xi^2\right)$$
$$Q\left(\xi\right) = \frac{L}{k}\left(\frac{\Omega_G - \Lambda}{\Omega_G + \Lambda}\right)\xi\left(1 - \overline{\Theta}\xi\right). \qquad (18)$$

Based on Euler's formula, we can get [20, 21]

$$\left(1 - \cos\left(Q\left(\xi\right)\kappa^2\right)\right) = \mathrm{Re}\left(1 - \exp\left(i\left|Q\left(\xi\right)\right|\kappa^2\right)\right). \qquad (19)$$

Thus, (10) can be rewritten as

$$\sigma_I^2\left(D_G\right) = 8\pi^2 k^2 L\int_0^1 d\xi \int_0^{+\infty} \kappa\Phi_n\left(\kappa, \alpha\right)$$
$$\cdot \exp\left(-P\left(\xi\right)\kappa^2\right) \qquad (20)$$
$$\cdot \mathrm{Re}\left(1 - \exp\left(-i\left|Q\left(\xi\right)\right|\kappa^2\right)\right)d\kappa.$$

Substituting (1) into (20), it follows that

$$\sigma_I^2\left(D_G\right) = 8\pi^2 k^2 L A\left(\alpha\right)C_n^2\int_0^1 d\xi$$
$$\cdot \int_0^{+\infty} \kappa^{1-\alpha}f\left(\kappa, \alpha, l_0, L_0\right)\exp\left(-P\left(\xi\right)\kappa^2\right) \qquad (21)$$
$$\cdot \mathrm{Re}\left(1 - \exp\left(-i\left|Q\left(\xi\right)\right|\kappa^2\right)\right)d\kappa.$$

To reduce the iterated integral in (21), it is necessary to expand the integrand by (6):

$$\kappa^{1-\alpha}f\left(\kappa, \alpha, l_0, L_0\right)\exp\left(-P\left(\xi\right)\kappa^2\right)$$
$$\cdot \mathrm{Re}\left(1 - \exp\left(-i\left|Q\left(\xi\right)\right|\kappa^2\right)\right) = \sum_{i=1}^{3}\sum_{j=1}^{2}(-1)^{j-1} \qquad (22)$$
$$\cdot c_i\kappa^{1-\alpha+p_i}\mathrm{Re}\left[\exp\left(-\left(P\left(\xi\right) + d_j^2\right)\kappa^2\right)\right.$$
$$\left. - \exp\left(-\left(P\left(\xi\right) + d_j^2 + i\left|Q\left(\xi\right)\right|\right)\kappa^2\right)\right].$$

Based on the equation for $p > -3$, $\mathrm{Re}(u) > 0$, and $\mathrm{Re}(v) > 0$ [21, 22]

$$\int_0^{+\infty} x^p\left(\exp\left(-ux^2\right) - \exp\left(-vx^2\right)\right)dx$$
$$= \frac{1}{p+1}\left(u^{-(p+1)/2} - v^{-(p+1)/2}\right)\Gamma\left(\frac{p+3}{2}\right), \qquad (23)$$

the improper integral of κ can be rewritten as

$$\int_0^{+\infty} \kappa^{1-\alpha}f\left(\kappa, \alpha, l_0, L_0\right)\exp\left(-P\left(\xi\right)\kappa^2\right)$$
$$\cdot \mathrm{Re}\left(1 - \exp\left(-i\left|Q\left(\xi\right)\right|\kappa^2\right)\right)d\kappa$$
$$= \sum_{i=1}^{3}\sum_{j=1}^{2}(-1)^{j-1}c_i\frac{1}{2-\alpha+p_i}\Gamma\left(\frac{4-\alpha+p_i}{2}\right) \qquad (24)$$
$$\cdot \mathrm{Re}\left(\left(P\left(\xi\right) + d_j^2\right)^{-(2-\alpha+p_i)/2}\right.$$
$$\left. - \left(P\left(\xi\right) + d_j^2 + i\left|Q\left(\xi\right)\right|\right)^{-(2-\alpha+p_i)/2}\right).$$

Based on Euler's formula and de Moivre's formula, we can get [20, 21]

$$\mathrm{Re}\left(P(\xi) + d_j^2 + i\left|Q(\xi)\right|\right)^{-(2-\alpha+p_i)/2}$$

$$= \left(\left(P(\xi) + d_j^2\right)^2 + Q^2(\xi)\right)^{-(2-\alpha+p_i)/4} \quad (25)$$

$$\cdot \cos\left(\frac{2-\alpha+p_i}{2}\arctan\frac{Q(\xi)}{P(\xi)+d_j^2}\right).$$

Thus, the iterated integral in (21) has been reduced to the definite integral of the real variable ξ in the bounded interval $[0, 1]$, which can be easily computed by methods of numerical integration with arbitrary precision.

4. Numerical Simulations

This section conducts numerical simulations to analyze the influences of α, l_0, L_0, W_0, and λ on the aperture-averaged irradiance scintillation index $\sigma_I^2(D_G)$ of different Gaussian beams. The simulations performed in this paper, however, should be only considered as arbitrary examples to indicate certain trend of results. Unless specified otherwise, all the numerical simulations are conducted with the default settings: $\lambda = 1.55 \times 10^{-6}$ m, $k \approx 4.0537 \times 10^6$ rad/m, $L = 8000$ m, $C_n = 1.0 \times 10^{-14}$ m$^{3-\alpha}$, $W_0 = 0.1$ m, and $\Lambda_0 = 0.0493$. Of course, other values which satisfy (12) can also be chosen.

Figure 1 depicts the aperture-averaged irradiance scintillation index $\sigma_I^2(D_G)$ for different Gaussian beams as a function of the spectral power law α for several pairs of the inner and outer scales l_0 and L_0. As shown for the convergent beam ($\Theta_0 = 1$) in Figure 1(c), the aperture-averaged irradiance scintillation index $\sigma_I^2(D_G)$ first increases and then decreases with an increase in the spectral power law α when other parameters are fixed, which acts in accordance with that under the generalized non-Kolmogorov spectrum and the generalized scale-dependent anisotropic spectrum, as presented in [8, 15]. Similar phenomena can be also found for the focused beam ($\Theta_0 = 0$) in Figure 1(a), the convergent beam ($\Theta_0 = 0.5$) in Figure 1(b), and the divergent beam ($\Theta_0 = 1.5$) in Figure 1(d), respectively. According to Figure 1, it also shows that the aperture-averaged irradiance scintillation index σ_I^2 is more sensitive to the outer scale L_0 than to the inner scale l_0, and an increase in the outer scale L_0 will lead to an increase in the aperture-averaged irradiance scintillation index $\sigma_I^2(D_G)$ when other optical parameters are fixed. These phenomena can be physically explained by the fact that the irradiance scintillation can be decomposed into radial and longitudinal components. The former is sensitive to the outer scale L_0, while the latter is sensitive to the inner scale l_0. The longitudinal component of the irradiance scintillation only exists at the beam center, and thus the radial component occupies the dominant position in the influences on the aperture-averaged irradiance scintillation index $\sigma_I^2(D_G)$. As the outer scale of turbulence L_0 increases,

the Gaussian beam will meet more turbulence eddies along its propagation.

For further discussions and analyses, the inner and outer scales of turbulence are assigned to constant values $l_0 = 0.01$ m and $L_0 = 1$ m, respectively. Some typical values of wavelength in the near infrared region $\lambda = 0.85 \times 10^{-6}$ m, $\lambda = 1.06 \times 10^{-6}$ m, and $\lambda = 1.55 \times 10^{-6}$ m are investigated in this simulation. Figure 2 depicts the aperture-averaged irradiance scintillation index $\sigma_I^2(D_G)$ for different Gaussian beams as a function of the radius W_0 for several pairs of the spectral power law α and the wavelength λ. The radius W_0 along the abscissa axis uses the logarithmic (10-base) scale. To avoid the mutual interferences, $\alpha = 3.6 < 11/3$ and $\alpha = 3.7 > 11/3$ are used. As shown for the convergent beam ($\Theta_0 = 1$) in Figure 2(c), the aperture-averaged irradiance scintillation index $\sigma_I^2(D_G)$ first decreases and then increases with an increase in the radius W_0 when other parameters are fixed. Physically, these turbulence eddies act as lenses to change the focusing properties of transmission medium, and the small turbulence eddies with scales less than $\sqrt{L/k}$ play a significant role in the fluctuations of irradiance. When the radius W_0 is small enough, the Gaussian beam is hypersensitively affected by the small turbulence eddies. When the radius W_0 is larger, the transversal surface of the Gaussian beam contains numerous small turbulence eddies, which makes the fluctuations of irradiance greater. Figure 2(c) also shows that the aperture-averaged irradiance scintillation index $\sigma_I^2(D_G)$ decreases with an increase in λ for a Gaussian wave if other optical parameters are fixed. This phenomenon may be caused by the fact that the longer the beam wavelength, the more pronounced the diffraction. Thus, a laser beam with longer wavelength can be less affected by turbulence eddies. Similar phenomena can be also found for the focused beam ($\Theta_0 = 0$) in Figure 2(a), the convergent beam ($\Theta_0 = 0.5$) in Figure 2(b), and the divergent beam ($\Theta_0 = 1.5$) in Figure 2(d), respectively.

To analyze the influence of the irradiance scintillation on the performance of point-to-point optical wireless communication system with intensity modulation/direct detection scheme, Figure 3 depicts the probability of fade P_I as a function of the aperture-averaged irradiance scintillation index $\sigma_I^2(D_G)$ and the fade parameter F_T, and Figure 4 depicts the mean signal-to-noise ratio $\langle SNR \rangle$ as a function of the aperture-averaged irradiance scintillation index $\sigma_I^2(D_G)$ and the signal-to-noise ratio in the absence of turbulence SNR_0, and Figure 5 depicts the mean bit-error rate $\langle BER \rangle$ as a function of the aperture-averaged irradiance scintillation index $\sigma_I^2(D_G)$ and the signal-to-noise ratio in the absence of turbulence SNR_0. Both the aperture-averaged irradiance scintillation index $\sigma_I^2(D_G)$ and the signal-to-noise ratio in the absence of turbulence SNR_0 are presented on a logarithmic (10-base) scale. Figures 3–5 indicate that the atmospheric turbulence would produce much more negative effects on the wireless optical communication system with an increase in the irradiance scintillation. To improve the performance of the wireless optical communication system, an alternative method is to raise the signal-to-noise ratio in the absence of turbulence SNR_0.

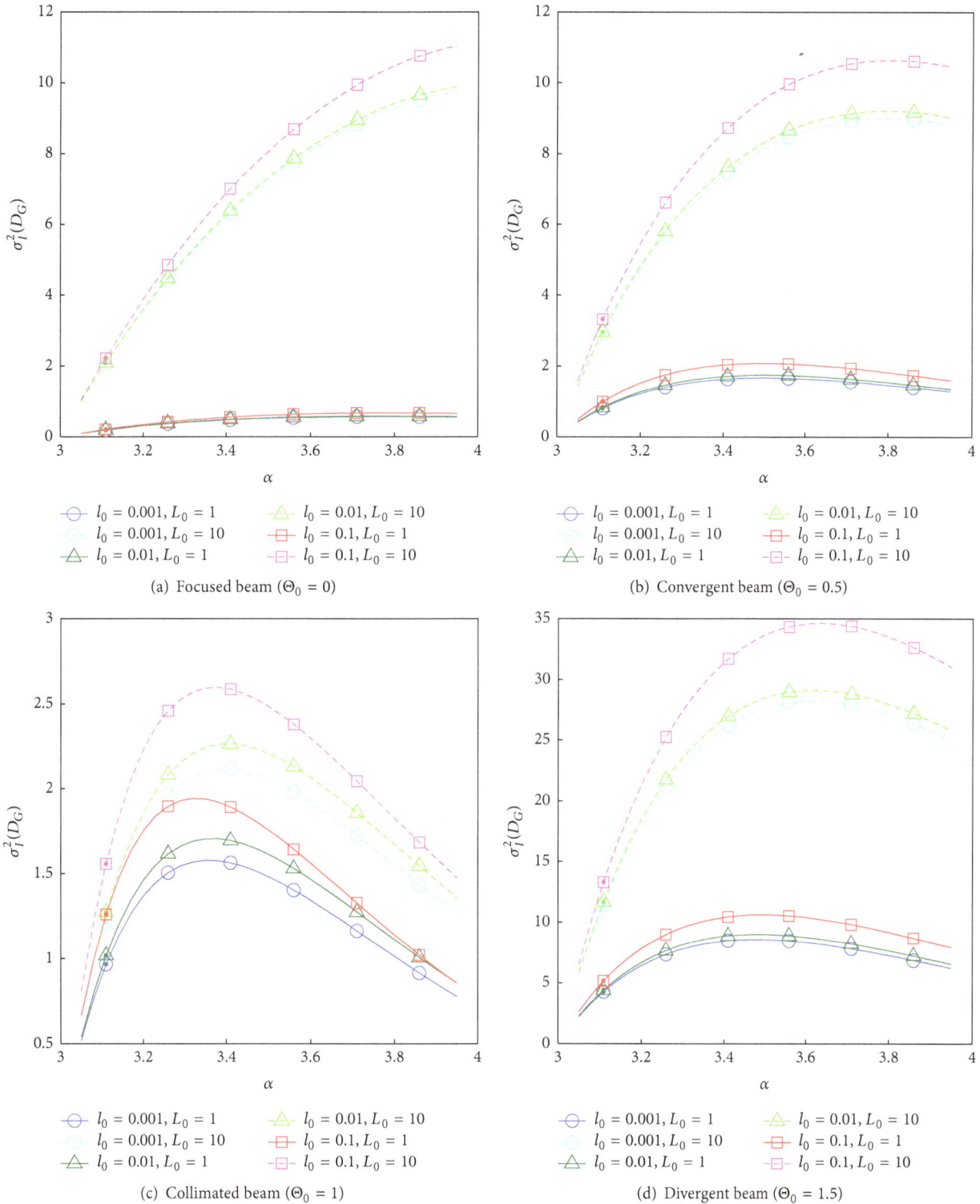

(a) Focused beam ($\Theta_0 = 0$)

(b) Convergent beam ($\Theta_0 = 0.5$)

(c) Collimated beam ($\Theta_0 = 1$)

(d) Divergent beam ($\Theta_0 = 1.5$)

FIGURE 1: σ_I^2 of different Gaussian beams as a function of α for several pairs of l_0 and L_0.

5. Conclusions

In this paper, the aperture-averaged irradiance scintillation derived for a Gaussian beam propagating through the weak non-Kolmogorov atmospheric turbulence along a horizontal path is investigated. The generalized modified atmospheric spectrum is utilized to take the variable spectral power law value, finite inner and outer scales of turbulence, and

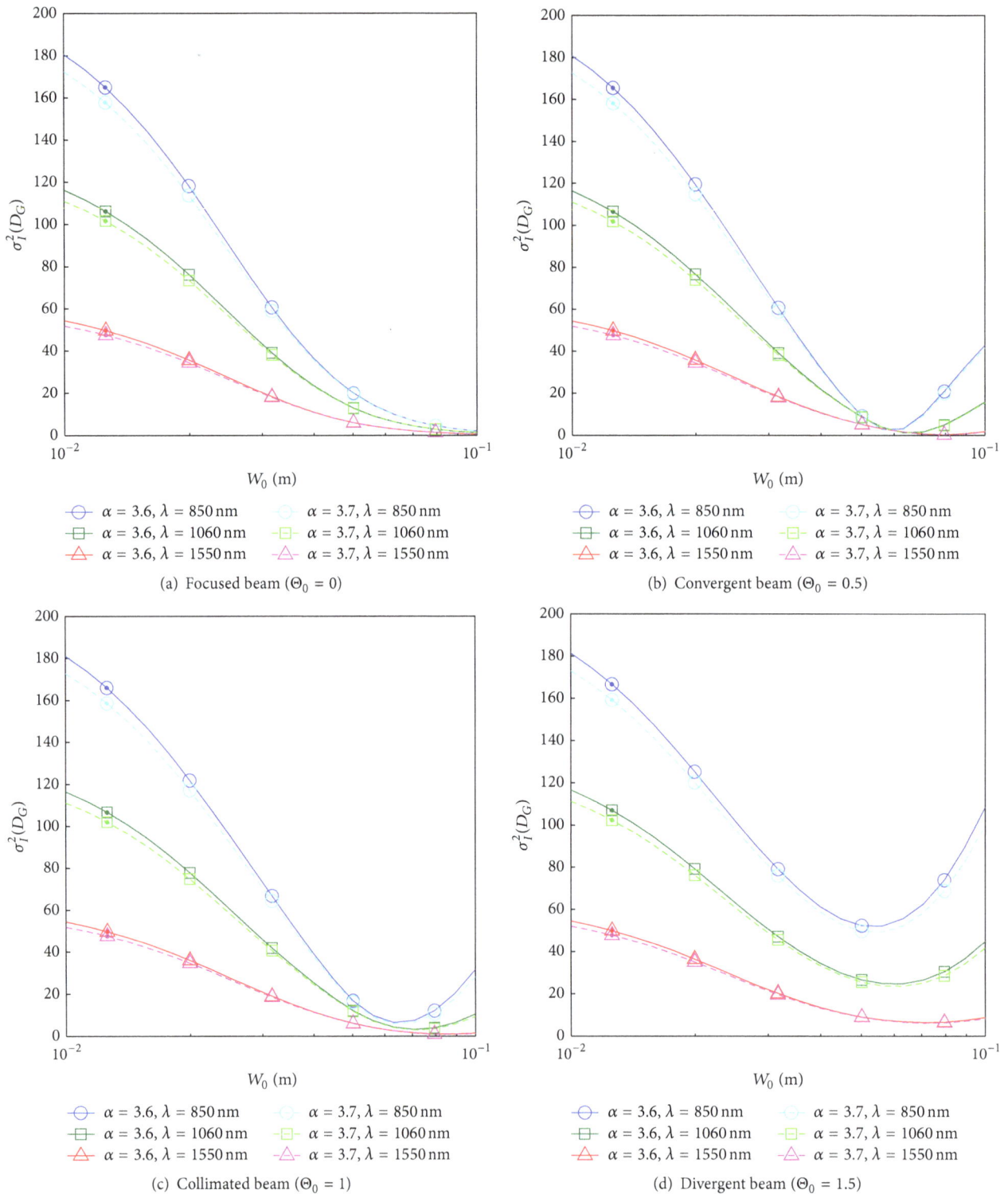

FIGURE 2: σ_I^2 of different Gaussian beams as a function of W_0 for several pairs of α and λ.

other important optical parameters of a Gaussian beam into account. The experimental results show the following:

(1) The irradiance scintillation index first increases and then decreases with an increase in the spectral power law.

(2) The aperture-averaged irradiance scintillation index is more sensitive to the outer scale than to the inner scale, and an increase in the outer scale will lead to an increase in the aperture-averaged irradiance scintillation index.

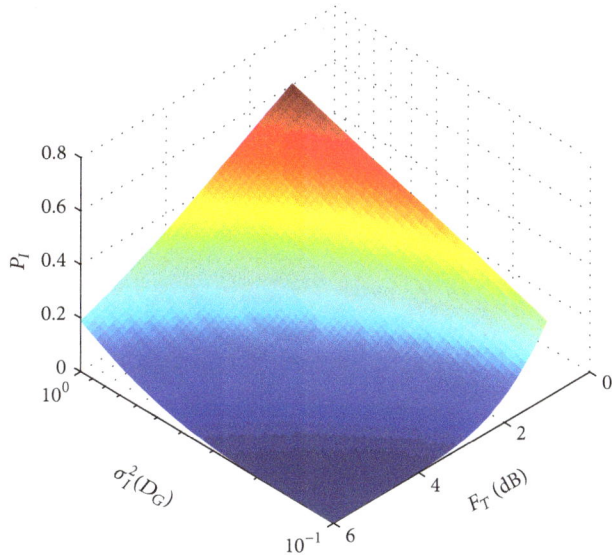

FIGURE 3: P_I as a function of σ_I^2 and F_T.

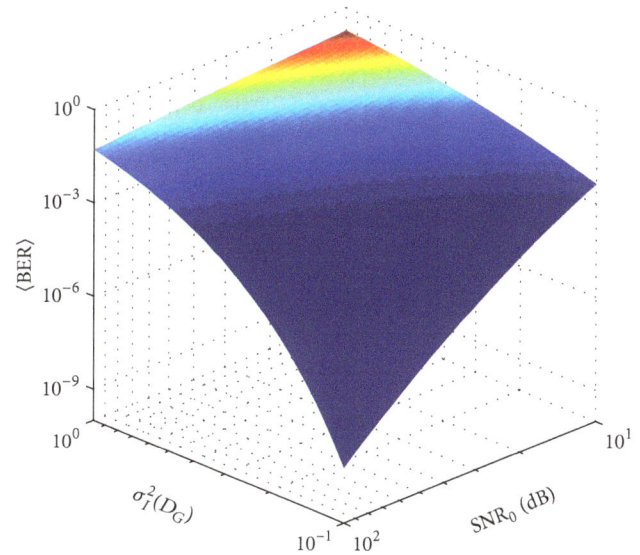

FIGURE 5: $\langle \text{BER} \rangle$ as a function of σ_I^2 and SNR_0.

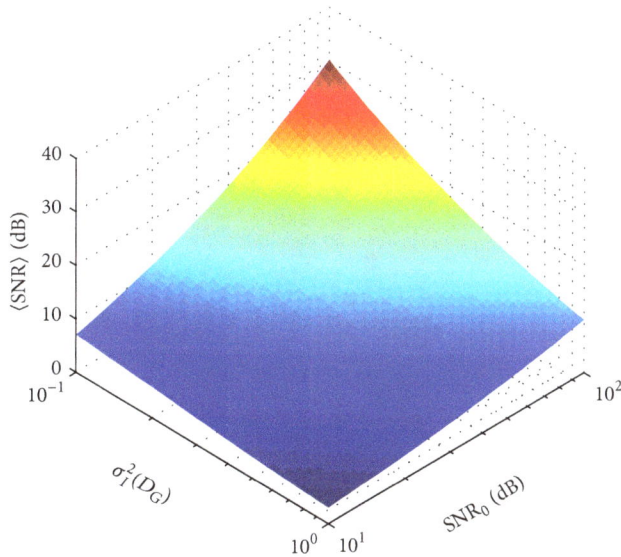

FIGURE 4: $\langle \text{SNR} \rangle$ as a function of σ_I^2 and SNR_0.

(3) The irradiance scintillation index first decreases and then increases with an increase in the radius.

(4) A longer beam wavelength will lead to a smaller irradiance scintillation index.

(5) The irradiance scintillation is harmful to the performance of the wireless optical communication system.

As mentioned above, the simulations performed in this paper should be only regarded as tentative examples to discover certain trend of results and mainly for the purpose of theoretical analyses. Due to the lack of adequate data and prior information about the non-Kolmogorov turbulence, the constant coefficients a, b, and β in (4) may fail to describe the actual situations of atmospheric turbulence. Once it is capable of determining the exact values of these coefficients

in the non-Kolmogorov turbulence, the expressions deduced in this paper could be generalized to the whole atmospheric layers. Besides, there are several literatures assuming the spectral power law value α to change in the range between 3 and 5, but in this paper the value of α is utilizing the smaller range between 3 and 4. For values of the spectral power law value α between 4 and 5, the condition in (23) is no more satisfied for the values of $1 - \alpha + p_i$ and thus the reduction fails. Therefore, the analysis is only valid for $\alpha \in (3, 4)$ in the derived model.

Conflict of Interests

The authors declare that there is no conflict of interests regarding the publication of this paper.

References

[1] M. A. Khalighi and M. Uysal, "Survey on free space optical communication: a communication theory perspective," *IEEE Communications Surveys and Tutorials*, vol. 16, no. 4, pp. 2231–2258, 2014.

[2] A. Malik and P. Singh, "Free space optics: current applications and future challenges," *International Journal of Optics*, vol. 2015, Article ID 945483, 7 pages, 2015.

[3] L. C. Andrews and R. L. Phillips, *Laser-Beam Propagation through Random Media*, SPIE Optical Engineering Press, Bellingham, Wash, USA, 2nd edition, 2005.

[4] M. Henriksson and L. Sjöqvist, "Scintillation index measurement using time-correlated single-photon counting laser radar," *Optical Engineering*, vol. 53, no. 8, Article ID 081902, 2014.

[5] W. B. Miller, J. C. Ricklin, and L. C. Andrews, "Effects of the refractive index spectral model on the irradiance variance of a Gaussian beam," *Journal of the Optical Society of America A*, vol. 11, no. 10, pp. 2719–2726, 1994.

[6] J. D. Shelton, "Turbulence-induced scintillation on Gaussian-beam waves: theoretical predictions and observations from

a laser-illuminated satellite," *Journal of the Optical Society of America A*, vol. 12, no. 10, p. 2172, 1995.

[7] L. C. Andrews, M. A. Al-Habash, C. Y. Hopen, and R. L. Phillips, "Theory of optical scintillation: Gaussian-beam wave model," *Waves Random Media*, vol. 11, no. 3, pp. 271–291, 2001.

[8] I. Toselli, L. C. Andrews, R. L. Phillips, and V. Ferrero, "Free space optical system performance for a gaussian beam propagating through non-kolmogorov weak turbulence," *IEEE Transactions on Antennas and Propagation*, vol. 57, no. 6, pp. 1783–1788, 2009.

[9] L. Cui, B. Xue, L. Cao et al., "Irradiance scintillation for Gaussian-beam wave propagating through weak non-Kolmogorov turbulence," *Optics Express*, vol. 19, no. 18, pp. 16872–16884, 2011.

[10] J. Cang and X. Liu, "Scintillation index and performance analysis of wireless optical links over non-Kolmogorov weak turbulence based on generalized atmospheric spectral model," *Optics Express*, vol. 19, no. 20, pp. 19067–19077, 2011.

[11] L. Cui, B. Xue, W. Xue, X. Bai, X. Cao, and F. Zhou, "Expressions of the scintillation index for optical waves propagating through weak non-Kolmogorov turbulence based on the generalized atmospheric spectral model," *Optics and Laser Technology*, vol. 44, no. 8, pp. 2453–2458, 2012.

[12] X. Yi, Z. Liu, and P. Yue, "Inner- and outer-scale effects on the scintillation index of an optical wave propagating through moderate-to-strong non-Kolmogorov turbulence," *Optics Express*, vol. 20, no. 4, pp. 4232–4247, 2012.

[13] X. Yi, Z. Liu, and P. Yue, "Optical scintillations and fade statistics for FSO communications through moderate-to-strong non-Kolmogorov turbulence," *Optics and Laser Technology*, vol. 47, pp. 199–207, 2013.

[14] L. Cui, "Temporal power spectra of irradiance scintillation for infrared optical waves' propagation through marine atmospheric turbulence," *Journal of the Optical Society of America A*, vol. 31, no. 9, pp. 2030–2037, 2014.

[15] I. Toselli and O. Korotkova, "General scale-dependent anisotropic turbulence and its impact on free space optical communication system performance," *Journal of the Optical Society of America A*, vol. 32, no. 6, pp. 1017–1025, 2015.

[16] I. Toselli, L. C. Andrews, and R. L. P. V. Ferrero, "Free-space optical system performance for laser beam propagation through non-Kolmogorov turbulence," *Optical Engineering*, vol. 47, no. 2, Article ID 026003, 2008.

[17] B. Xue, L. Cui, W. Xue, X. Bai, and F. Zhou, "Generalized modified atmospheric spectral model for optical wave propagating through non-Kolmogorov turbulence," *Journal of the Optical Society of America A: Optics and Image Science, and Vision*, vol. 28, no. 5, pp. 912–916, 2011.

[18] R. M. Williams and C. A. Paulson, "Microscale temperature and velocity spectra in the atmospheric boundary layers," *Journal of Fluid Mechanics*, vol. 83, no. 3, pp. 547–567, 1977.

[19] C. Gao, L. Su, and W. Yu, "Long-term spreading of Gaussian beam using generalized modified atmospheric spectrum," in *Proceedings of the IEEE International Conference on Mechatronics and Automation (ICMA'15)*, pp. 2375–2380, Beijing, China, August 2015.

[20] F. W. J. Olver, D. W. Lozier, R. F. Boisvert, and C. W. Clark, *NIST Handbook of Mathematical Functions*, Cambridge University Press, New York, NY, USA, 2010.

[21] I. S. Gradshteyn and I. M. Ryzhik, *Table of Integrals, Series, and Products*, Academic Press, Waltham, Mass, USA, 8th edition, 2014.

[22] A. Erdelyi, W. Magnus, and F. Oberhettinger, *Tables of Integral Transforms*, McGraw-Hill, New York, NY, USA, 1954.

Permissions

All chapters in this book were first published in IJO, by Hindawi Publishing Corporation; hereby published with permission under the Creative Commons Attribution License or equivalent. Every chapter published in this book has been scrutinized by our experts. Their significance has been extensively debated. The topics covered herein carry significant findings which will fuel the growth of the discipline. They may even be implemented as practical applications or may be referred to as a beginning point for another development.

The contributors of this book come from diverse backgrounds, making this book a truly international effort. This book will bring forth new frontiers with its revolutionizing research information and detailed analysis of the nascent developments around the world.

We would like to thank all the contributing authors for lending their expertise to make the book truly unique. They have played a crucial role in the development of this book. Without their invaluable contributions this book wouldn't have been possible. They have made vital efforts to compile up to date information on the varied aspects of this subject to make this book a valuable addition to the collection of many professionals and students.

This book was conceptualized with the vision of imparting up-to-date information and advanced data in this field. To ensure the same, a matchless editorial board was set up. Every individual on the board went through rigorous rounds of assessment to prove their worth. After which they invested a large part of their time researching and compiling the most relevant data for our readers.

The editorial board has been involved in producing this book since its inception. They have spent rigorous hours researching and exploring the diverse topics which have resulted in the successful publishing of this book. They have passed on their knowledge of decades through this book. To expedite this challenging task, the publisher supported the team at every step. A small team of assistant editors was also appointed to further simplify the editing procedure and attain best results for the readers.

Apart from the editorial board, the designing team has also invested a significant amount of their time in understanding the subject and creating the most relevant covers. They scrutinized every image to scout for the most suitable representation of the subject and create an appropriate cover for the book.

The publishing team has been an ardent support to the editorial, designing and production team. Their endless efforts to recruit the best for this project, has resulted in the accomplishment of this book. They are a veteran in the field of academics and their pool of knowledge is as vast as their experience in printing. Their expertise and guidance has proved useful at every step. Their uncompromising quality standards have made this book an exceptional effort. Their encouragement from time to time has been an inspiration for everyone.

The publisher and the editorial board hope that this book will prove to be a valuable piece of knowledge for researchers, students, practitioners and scholars across the globe.

List of Contributors

Naresh Kumar Reddy Andra
Department of Physics, CMR Institute of Technology, Medchal Road, Kandlakoya, Hyderabad, Telangana 501401, India

Karuna Sagar Dasari
Optics Research Group, Department of Physics, University College of Science, Osmania University, No. 49, Hyderabad, Telangana 500007, India

Udaya Laxmi Sriperumbudur
Department of Physics, Keshav Memorial Institute of Technology, Narayanguda, Hyderabad, Telangana 500029, India

Xiongbin Chen
State Key Laboratory of Integrated Optoelectronics, Institute of Semiconductors, Chinese Academy of Sciences, Beijing, China
School of Electronic, Electrical and Communication Engineering, University of Chinese Academy of Sciences, Beijing, China

Chengyu Min and Junqing Guo
State Key Laboratory of Integrated Optoelectronics, Institute of Semiconductors, Chinese Academy of Sciences, Beijing, China

Nadia Anam and Ebad Zahir
American International University-Bangladesh (AIUB), Kemal Ataturk Avenue, Banani, Dhaka 1213, Bangladesh

Chunyang Ma, Ge Wu, Tian Zhang and Xiaojian Tian
College of Electronic Science and Engineering, Jilin University, Changchun 130012, China

Bo Gao
College of Communication Engineering, Jilin University, Changchun 130012, China

Tomasz Moscicki
Institute of Fundamental Technological Research, PAS, Pawinskiego 5B, 02-106Warsaw, Poland

Hui Zeng and Yu Gu
School of Automation and Electrical Engineering, University of Science and Technology Beijing, Beijing 100083, China

XiuqingWang
Vocational & Technical Institute, Hebei Normal University, Shijiazhuang 050031, China

Wang Xun, Huang Kelin, Liu Zhirong and Zhao Kangyi
Department of Applied Physics, East China Jiaotong University, Nanchang, Jiangxi 330013, China

Yury Fedotov, Olga Bullo, Michael Belov and Viktor Gorodnichev
Faculty of Radioelectronics and Laser Techniques, Department of Laser and Optoelectronics Systems, Bauman Moscow State Technical University, 2-ya Baumanskaya Ulitsa, Moscow 105005, Russia

Zhao Peng, Li Yue and Ning Xiao
Information and Computer Engineering College, Northeast Forestry University, Harbin 150040, China

Chao Gao and Xiaofeng Li
School of Astronautics and Aeronautics, University of Electronic Science and Technology of China, 2006 Xiyuan Ave, West Hi-Tech Zone, Chengdu 611731, China

Mohamed Abdel-Nasser, AntonioMoreno and Domenec Puig
Departament d'Enginyeria Informàtica i Matemàtiques, Universitat Rovira i Virgili, Avinguda Paisos Catalans 26, 43007 Tarragona, Spain

Jaime Melendez
Department of Radiology, Radboud University Medical Center, 6525 GA Nijmegen, Netherlands

Shuanghong Wu and Qi Qiu
1School of Optoelectronic Information, University of Electronic Science and Technology of China, Chengdu 610054, China

Xiangru Wang
School of Optoelectronic Information, University of Electronic Science and Technology of China, Chengdu 610054, China
Science and Technology on Electro-Optical Information Security Control Laboratory, Sanhe 065201, China

Man Li
Science and Technology on Electro-Optical Information Security Control Laboratory, Sanhe 065201, China

Liang Wu
School of Physical Electronics, University of Electronic Science and Technology of China, Chengdu 610054, China

Jiyang Shang
Shanghai Aerospace Electronic Technology Institute, Shanghai 201109, China

Rajesh Yadav and Gurjit Kaur
Department of ECE, School of ICT, Gautam Buddha University, Greater Noida, India

Ali Kadhim and Alaa Al-Mebir
Department of Physics and Astronomy, University of Kansas, Lawrence, KS 66046, USA
Departments of Physics, College of Science, University of Thi-Qar, Nasiriya,Thi-Qar, Iraq

Paul Harrison and Judy Wu
Department of Physics and Astronomy, University of Kansas, Lawrence, KS 66046, USA

Jake Meeth
Department of Physics and Astronomy, University of Kansas, Lawrence, KS 66046, USA
Electrical Engineering Division, Department of Engineering, University of Cambridge, Cambridge CB3 OFA, UK

Guanggen Zeng
Department of Physics and Astronomy, University of Kansas, Lawrence, KS 66046, USA
College of Materials Science and Engineering, Sichuan University, Chengdu 610064, China

Mei-Ling Wang, Guo-Qing Zhong and Ling Chen
School of Material Science and Engineering, Southwest University of Science and Technology, Mianyang 621010, China

Jian Wang
Graduate School, Yanshan University, Qinhuangdao 066004, China
School of Information Science and Engineering, Yanshan University, Qinhuangdao 066004, China
The First Hospital of Qinhuangdao, Qinhuangdao 066000, China

Xiao Bing Li, Hong Ju Xu, Wei Bing Lu and JianWang
State Key Laboratory ofMillimeterWaves, School of Information Science and Engineering, SoutheastUniversity,Nanjing 210096, China

Yanpeng Shi
School of Physics, Shandong University, Jinan 250100, China

Xiaodong Wang and Fuhua Yang
Engineering Research Center for Semiconductor Integrated Technology, Institute of Semiconductors, Chinese Academy of Sciences, Beijing 100083, China

H. Abelman
School of Electrical and Information Engineering, University of theWitwatersrand, Johannesburg Private Bag 3,Wits 2050, South Africa

S. Abelman
School of Computer Science and Applied Mathematics, University of the Witwatersrand, Johannesburg, Private Bag 3,Wits 2050, South Africa

Manal Alshami,Mohamed Fawaz Mousselly and Anas Wabby
Higher Institute for Applied Sciences and Technology, Damascus, Syria

K. Rajesh
Department of Physics, Dhanalakshmi College of Engineering, Chennai 601301, India

A. Mani, V. Thayanithi and P. Praveen Kumar
Department of Physics, Presidency College, Chennai 5, India

Cristian Acevedo and Yezid Torres Moreno
Grupo de Óptica y Tratamiento de Señales (GOTS), Escuela de Física, Universidad Industrial de Santander, Carrera 27 Calle 9, A.A. 678, Bucaramanga, Colombia

Angela Guzmán and Aristide Dogariu
CREOL, The College of Optics and Photonics, University of Central Florida, P.O. Box 162700, 4000 Central Florida Boulevard, Orlando, FL 32816-2700, USA

Vinod K.Mishra
US Army Research Laboratory, Aberdeen, MD 21005, USA

Thai-Chien Bui and Suwit Kiravittaya
Advanced Optical Technology (AOT) Laboratory, Department of Electrical and Computer Engineering, Faculty of Engineering, Naresuan University, Muang, Phitsanulok,Thailand

Keattisak Sripimanwat
ECTI Association, Klong Luang, Pathumthani,Thailand

Nam-Hoang Nguyen
Faculty of Electronics and Telecommunications, VNU University of Engineering and Technology, Hanoi, Vietnam

Mohd Taufiq Mohd Khairi, Sallehuddin Ibrahim, Mohd Amri Md Yunus and Mahdi Faramarzi
Department of Control and Mechatronics Engineering, Faculty of Electrical Engineering, Universiti Teknologi Malaysia, 81310 Skudai, Johor, Malaysia

Chao Gao, Yiming Li and Xiaofeng Li
School of Astronautics and Aeronautic, University of Electronic Science and Technology of China, 2006 Xiyuan Avenue, Chengdu 611731, China

Yang Li
School of Accounting, Southwestern University of Finance and Economics, 555 Liutai Avenue, Chengdu 611130, China

Index

www.ingramcontent.com/pod-product-compliance
Lightning Source LLC
Chambersburg PA
CBHW070153240326
41458CB00126B/4538

9 781632 385840